T0350979

Dynamics of Water Surface Flows and Waves

Dynamics of Water Surface Flows and Waves

Yasunori Watanabe

CRC Press is an imprint of the
Taylor & Francis Group, an **informa** business
A CHAPMAN & HALL BOOK

First edition published 2022
by CRC Press
6000 Broken Sound Parkway NW, Suite 300, Boca Raton, FL 33487-2742

and by CRC Press
4 Park Square, Milton Park, Abingdon, Oxon, OX14 4RN

CRC Press is an imprint of Taylor & Francis Group, LLC

ISBN: 978-0-367-69042-7 (hbk)
ISBN: 978-0-367-69043-4 (pbk)
ISBN: 978-1-003-14016-0 (ebk)

DOI: 10.1201/9781003140160

Typeset in CMR10
by KnowledgeWorks Global Ltd.

Contents

Preface

Evaporation of seawater owing to solar radiation creates ascending air current, which varies local atmospheric pressure that causes winds. The winds generate ocean waves and current as a result of momentum transfer from air to ocean. Waves travel over ocean through various interactions with currents and bathymetry and arrive at shoaling coastal water where the long travel culminates in wave breaking. The wave breaking dissipates the wave energy and produces a huge number of air bubbles, entrained in coastal water, and sea spray dispersed in the air. While the air bubbles dissolving gas into seawater contribute to managing the ecosystem in the ocean, sea spray enhances evaporation and influences weather system through moisture and heat transfers. Ocean waves have important roles to maintain the global water cycle and environment.

This book is designed for theoretically understanding various aspects of sea waves over the lifetime. While the established linear potential wave theory well describes the evolution of waves in the normal state, multidisciplinary understandings of fluid dynamics are required to interpret the final stage of the lifetimes of waves, wave breaking, which causes a variety of physical processes. Fundamental theories and their interconnections over the diverse subjects, such as turbulence, vorticity dynamics, capillary dynamics, thermal mechanics, diffusion theory, multi-phase flow analysis, and stability analysis, are compiled in this book with the aim to enhance comprehensive understanding of fluid mechanics associated with water surface flows and waves.

I have had courses of fluid mechanics, coastal and ocean engineering at Hokkaido University for several decades. In the courses, I have always told my students that any mathematical formula exists not for looking at but for deriving and solving. I believe the derivation procedure is most important for properly understanding physics, especially fluid mechanics. As I know students have trouble in deriving equations during my teaching experiences in Hokkaido University, I have carefully recorded as much of development of formulas, coming to the solution, in this book. I hope readers derive the formulas following the procedures given in this book, not just look at the final solutions.

This book is organized for promoting step-by-step understanding. Firstly the fundamentals of fluid dynamics, mathematically describing potential and viscous flows are introduced in Chapter 1. Chapter 2 provides theories of turbulence and diffusion, which is required for understanding turbulence in breaking waves, marine boundary layer flows, and gas transfer across sea surfaces, introduced in Chapter 8. Chapter 3 theoretically explains the dynamics of free-surface/interface flows, capillary effects, and interactions with vortices, which is associated with any applications of surface flows and waves introduced in the following chapters. In Chapter 4, mathematical procedures to solve a Laplace equation under boundary conditions describing various wave motions, including progressive waves, deformations of thin films and jets, standing waves in rectangular and cylindrical domains, and the waves of another progressive modes (evanescent and edge waves). A shallow water equation, approximating surface flows and waves in thin sheet, open channel, tides and currents in ocean, is derived and applied to coastal and ocean waves in Chapter 5. Evolution of tsunami is also introduced in this chapter. In Chapter 6, the flow theory extends to analyze stability of water surfaces via well-known Kelvin-Helmholtz, Rayleigh-Taylor, and Rayleigh-Plateau instabilities, which explains the underlying mechanisms of surface deformation and breakup, and generation of surface waves. Chapter 7 describes various features of ocean waves observed in open ocean and shallower coastal regions, such as wave shoaling, evolution of wave packets, interactions with currents and bathymetry, and statistical and spectrum representations of irregular waves as a theoretical extension of Chapters 4 and 6. Fully developed ocean waves, through gravity-governed evolution process, noted in Chapter 7, culminate in wave breaking in which small-scale flow dynamics, governed by turbulence, vortices and capillary effects (Chapters 2, 3, 6), become more important; that is, multi-scale physics with different governing forcing is required to understand the wave breaking process. Chapter 8 describes mechanisms and contributions of wave breaking in coastal surf zone are described in terms of surface dynamics; that is, the formation of vorticity on the surfaces of overturning jets, the generation of sub-surface counter-rotating vortices causing scars, the development of secondary finger jets, and fragmentation into sea sprays. Capillary dynamics in the whitecapping process, namely entrained bubbles, residual foams, and sea sprays created by bursting bubbles, is also explained as theoretical extensions of the surface-vortex interaction (Chapter 3) and flow stability (Chapter 6). The main Chapters 4–8 are intended to promote a comprehensive understanding of the entire lives of ocean waves; as the growth of surface instability creates initial capillary waves, which evolve into gravity-dominant ocean waves with nonlinearity, and which in turn experience various interactions in the shallower coastal region, until wave energy is dissipated through turbulence generation and finally spent by creating air bubbles and sea sprays during the wave breaking process. All the chapters are designed to be interconnected with cross-references. The mathematical definitions required in the derivation procedures throughout the book are provided in Appendix.

In order to turn the various aspects of fluid flows and waves into a book, I have drawn my own research undertaken at the Laboratory of Coastal and Offshore Engineering, School of Engineering, Hokkaido University Japan. I would like to first thank my former students and colleagues who have joined the research and worked with me. In particular, the research has much progressed in collaboration with Prof. David Ingram, the University of Edinburgh, Prof. Nobuhito Mori, Kyoto University, Dr. Junichi Otsuka, Civil Engineering Research Institute, Dr. Ayumi Saruwatari, Hokkaido University, and Dr. Yasuo Niida, Central Research Institute of Electric Power Industry. I also wish to acknowledge encouragement and support from my friends and colleagues from the Coastal Engineering Community, particularly the JSCE Coastal Engineering Committee. My special thanks to Dr. Junichi Otsuka, Dr. Ayumi Saruwatari and Dr. Yasuo Niida for checking formulas on the draft of this book and advising on the presentation.

Lastly, I wish to express my gratitude to Miho, Kosei, and σ for their patience and understanding.

Yasunori Watanabe
Sapporo, Japan
October 2021

Author

Yasunori Watanabe is currently a Professor of Engineering at Hokkaido University Japan. He was Editor-in-Chief of *Coastal Engineering Journal* from 2013 to 2019. He has studied various topics in surface waves within the broad research area of fluid mechanics, such as capillary dynamics on a drop, bubble, and a thin film; flow stabilities in breaking waves; wave-structure interactions; and ocean wave dynamics. His research interests also cover coastal disasters associated with tsunamis, storm surges and extreme waves.

1

Introduction of Fluid Mechanics

Fundamentals of fluid mechanics, which are required to acquire understanding of practical issues in the following chapters, are described in this chapter. All the issues coming up in this chapter are also important to interpret problems raised in various flows. Readers who have basic knowledge of fluid mechanics may skip this chapter and come back if they require to know it in the following chapters.

1.1 Mathematical representation of flows

The preliminary study of fluid mechanics starting from the definitions of coordinates to representations of flows and stresses by vector and index is introduced in this section.

1.1.1 Coordinates and vector calculus

Three different coordinate systems, Cartesian, cylindrical, and spherical coordinates, are used in this book (Fig. 1.1*a*).

1.1.1.1 Cartesian coordinates

A Cartesian orthogonal coordinate system (x, y, z) is defined by perpendicular pairs of horizontal axes x, y, and vertical axis z; the unit vectors directing the axes have relations $\boldsymbol{e}_x \times \boldsymbol{e}_y = \boldsymbol{e}_z$, $\boldsymbol{e}_y \times \boldsymbol{e}_z = \boldsymbol{e}_x$, and $\boldsymbol{e}_z \times \boldsymbol{e}_x = \boldsymbol{e}_y$, where $\boldsymbol{e}_x = (1, 0, 0)$, $\boldsymbol{e}_y = (0, 1, 0)$, and $\boldsymbol{e}_z = (0, 0, 1)$.

An arbitrary vector $\boldsymbol{A} = (a_x, a_y, a_z)$ is described in terms of the unit vector as $\boldsymbol{A} = \boldsymbol{e}_x a_x + \boldsymbol{e}_y a_y + \boldsymbol{e}_z a_z$. A vector differential operator, $\boldsymbol{\nabla}$, is defined in the Cartesian coordinates as

$$\boldsymbol{\nabla} = \left(\frac{\partial}{\partial x}, \frac{\partial}{\partial y}, \frac{\partial}{\partial z}\right) = \boldsymbol{e}_x \frac{\partial}{\partial x} + \boldsymbol{e}_y \frac{\partial}{\partial y} + \boldsymbol{e}_z \frac{\partial}{\partial z} \tag{1.1}$$

Gradient of a scalar ϕ is thus expressed by

$$\boldsymbol{\nabla}\phi = \left(\frac{\partial \phi}{\partial x}, \frac{\partial \phi}{\partial y}, \frac{\partial \phi}{\partial z}\right) = \boldsymbol{e}_x \frac{\partial \phi}{\partial x} + \boldsymbol{e}_y \frac{\partial \phi}{\partial y} + \boldsymbol{e}_z \frac{\partial \phi}{\partial z} \tag{1.2}$$

DOI: 10.1201/9781003140160-1

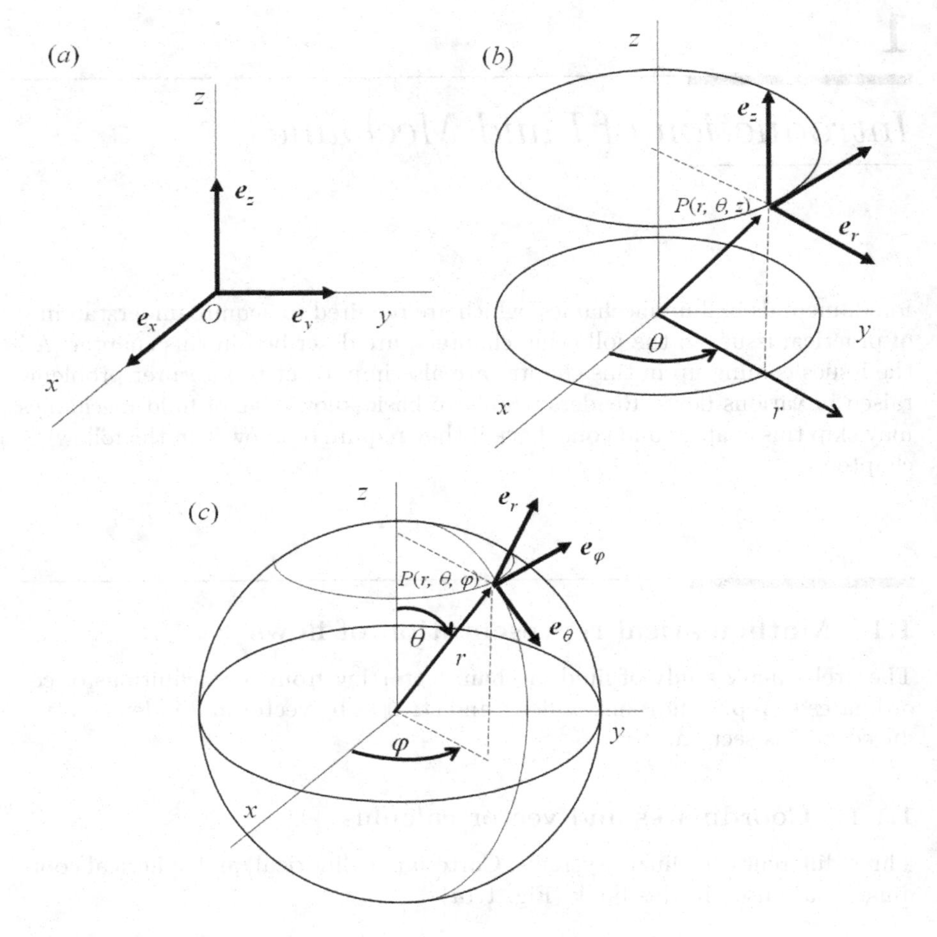

FIGURE 1.1
(*a*) Cartesian, (*b*) cylindrical, and (*c*) spherical coordinate systems.

Divergence of the vector $\boldsymbol{A}$ is defined by

$$\begin{aligned}\nabla \cdot \boldsymbol{A} &= \left(\boldsymbol{e}_x \frac{\partial}{\partial x} + \boldsymbol{e}_y \frac{\partial}{\partial y} + \boldsymbol{e}_z \frac{\partial}{\partial z}\right) \cdot (\boldsymbol{e_x} a_x + \boldsymbol{e_y} a_y + \boldsymbol{e_z} a_z) \\ &= \frac{\partial a_x}{\partial x} + \frac{\partial a_y}{\partial y} + \frac{\partial a_z}{\partial z} \end{aligned} \tag{1.3}$$

Rotation of the vector $\boldsymbol{A}$:

$$\begin{aligned}\nabla \times \boldsymbol{A} &= \left(\boldsymbol{e}_x \frac{\partial}{\partial x} + \boldsymbol{e}_y \frac{\partial}{\partial y} + \boldsymbol{e}_z \frac{\partial}{\partial z}\right) \times (\boldsymbol{e_x} a_x + \boldsymbol{e_y} a_y + \boldsymbol{e_z} a_z) \\ &= \boldsymbol{e}_x \left(\frac{\partial a_z}{\partial y} - \frac{\partial a_y}{\partial z}\right) + \boldsymbol{e}_y \left(\frac{\partial a_x}{\partial z} - \frac{\partial a_z}{\partial x}\right) + \boldsymbol{e}_z \left(\frac{\partial a_y}{\partial x} - \frac{\partial a_x}{\partial y}\right) \end{aligned} \tag{1.4}$$

The Laplace operator or Laplacian is given by

$$\nabla^2\phi = \nabla\cdot\nabla\phi = \left(\boldsymbol{e}_x\frac{\partial}{\partial x}+\boldsymbol{e}_y\frac{\partial}{\partial y}+\boldsymbol{e}_z\frac{\partial}{\partial z}\right)\cdot\left(\boldsymbol{e}_x\frac{\partial\phi}{\partial x}+\boldsymbol{e}_y\frac{\partial\phi}{\partial y}+\boldsymbol{e}_z\frac{\partial\phi}{\partial z}\right)$$
$$= \frac{\partial^2\phi}{\partial x^2}+\frac{\partial^2\phi}{\partial y^2}+\frac{\partial^2\phi}{\partial z^2} \tag{1.5}$$

Useful vector formulas are listed in Section A.1 of Appendix.

1.1.1.2 Cylindrical coordinates

A cylindrical coordinate system (r,θ,z) has the radial r, azimuthal θ, and perpendicular vertical z-axis. The system (r,θ,z) has relations to the Cartesian coordinates; $r^2 = x^2+y^2$, $x = r\cos\theta$, $y = r\sin\theta$, i.e. $\theta = \tan^{-1}(y/x)$, and $z = z$ (see Fig. 1.1c). The unit vectors $\boldsymbol{e}_r$, $\boldsymbol{e}_\theta$, and $\boldsymbol{e}_z$ have the orthogonal conditions; $\boldsymbol{e}_r\times\boldsymbol{e}_\theta = \boldsymbol{e}_z$, $\boldsymbol{e}_\theta\times\boldsymbol{e}_z = \boldsymbol{e}_r$, and $\boldsymbol{e}_z\times\boldsymbol{e}_r = \boldsymbol{e}_\theta$, where $\boldsymbol{e}_r = \cos\theta\boldsymbol{e}_x+\sin\theta\boldsymbol{e}_y$ and $\boldsymbol{e}_\theta = -\sin\theta\boldsymbol{e}_x+\cos\theta\boldsymbol{e}_y$.

In the cylindrical coordinates, ∇ is defined by

$$\nabla = \boldsymbol{e}_r\frac{\partial}{\partial r}+\boldsymbol{e}_\theta\frac{1}{r}\frac{\partial}{\partial\theta}+\boldsymbol{e}_z\frac{\partial}{\partial z} \tag{1.6}$$

Gradient, divergence, rotation, and Laplacian in this coordinates are

$$\nabla\phi = \boldsymbol{e}_r\frac{\partial\phi}{\partial r}+\boldsymbol{e}_\theta\frac{1}{r}\frac{\partial\phi}{\partial\theta}+\boldsymbol{e}_z\frac{\partial\phi}{\partial z} \tag{1.7}$$

$$\nabla\cdot\boldsymbol{A} = \frac{1}{r}\frac{\partial}{\partial r}(ra_r)+\frac{1}{r}\frac{\partial a_\theta}{\partial\theta}+\frac{\partial a_z}{\partial z} \tag{1.8}$$

$$\nabla\times\boldsymbol{A} = \boldsymbol{e}_r\left(\frac{1}{r}\frac{\partial a_z}{\partial\theta}-\frac{\partial a_\theta}{\partial z}\right)+\boldsymbol{e}_\theta\left(\frac{\partial a_r}{\partial z}-\frac{\partial a_z}{\partial r}\right)+\boldsymbol{e}_z\frac{1}{r}\left(\frac{\partial}{\partial r}(ra_\theta)-\frac{\partial a_r}{\partial\theta}\right) \tag{1.9}$$

$$\nabla^2\phi = \frac{1}{r}\frac{\partial}{\partial r}\left(r\frac{\partial\phi}{\partial r}\right)+\frac{1}{r^2}\frac{\partial^2\phi}{\partial\theta^2}+\frac{\partial^2\phi}{\partial z^2} \tag{1.10}$$

1.1.1.3 Spherical coordinates

In a spherical coordinate (r,θ,φ), the radial r, the polar θ, and azimuthal φ axis are described in the Cartesian coordinates by $r^2 = x^2+y^2+z^2$, $x = r\sin\theta\cos\varphi$, $y = r\sin\varphi\sin\theta$, and $z = r\cos\theta$, i.e. $\varphi = \tan^{-1}(y/x)$ and $\theta = \tan^{-1}\left(\sqrt{x^2+y^2}/z\right)$ (see Fig. 1.1c). The orthogonal unit vectors with the relations of $\boldsymbol{e}_r\times\boldsymbol{e}_\theta = \boldsymbol{e}_\varphi$, $\boldsymbol{e}_\theta\times\boldsymbol{e}_\varphi = \boldsymbol{e}_r$, and $\boldsymbol{e}_\varphi\times\boldsymbol{e}_r = \boldsymbol{e}_\theta$ are given as $\boldsymbol{e}_r = \sin\theta\ \cos\varphi\ \boldsymbol{e}_x+\sin\theta\ \sin\varphi\ \boldsymbol{e}_y+\cos\theta\ \boldsymbol{e}_z$, $\boldsymbol{e}_\theta = \cos\theta\ \cos\varphi\ \boldsymbol{e}_x+\cos\theta\ \sin\varphi\ \boldsymbol{e}_y-\sin\theta\ \boldsymbol{e}_z$, and $\boldsymbol{e}_\varphi = -\sin\theta\ \boldsymbol{e}_x+\cos\varphi\ \boldsymbol{e}_y$.

∇ in the spherical coordinates is defined by

$$\nabla = \boldsymbol{e}_r\frac{\partial}{\partial r}+\boldsymbol{e}_\theta\frac{1}{r}\frac{\partial}{\partial\theta}+\boldsymbol{e}_\varphi\frac{1}{r\sin\theta}\frac{\partial}{\partial\varphi} \tag{1.11}$$

The vector operations in these coordinates are given by

$$\nabla\phi = \boldsymbol{e}_r \frac{\partial \phi}{\partial r} + \boldsymbol{e}_\theta \frac{1}{r}\frac{\partial \phi}{\partial \theta} + \boldsymbol{e}_\varphi \frac{1}{r\sin\theta}\frac{\partial \phi}{\partial \varphi} \tag{1.12}$$

$$\nabla\cdot\boldsymbol{A} = \frac{1}{r^2}\frac{\partial}{\partial r}\left(r^2 a_r\right) + \frac{1}{r\sin\theta}\frac{\partial}{\partial \theta}\left(a_\theta \sin\theta\right) + \frac{1}{r\sin\theta}\frac{\partial a_\varphi}{\partial \varphi} \tag{1.13}$$

$$\begin{aligned}\nabla\times\boldsymbol{A} = {} & \boldsymbol{e}_r \frac{1}{r\sin\theta}\left(\frac{\partial}{\partial\theta}\left(a_\varphi\sin\theta\right) - \frac{\partial a_\theta}{\partial\varphi}\right) \\ & + \boldsymbol{e}_\theta \frac{1}{r}\left(\frac{1}{\sin\theta}\frac{\partial a_r}{\partial\varphi} - \frac{\partial}{\partial r}\left(r a_\varphi\right)\right) + \boldsymbol{e}_\varphi \frac{1}{r}\left(\frac{\partial}{\partial r}\left(r a_\theta\right) - \frac{\partial a_r}{\partial\theta}\right)\end{aligned} \tag{1.14}$$

$$\nabla^2\phi = \frac{1}{r^2}\frac{\partial}{\partial r}\left(r^2\frac{\partial\phi}{\partial r}\right) + \frac{1}{r^2\sin\theta}\frac{\partial}{\partial\theta}\left(\sin\theta\frac{\partial\phi}{\partial\theta}\right) + \frac{1}{r^2\sin^2\theta}\frac{\partial^2\phi}{\partial\varphi^2} \tag{1.15}$$

1.1.2 Index notation

Index representations of vectors and tensors are also used in this book. In this notation, for instance, a position vector $\boldsymbol{x}$ in the Cartesian coordinates is expressed in terms of an indexed variable, x_i, where $i = 1, 2, 3$:

$$x_i = (x_1, x_2, x_3) = (x, y, z) = \boldsymbol{x} \tag{1.16}$$

Similarly, for a velocity vector, $u_i = (u, v, w) = \boldsymbol{u}$.

When the same index appears twice in a single term, like $a_i b_i$, we follow the Einstein summation convention:

$$a_i b_i = \sum_{j=1}^{3} a_j b_j = a_1 b_1 + a_2 b_2 + a_3 b_3 \tag{1.17}$$

An arbitrary vector $\boldsymbol{u}$ is then expressed in terms of the unit vector $\boldsymbol{e}_i$ as

$$\boldsymbol{u} = u_i \boldsymbol{e}_i, \quad \text{and} \quad u_i = \boldsymbol{u}\cdot\boldsymbol{e}_i \tag{1.18}$$

The derivative operations of vectors are expressed by

$$\nabla\phi = \frac{\partial\phi}{\partial x_i} \tag{1.19}$$

$$\nabla\cdot\boldsymbol{u} = \frac{\partial u_i}{\partial x_i} \tag{1.20}$$

$$\nabla^2\phi = \frac{\partial^2\phi}{\partial x_i \partial x_i} \tag{1.21}$$

1.1.3 Trigonometric and hyperbolic functions

Oscillatory features of flows and surface forms are most important in the issues dealt with in this book. The exponential form, following Euler's formula, is preferably used to describe the oscillatory motion throughout the book:

$$e^{ix} = \cos x + i \sin x \tag{1.22}$$

where the trigonometry is expressed by

$$\sin x = \frac{e^{ix} - e^{-ix}}{2i} \tag{1.23}$$

$$\cos x = \frac{e^{ix} + e^{-ix}}{2} \tag{1.24}$$

The exponential representation of the oscillation is often convenient for solving differential equations, rather than the trigonometric form, as differentiation is straightforward (no need to replace sin and cos, and change a sign during differentiation). When the trigonometric form is required for complex solution $f = (a+ib)e^{ix}$, the real part of f provides the trigonometric solution:

$$Re[f] = Re\left[(a + ib)(\cos x + i \sin x)\right] = a \cos x - b \sin x$$

If x in Eqs. (1.23) and (1.24) is replaced with ix, the hyperbolic functions are defined as (Fig. 1.2)

$$\sinh x = \frac{e^{x} - e^{-x}}{2} = -i \sin(ix) \tag{1.25}$$

$$\cosh x = \frac{e^{x} + e^{-x}}{2} = \cos(ix) \tag{1.26}$$

$$\tanh x = \frac{\sinh x}{\cosh x} \tag{1.27}$$

Considering approximate values of the hyperbolic function at the limit $x \to 0$, e^x and e^{-x} are expanded to $x = \xi$ in a Taylor series about zero:

$$\begin{aligned} e^{\xi} &= e^{0} + \xi \frac{d}{dx} e^{x} \mid_{x=0} + \frac{\xi^2}{2} \frac{d^2}{dx^2} e^{x} \mid_{x=0} + \cdots \\ &= 1 + \xi + \frac{\xi^2}{2} + \cdots \\ e^{-\xi} &= e^{0} - \xi \frac{d}{dx} e^{x} \mid_{x=0} + \frac{\xi^2}{2} \frac{d^2}{dx^2} e^{x} \mid_{x=0} + \cdots \\ &= 1 - \xi + \frac{\xi^2}{2} + \cdots \end{aligned}$$

If we neglect the terms of the second order (ξ^2) or higher, the approximated

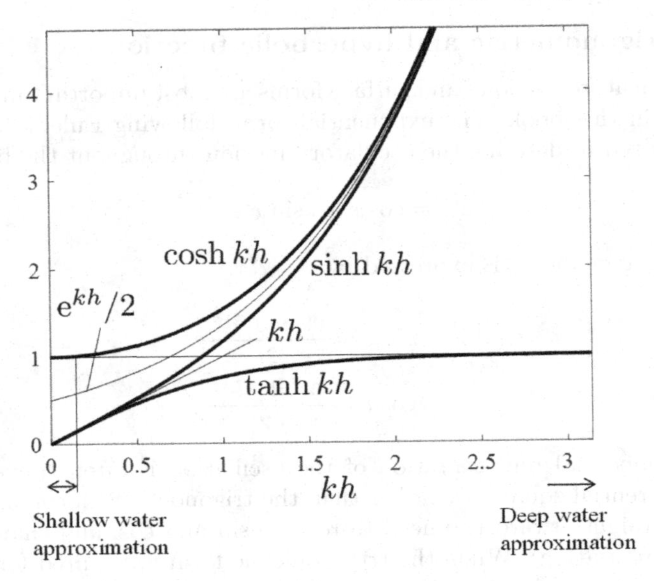

FIGURE 1.2
Hyperbolic functions and their approximations at the limit $kh \to 0$ (shallow water regime) and $kh \to \infty$ (deep water regime).

values of $e^{\xi} \approx 1 + \xi$ and $e^{-\xi} \approx 1 - \xi$ give

$$\sinh \xi = \frac{e^{\xi} - e^{-\xi}}{2} = \frac{1}{2}(2\xi + \cdots) \approx \xi \tag{1.28}$$

$$\cosh \xi = \frac{e^{\xi} + e^{-\xi}}{2} = \frac{1}{2}(2 + \cdots) \approx 1 \tag{1.29}$$

$$\tanh \xi \approx \xi \tag{1.30}$$

Solutions of surface waves, introduced in the later chapters, are described in terms of hyperbolic functions with $\xi = kh$, where h is the water depth and $k = 2\pi/L$ is the wave number. If h is much smaller than the wavelength L, i.e. very small kh, Eqs. (1.28)–(1.30) with $\xi = kh$ give

$$\sinh kh \approx kh \tag{1.31}$$

$$\cosh kh \approx 1 \tag{1.32}$$

$$\tanh kh \approx kh \tag{1.33}$$

which is called the shallow water approximation (see Fig. 1.2). When h is much larger than L, $kh \to \infty$ and thus $e^{-kh} \to 0$, termed deep water approximation (see Fig. 1.2), the approximate forms of the hyperbolic functions in this limit are given

$$\sinh kh = \frac{e^{kh} - \cancel{e^{-kh}}}{2} \approx \frac{e^{kh}}{2} \tag{1.34}$$

$$\cosh kh = \frac{e^{kh} + e^{-kh}}{2} \approx \frac{e^{kh}}{2} \tag{1.35}$$

$$\tanh kh \approx 1 \tag{1.36}$$

1.1.4 Stresses acting on fluid

Motions of any object and matter, including gas and liquid, are governed universally by the Newton's second law of motion that states an object accelerates if there is a net force acting upon the object. In this section, we consider, as the forces acting on a fluid element, pressure, viscous stress and gravity, which balances with the inertial force of the element.

1.1.4.1 Inertial force

Consider physical quantity $f(x, y, z, t)$ characterizing flow, which is passively transported in the flow with fluid velocity (u, v, w). The variation of f in the transport during the short interval Δt is approximated in terms of Taylor series by

$$\begin{aligned} \Delta f &= f(x + u\Delta t, y + v\Delta t, z + w\Delta t, t + \Delta t) - f(x, y, z, t) \\ &= \frac{\partial f}{\partial t}\Delta t + u\Delta t\frac{\partial f}{\partial x} + v\Delta t\frac{\partial f}{\partial y} + w\Delta t\frac{\partial f}{\partial z} \end{aligned} \tag{1.37}$$

We can define from Eq. (1.37) the material derivative (also termed the Lagrangian or substantial derivative) as a temporal variation of f at the limit of $\Delta t \to 0$:

$$\frac{Df}{Dt} = \lim_{\Delta t \to 0} \frac{\Delta f}{\Delta t} = \frac{\partial f}{\partial t} + u\frac{\partial f}{\partial x} + v\frac{\partial f}{\partial y} + w\frac{\partial f}{\partial z} \tag{1.38}$$

where the derivative operator expressed by

$$\frac{D}{Dt} = \frac{\partial}{\partial t} + u\frac{\partial}{\partial x} + v\frac{\partial}{\partial y} + w\frac{\partial}{\partial z} = \frac{\partial}{\partial t} + (\boldsymbol{u} \cdot \boldsymbol{\nabla}) \tag{1.39}$$

If f is the fluid velocity $\boldsymbol{u} = (u, v, w)$, acceleration of a fluid can be expressed by

$$\frac{D\boldsymbol{u}}{Dt} = \frac{\partial \boldsymbol{u}}{\partial t} + (\boldsymbol{u} \cdot \boldsymbol{\nabla})\boldsymbol{u} \tag{1.40}$$

or in index notation,

$$\frac{Du_i}{Dt} = \frac{\partial u_i}{\partial t} + u_j\frac{\partial u_i}{\partial x_j} \tag{1.41}$$

The acceleration in each axis direction of the Cartesian coordinates is given by

$$\frac{Du}{Dt} = \frac{\partial u}{\partial t} + u\frac{\partial u}{\partial x} + v\frac{\partial u}{\partial y} + w\frac{\partial u}{\partial z} \tag{1.42}$$

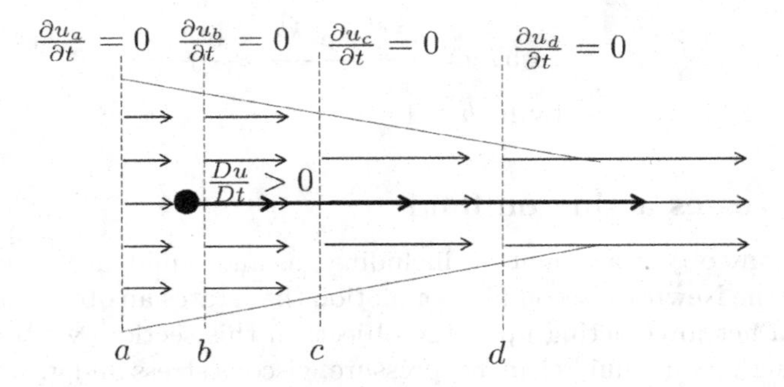

FIGURE 1.3
A fluid particle is accelerated in a steady flow in a narrowing channel; while $\partial u/\partial t = 0$ at any cross sections a–d, as the advection $u\partial u/\partial x > 0$, the fluid is accelerated at positive Du/Dt.

$$\frac{Dv}{Dt} = \frac{\partial v}{\partial t} + u\frac{\partial v}{\partial x} + v\frac{\partial v}{\partial y} + w\frac{\partial v}{\partial z} \tag{1.43}$$

$$\frac{Dw}{Dt} = \frac{\partial w}{\partial t} + u\frac{\partial w}{\partial x} + v\frac{\partial w}{\partial y} + w\frac{\partial w}{\partial z} \tag{1.44}$$

The first terms on right-hand sides of these equations are called unsteady terms, and the following nonlinear terms are called advection (or convection) terms. While the material derivative of fluid velocity, the left-hand side, describes the Lagrangian acceleration of a fluid particle transported in a flow, in Eulerian representation, the advection terms describe the Lagrangian variation of fluid velocity during transport of a fluid particle. For instance, as shown in Fig. 1.3, considering a fluid particle (black circle) being transported in narrowing steady channel flow, because of steady flow, the unsteady term $\partial u/\partial t$ is zero at any cross-sections. Since fluid velocity increases in the flow direction as the cross-sectional area decreases, $\partial u/\partial x > 0$, the particle must accelerate during the transport, i.e. $Du/Dt > 0$, since the advection term, $u\,(\partial u/\partial x)$, in Eq. (1.42) takes a positive value.

1.1.4.2 Pressure and viscous stress

Fluid forces acting on a fluid element to balance with the fluid inertia, i.e. pressure and viscous stress, are considered. The pressure is independent of orientation, i.e. isotropic, and it acts only normal to a fluid element. In this case, the stress tensor, $\boldsymbol{\Sigma}$, defining nine components of stress on the element (see Fig. 1.4), has only diagonal components, Σ_{xx}, Σ_{yy}, and Σ_{zz} in Fig. 1.4:

$$\boldsymbol{\Sigma} = -p\boldsymbol{I} = -p\begin{pmatrix} 1 & 0 & 0 \\ 0 & 1 & 0 \\ 0 & 0 & 1 \end{pmatrix} \tag{1.45}$$

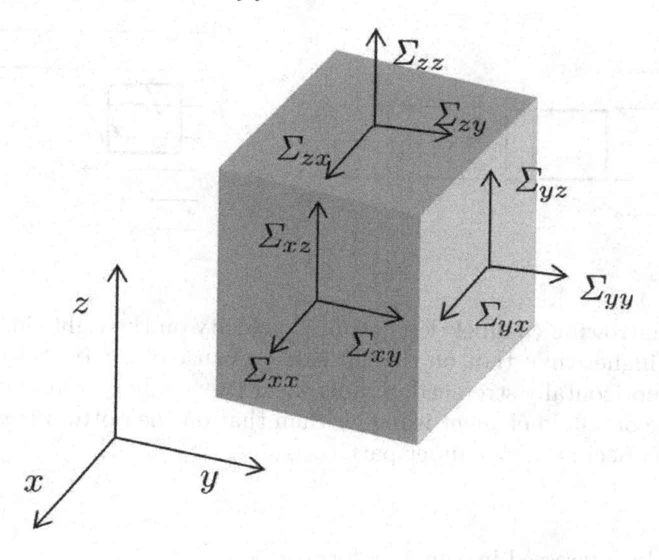

FIGURE 1.4
Components of a stress tensor on a rectangle fluid element.

where $\boldsymbol{I}$ is the unit diagonal matrix. In the index notation, Eq. (1.45) is expressed by

$$\Sigma_{ij} = -p\delta_{ij} \tag{1.46}$$

where the Kronecker delta is defined by

$$\delta_{ij} = \begin{cases} 1 & (i = j) \\ 0 & (i \neq j) \end{cases} \tag{1.47}$$

When a fluid is strained, internal stress occurs in the element, like an elastic material. The fluid strain occurs where there is a velocity difference around the fluid element (see Fig. 1.5). The velocity difference, $\Delta\boldsymbol{u}(x, y, z)$, approximated by Taylor series, can be written

$$\Delta u = \frac{\partial u}{\partial x}\Delta x + \frac{\partial u}{\partial y}\Delta y + \frac{\partial u}{\partial z}\Delta z$$
$$\Delta v = \frac{\partial v}{\partial x}\Delta x + \frac{\partial v}{\partial y}\Delta y + \frac{\partial v}{\partial z}\Delta z$$
$$\Delta w = \frac{\partial w}{\partial x}\Delta x + \frac{\partial w}{\partial y}\Delta y + \frac{\partial w}{\partial z}\Delta z$$

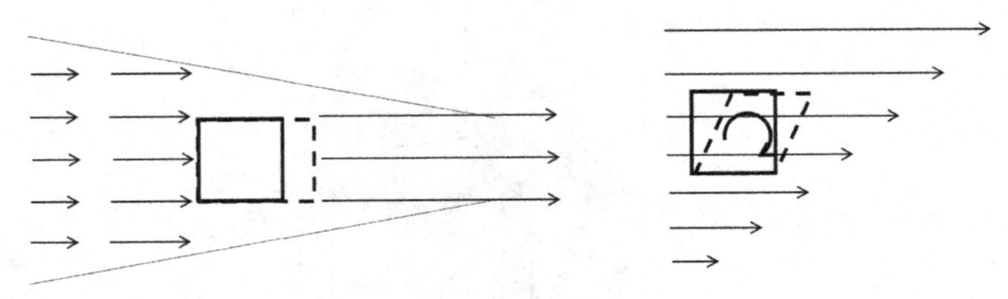

FIGURE 1.5
Flow in a narrowing channel (left): as fluid velocity on the right side of a fluid element is higher than that on the left side (because of $\partial u/\partial x > 0$), the fluid element is horizontally stretched. Simple shear flow (right): as fluid velocity on the top side of a fluid element is higher than that on the bottom ($\partial u/\partial z > 0$), deformation occurs in the upper part.

which can be expressed in a matrix form:

$$\begin{pmatrix} \Delta u \\ \Delta v \\ \Delta w \end{pmatrix} = \begin{pmatrix} \frac{\partial u}{\partial x} & \frac{\partial u}{\partial y} & \frac{\partial u}{\partial z} \\ \frac{\partial v}{\partial x} & \frac{\partial v}{\partial y} & \frac{\partial v}{\partial z} \\ \frac{\partial w}{\partial x} & \frac{\partial w}{\partial y} & \frac{\partial w}{\partial z} \end{pmatrix} \begin{pmatrix} \Delta x \\ \Delta y \\ \Delta z \end{pmatrix} \tag{1.48}$$

On the one hand, an arbitrary square matrix $\boldsymbol{A}$ is decomposed into symmetric and antisymmetric matrices, $\boldsymbol{A} = \frac{1}{2}\left(\boldsymbol{A} + \boldsymbol{A}^T\right) + \frac{1}{2}\left(\boldsymbol{A} - \boldsymbol{A}^T\right)$. The matrix of the velocity gradient in Eq. (1.48) is decomposed as $\boldsymbol{\nabla u} = \frac{1}{2}\left(\boldsymbol{\nabla u} + \boldsymbol{\nabla u}^T\right) + \frac{1}{2}\left(\boldsymbol{\nabla u} - \boldsymbol{\nabla u}^T\right)$

$$\boldsymbol{\nabla u} = \begin{pmatrix} \frac{\partial u}{\partial x} & \frac{\partial u}{\partial y} & \frac{\partial u}{\partial z} \\ \frac{\partial v}{\partial x} & \frac{\partial v}{\partial y} & \frac{\partial v}{\partial z} \\ \frac{\partial w}{\partial x} & \frac{\partial w}{\partial y} & \frac{\partial w}{\partial z} \end{pmatrix} = \boldsymbol{S} + \boldsymbol{\Omega} \tag{1.49}$$

where the symmetric one is termed the rate of strain:

$$\boldsymbol{S} = \frac{1}{2}\left(\boldsymbol{\nabla u} + \boldsymbol{\nabla u}^T\right) = \frac{1}{2} \begin{pmatrix} 2\frac{\partial u}{\partial x} & \frac{\partial u}{\partial y} + \frac{\partial v}{\partial x} & \frac{\partial u}{\partial z} + \frac{\partial w}{\partial x} \\ \frac{\partial v}{\partial x} + \frac{\partial u}{\partial y} & 2\frac{\partial v}{\partial y} & \frac{\partial v}{\partial z} + \frac{\partial w}{\partial y} \\ \frac{\partial w}{\partial x} + \frac{\partial u}{\partial z} & \frac{\partial w}{\partial y} + \frac{\partial v}{\partial z} & 2\frac{\partial w}{\partial z} \end{pmatrix} \tag{1.50}$$

and the antisymmetric, rotation tensor:

$$\boldsymbol{\Omega} = \frac{1}{2}\left(\boldsymbol{\nabla u} - \boldsymbol{\nabla u}^T\right) = \frac{1}{2}\begin{pmatrix} 0 & \dfrac{\partial u}{\partial y} - \dfrac{\partial v}{\partial x} & \dfrac{\partial u}{\partial z} - \dfrac{\partial w}{\partial x} \\ \dfrac{\partial v}{\partial x} - \dfrac{\partial u}{\partial y} & 0 & \dfrac{\partial v}{\partial z} - \dfrac{\partial w}{\partial y} \\ \dfrac{\partial w}{\partial x} - \dfrac{\partial u}{\partial z} & \dfrac{\partial w}{\partial y} - \dfrac{\partial v}{\partial z} & 0 \end{pmatrix} \tag{1.51}$$

While $\boldsymbol{S}$ contributes to deformation of a fluid element (see Fig. 1.5), $\boldsymbol{\Omega}$ does not, it works for rotating a fluid element. The stress in proportion to the rate of strain defines viscous stress:

$$\boldsymbol{\tau} = \mu\,(2\boldsymbol{S}) = \begin{pmatrix} \tau_{xx} & \tau_{xy} & \tau_{xz} \\ \tau_{yx} & \tau_{yy} & \tau_{yz} \\ \tau_{zx} & \tau_{zy} & \tau_{zz} \end{pmatrix} = \mu \begin{pmatrix} 2\dfrac{\partial u}{\partial x} & \dfrac{\partial u}{\partial y} + \dfrac{\partial v}{\partial x} & \dfrac{\partial u}{\partial z} + \dfrac{\partial w}{\partial x} \\ \dfrac{\partial v}{\partial x} + \dfrac{\partial u}{\partial y} & 2\dfrac{\partial v}{\partial y} & \dfrac{\partial v}{\partial z} + \dfrac{\partial w}{\partial y} \\ \dfrac{\partial w}{\partial x} + \dfrac{\partial u}{\partial z} & \dfrac{\partial w}{\partial y} + \dfrac{\partial v}{\partial z} & 2\dfrac{\partial w}{\partial z} \end{pmatrix} \tag{1.52}$$

where the constant of proportion μ is called viscosity. The fluid whose viscous stress is in proportion to the rate of strain is called Newtonian fluid. The stress tensor $\boldsymbol{\Sigma}$ of viscous fluid, Fig. 1.4, is thus defined by

$$\begin{aligned} \boldsymbol{\Sigma} &= \begin{pmatrix} \Sigma_{xx} & \Sigma_{xy} & \Sigma_{xz} \\ \Sigma_{yx} & \Sigma_{yy} & \Sigma_{yz} \\ \Sigma_{zx} & \Sigma_{zy} & \Sigma_{zz} \end{pmatrix} = -p\boldsymbol{I} + \boldsymbol{\tau} \\ &= -p\begin{pmatrix} 1 & 0 & 0 \\ 0 & 1 & 0 \\ 0 & 0 & 1 \end{pmatrix} + \mu \begin{pmatrix} 2\dfrac{\partial u}{\partial x} & \dfrac{\partial u}{\partial y} + \dfrac{\partial v}{\partial x} & \dfrac{\partial u}{\partial z} + \dfrac{\partial w}{\partial x} \\ \dfrac{\partial v}{\partial x} + \dfrac{\partial u}{\partial y} & 2\dfrac{\partial v}{\partial y} & \dfrac{\partial v}{\partial z} + \dfrac{\partial w}{\partial y} \\ \dfrac{\partial w}{\partial x} + \dfrac{\partial u}{\partial z} & \dfrac{\partial w}{\partial y} + \dfrac{\partial v}{\partial z} & 2\dfrac{\partial w}{\partial z} \end{pmatrix} \end{aligned} \tag{1.53}$$

which is expressed by the index as

$$\Sigma_{ij} = -p\delta_{ij} + \mu\left(\frac{\partial u_i}{\partial x_j} + \frac{\partial u_j}{\partial x_i}\right) \tag{1.54}$$

With the same operation following Section 1.1.1.2, the viscous stress in cylindrical coordinates is given by

$$\boldsymbol{\tau} = \mu \begin{pmatrix} 2\dfrac{\partial u_r}{\partial r} & \dfrac{1}{r}\dfrac{\partial u_r}{\partial \theta} + \dfrac{\partial u_\theta}{\partial r} - \dfrac{u_\theta}{r} & \dfrac{\partial u_r}{\partial z} + \dfrac{\partial u_z}{\partial r} \\ \dfrac{\partial u_\theta}{\partial r} - \dfrac{u_\theta}{r} + \dfrac{1}{r}\dfrac{\partial u_r}{\partial \theta} & \dfrac{2}{r}\left(\dfrac{\partial u_\theta}{\partial \theta} + u_r\right) & \dfrac{\partial u_\theta}{\partial z} + \dfrac{1}{r}\dfrac{\partial u_z}{\partial \theta} \\ \dfrac{\partial u_z}{\partial r} + \dfrac{\partial u_r}{\partial z} & \dfrac{1}{r}\dfrac{\partial u_z}{\partial \theta} + \dfrac{\partial u_\theta}{\partial z} & 2\dfrac{\partial u_z}{\partial z} \end{pmatrix} \tag{1.55}$$

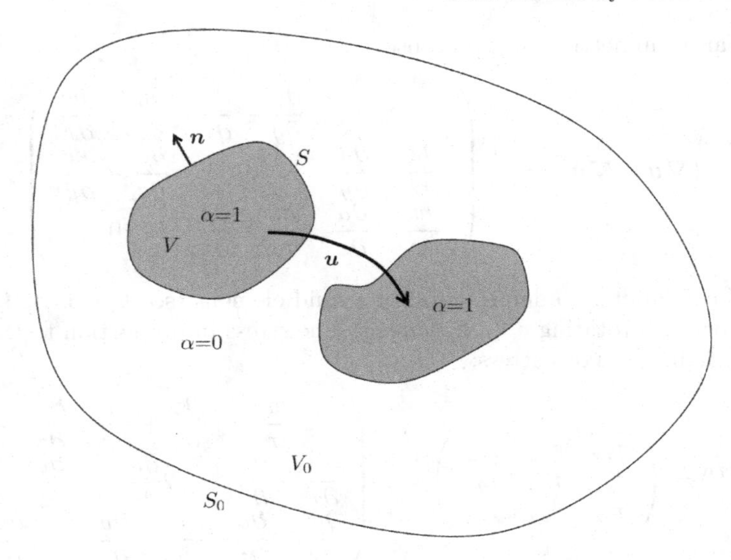

FIGURE 1.6
Illustration of a control volume transported in the flow domain.

1.1.5 Reynolds transport theorem

Consider physical quantity $q(\boldsymbol{x}, t)$ in a control volume (V), surrounded by surface area (S), transported in a flow with velocity $\boldsymbol{u}$ in a fluid domain V_0 with surface S_0 (see Fig. 1.6). In order to define the transported control volume, we introduce a color function $\alpha(\boldsymbol{x}, t)$ indicating 1 in an inner region across S and 0 for else. The variation of α is followed by the advection equation, Eq. (1.38),

$$\frac{D\alpha}{Dt} = \frac{\partial \alpha}{\partial t} + \boldsymbol{u} \cdot \boldsymbol{\nabla} \alpha = 0 \tag{1.56}$$

which defines an arbitrary form of the control volume transported in the flow. Note that as the control volume is located far from the boundary of the fluid domain, $\alpha = 0$ on S_0. Using Eq. (1.56), the temporal change of q in V is given as

$$\begin{aligned} \frac{d}{dt}\int_V q dv &= \frac{d}{dt}\int_{V_0} q\alpha dv = \int_{V_0} \frac{\partial}{\partial t}(q\alpha)\, dv = \int_{V_0}\left(\frac{\partial q}{\partial t}\alpha + q\frac{\partial \alpha}{\partial t}\right) dv \\ &= \int_{V_0}\left(\frac{\partial q}{\partial t}\alpha - q\boldsymbol{u} \cdot \boldsymbol{\nabla}\alpha\right) dv \end{aligned} \tag{1.57}$$

As $\boldsymbol{\nabla} \cdot (q\boldsymbol{u}\alpha) = q\boldsymbol{u} \cdot \boldsymbol{\nabla}\alpha + \alpha \boldsymbol{\nabla} \cdot (q\boldsymbol{u})$, following the chain rule (see Eq. (A.3) in Appendix), the second term on the right-hand side of Eq. (1.57) is expressed by

$$\int_{V_0} q\boldsymbol{u}\cdot\boldsymbol{\nabla}\alpha dv = \int_{V_0} \boldsymbol{\nabla}\cdot(q\boldsymbol{u}\alpha)\,dv - \int_{V_0} \alpha\boldsymbol{\nabla}\cdot(q\boldsymbol{u})\,dv \tag{1.58}$$

Using Gauss' divergent theorem, Eq. (A.29) in Appendix, for the right-hand side, $\int_{V_0} \boldsymbol{\nabla}\cdot(q\boldsymbol{u}\alpha)\,dv = \int_{S_0} (\alpha q\boldsymbol{u})\cdot\boldsymbol{n}ds = 0$ as $\alpha = 0$ on the end of fluid domain S_0, where $\boldsymbol{n}$ is the unit normal vector on S_0. Eq. (1.58) is then reduced to

$$\int_{V_0} q\boldsymbol{u}\cdot\boldsymbol{\nabla}\alpha dv = -\int_{V_0} \alpha\boldsymbol{\nabla}\cdot(q\boldsymbol{u})\,dv \tag{1.59}$$

Eq. (1.57) is rewritten by

$$\frac{d}{dt}\int_{V_0} q\alpha dv = \int_{V_0} \alpha\left(\frac{\partial q}{\partial t} + \boldsymbol{\nabla}\cdot(q\boldsymbol{u})\right)dv \tag{1.60}$$

Since $\alpha = 1$ in V, the Reynolds transport theorem is given

$$\frac{d}{dt}\int_{V} q dv = \int_{V} \left(\frac{\partial q}{\partial t} + \boldsymbol{\nabla}\cdot(q\boldsymbol{u})\right)dv \tag{1.61}$$

1.2 Mass conservation

Mass within a certain control volume V is conserved during the transport process

$$\frac{d}{dt}\int_{V} \rho dv = 0 \tag{1.62}$$

Substituting $q = \rho$ into Eq. (1.61), the mass conservation Eq. (1.62) can be expressed in terms of the Reynolds transport theorem as

$$\int_{V} \left(\frac{\partial \rho}{\partial t} + \boldsymbol{\nabla}\cdot(\rho\boldsymbol{u})\right)dv = 0 \tag{1.63}$$

The condition to identically satisfy Eq. (1.63) for arbitrary control volume V is

$$\frac{\partial \rho}{\partial t} + \boldsymbol{\nabla}\cdot(\rho\boldsymbol{u}) = 0 \tag{1.64}$$

In the Cartesian coordinates, component form, Eq. (1.64) is specifically expressed by

$$\frac{\partial \rho}{\partial t} + \frac{\partial \rho u}{\partial x} + \frac{\partial \rho v}{\partial y} + \frac{\partial \rho w}{\partial z} = 0 \tag{1.65}$$

Following a chain rule, the equation is rewritten as

$$\frac{\partial \rho}{\partial t} + u\frac{\partial \rho}{\partial x} + v\frac{\partial \rho}{\partial y} + w\frac{\partial \rho}{\partial z} + \rho\left(\frac{\partial u}{\partial x} + \frac{\partial v}{\partial y} + \frac{\partial w}{\partial z}\right) = 0 \tag{1.66}$$

Assuming the density of a fluid particle is unchanged during the transport, we may state

$$\frac{D\rho}{Dt} = \frac{\partial \rho}{\partial t} + u\frac{\partial \rho}{\partial x} + v\frac{\partial \rho}{\partial y} + w\frac{\partial \rho}{\partial z} = 0 \tag{1.67}$$

The substitution of Eq. (1.67) into Eq. (1.66) gives the incompressible mass conservation termed a continuity equation:

$$\frac{\partial u}{\partial x} + \frac{\partial v}{\partial y} + \frac{\partial w}{\partial z} = 0 \tag{1.68}$$

which is also expressed as

$$\boldsymbol{\nabla} \cdot \boldsymbol{u} = 0 \quad \text{or} \quad \frac{\partial u_j}{\partial x_j} = 0 \tag{1.69}$$

In cylindrical coordinates, the continuity equation is given by

$$\frac{1}{r}\frac{\partial (r u_r)}{\partial r} + \frac{1}{r}\frac{\partial u_\theta}{\partial \theta} + \frac{\partial u_z}{\partial z} = 0 \tag{1.70}$$

1.3 Momentum conservation

A rate of change of momentum in a control volume V bounded by a surface S balances with stress acting upon the surface and external body force:

$$\frac{d}{dt}\int_V \rho \boldsymbol{u} dV = \int_S (\boldsymbol{\Sigma} \cdot \boldsymbol{n}) dS + \int_V \rho \boldsymbol{K} dV \tag{1.71}$$

where $\boldsymbol{\Sigma}$ is the stress tensor, Eq. (1.53), and $\boldsymbol{K}$ is the external force. If only the gravity acts as the force, $\boldsymbol{K} = (0, 0, -g)$, in the Cartesian system. The surface integral of the first term on the right-hand side can be replaced with the volume integral by using Gauss' divergence theorem (see Eq. (A.29) in Appendix):

$$\int_S (\boldsymbol{\Sigma} \cdot \boldsymbol{n}) dS = \int_V \boldsymbol{\nabla} \cdot \boldsymbol{\Sigma} dV \tag{1.72}$$

Substituting $q = \rho \boldsymbol{u}$ into the Reynolds transport theorem, Eq. (1.61), to replace with the left-hand side of Eq. (1.71), we have

$$\int_V \left[\frac{\partial \rho \boldsymbol{u}}{\partial t} + \boldsymbol{\nabla} \cdot (\rho \boldsymbol{u} \otimes \boldsymbol{u}) - \boldsymbol{\nabla} \cdot \boldsymbol{\Sigma} - \rho \boldsymbol{K}\right] dV = 0 \tag{1.73}$$

where the tensor product

$$\boldsymbol{u} \otimes \boldsymbol{u} = \begin{pmatrix} u^2 & uv & uw \\ uv & v^2 & vw \\ uw & uv & w^2 \end{pmatrix} \tag{1.74}$$

As V is arbitrary, the condition to satisfy Eq. (1.73) is given by

$$\frac{\partial \rho \boldsymbol{u}}{\partial t} + \boldsymbol{\nabla} \cdot (\rho \boldsymbol{u} \otimes \boldsymbol{u}) = \boldsymbol{\nabla} \cdot \boldsymbol{\Sigma} + \rho \boldsymbol{K} \tag{1.75}$$

Since the chain rule, $\boldsymbol{\nabla} \cdot (\rho \boldsymbol{u} \otimes \boldsymbol{u}) = \rho \boldsymbol{u} \cdot \boldsymbol{\nabla} \boldsymbol{u} + \boldsymbol{u} \boldsymbol{\nabla} \cdot (\rho \boldsymbol{u})$ and $\boldsymbol{u} \cdot \boldsymbol{\nabla} \boldsymbol{u} = (\boldsymbol{u} \cdot \boldsymbol{\nabla}) \boldsymbol{u}$ (see Eqs. (A.4) and (A.5) in Appendix), Eq. (1.75) is rewritten as

$$\rho \frac{\partial \boldsymbol{u}}{\partial t} + \rho (\boldsymbol{u} \cdot \boldsymbol{\nabla}) \boldsymbol{u} + \boldsymbol{u} \left(\frac{\partial \rho}{\partial t} + \boldsymbol{\nabla} \cdot (\rho \boldsymbol{u}) \right) = \boldsymbol{\nabla} \cdot \boldsymbol{\Sigma} + \rho \boldsymbol{K} \tag{1.76}$$

The mass conservation, Eq. (1.64), removes the third term on the left-hand side of Eq. (1.76), yielding a convective form of momentum equation:

$$\frac{\partial \boldsymbol{u}}{\partial t} + (\boldsymbol{u} \cdot \boldsymbol{\nabla}) \boldsymbol{u} = \frac{1}{\rho} \boldsymbol{\nabla} \cdot \boldsymbol{\Sigma} + \boldsymbol{K} \tag{1.77}$$

In the case of inviscid fluid, the stress tensor Eq. (1.45) is substituted into Eq. (1.77), giving the well-known Euler equation:

$$\frac{\partial \boldsymbol{u}}{\partial t} + (\boldsymbol{u} \cdot \boldsymbol{\nabla}) \boldsymbol{u} = -\frac{1}{\rho} \boldsymbol{\nabla} p + \boldsymbol{K}, \quad \frac{\partial u_i}{\partial t} + u_j \frac{\partial u_i}{\partial x_j} = -\frac{1}{\rho} \frac{\partial p}{\partial x_i} + K_i \tag{1.78}$$

The components of Eq. (1.78) with $\boldsymbol{K} = (0, 0, -g)$ in the Cartesian coordinates are expressed by

$$\frac{\partial u}{\partial t} + u \frac{\partial u}{\partial x} + v \frac{\partial u}{\partial y} + w \frac{\partial u}{\partial z} = -\frac{1}{\rho} \frac{\partial p}{\partial x} \tag{1.79}$$

$$\frac{\partial v}{\partial t} + u \frac{\partial v}{\partial x} + v \frac{\partial v}{\partial y} + w \frac{\partial v}{\partial z} = -\frac{1}{\rho} \frac{\partial p}{\partial y} \tag{1.80}$$

$$\frac{\partial w}{\partial t} + u \frac{\partial w}{\partial x} + v \frac{\partial w}{\partial y} + w \frac{\partial w}{\partial z} = -\frac{1}{\rho} \frac{\partial p}{\partial z} - g \tag{1.81}$$

where g is the gravitational acceleration.

The Euler equation in cylindrical coordinates (r, θ, z) is given by

$$\frac{\partial u_r}{\partial t} + u_r \frac{\partial u_r}{\partial r} + \frac{u_\theta}{r} \frac{\partial u_r}{\partial \theta} + u_z \frac{\partial u_r}{\partial z} - \frac{u_\theta^2}{r} = -\frac{1}{\rho} \frac{\partial p}{\partial r} + K_r \tag{1.82}$$

$$\frac{\partial u_\theta}{\partial t} + u_r \frac{\partial u_\theta}{\partial r} + \frac{u_\theta}{r} \frac{\partial u_\theta}{\partial \theta} + u_z \frac{\partial u_\theta}{\partial z} + \frac{u_r u_\theta}{r} = -\frac{1}{\rho r} \frac{\partial p}{\partial \theta} + K_\theta \tag{1.83}$$

$$\frac{\partial u_z}{\partial t} + u_r \frac{\partial u_z}{\partial r} + \frac{u_\theta}{r} \frac{\partial u_z}{\partial \theta} + u_z \frac{\partial u_z}{\partial z} = -\frac{1}{\rho} \frac{\partial p}{\partial z} + K_z \tag{1.84}$$

Substituting the stress tensor of viscous fluid, Eq. (1.53) into Eq. (1.77), the well-known Navier-Stokes equation is given as

$$\frac{\partial \boldsymbol{u}}{\partial t} + (\boldsymbol{u} \cdot \boldsymbol{\nabla})\boldsymbol{u} = -\frac{1}{\rho}\boldsymbol{\nabla} p + \frac{1}{\rho}\boldsymbol{\nabla} \cdot \boldsymbol{\tau} + \boldsymbol{K} \tag{1.85}$$

$$\frac{\partial u_i}{\partial t} + u_j \frac{\partial u_i}{\partial x_j} = -\frac{1}{\rho}\frac{\partial p}{\partial x_i} + \frac{\mu}{\rho}\frac{\partial}{\partial x_j}\left(\frac{\partial u_i}{\partial x_j} + \frac{\partial u_j}{\partial x_i}\right) + K_i \tag{1.86}$$

If incompressible fluid is assumed, Eq. (1.69), the second term of the right-hand side can be reduced to

$$\frac{\mu}{\rho}\frac{\partial}{\partial x_j}\left(\frac{\partial u_i}{\partial x_j} + \frac{\partial u_j}{\partial x_i}\right) = \frac{\mu}{\rho}\left(\frac{\partial^2 u_i}{\partial x_j \partial x_j} + \frac{\partial}{\partial x_i}\frac{\partial u_j}{\partial x_j}\right) = \frac{\mu}{\rho}\frac{\partial^2 u_i}{\partial x_j \partial x_j}$$

The Navier-Stokes equation for incompressible fluid is thus given by

$$\frac{\partial \boldsymbol{u}}{\partial t} + (\boldsymbol{u} \cdot \boldsymbol{\nabla})\boldsymbol{u} = -\frac{1}{\rho}\boldsymbol{\nabla} p + \nu \boldsymbol{\nabla}^2 \boldsymbol{u} + \boldsymbol{K} \tag{1.87}$$

$$\frac{\partial u_i}{\partial t} + u_j \frac{\partial u_i}{\partial x_j} = -\frac{1}{\rho}\frac{\partial p}{\partial x_i} + \nu \frac{\partial^2 u_i}{\partial x_j \partial x_j} + K_i \tag{1.88}$$

where the kinematic viscosity $\nu = \mu/\rho$. In the Cartesian coordinates, with $\boldsymbol{K} = (0, 0, -g)$, the vector components of Eq. (1.87) are given as

$$\frac{\partial u}{\partial t} + u\frac{\partial u}{\partial x} + v\frac{\partial u}{\partial y} + w\frac{\partial u}{\partial z} = -\frac{1}{\rho}\frac{\partial p}{\partial x} + \nu\left(\frac{\partial^2 u}{\partial x^2} + \frac{\partial^2 u}{\partial y^2} + \frac{\partial^2 u}{\partial z^2}\right) \tag{1.89}$$

$$\frac{\partial v}{\partial t} + u\frac{\partial v}{\partial x} + v\frac{\partial v}{\partial y} + w\frac{\partial v}{\partial z} = -\frac{1}{\rho}\frac{\partial p}{\partial y} + \nu\left(\frac{\partial^2 v}{\partial x^2} + \frac{\partial^2 v}{\partial y^2} + \frac{\partial^2 v}{\partial z^2}\right) \tag{1.90}$$

$$\frac{\partial w}{\partial t} + u\frac{\partial w}{\partial x} + v\frac{\partial w}{\partial y} + w\frac{\partial w}{\partial z} = -\frac{1}{\rho}\frac{\partial p}{\partial z} + \nu\left(\frac{\partial^2 w}{\partial x^2} + \frac{\partial^2 w}{\partial y^2} + \frac{\partial^2 w}{\partial z^2}\right) - g \tag{1.91}$$

The Navier-Stokes equation in cylindrical coordinates (r, θ, z) is expressed by

$$\begin{aligned}
&\frac{\partial u_r}{\partial t} + u_r\frac{\partial u_r}{\partial r} + \frac{u_\theta}{r}\frac{\partial u_r}{\partial \theta} + u_z\frac{\partial u_r}{\partial z} - \frac{u_\theta^2}{r} \\
&\quad = -\frac{1}{\rho}\frac{\partial p}{\partial r} + \nu\left(\boldsymbol{\nabla}^2 u_r - \frac{2}{r^2}\frac{\partial u_\theta}{\partial \theta} - \frac{u_r}{r^2}\right) + K_r
\end{aligned} \tag{1.92}$$

$$\begin{aligned}
&\frac{\partial u_\theta}{\partial t} + u_r\frac{\partial u_\theta}{\partial r} + \frac{u_\theta}{r}\frac{\partial u_\theta}{\partial \theta} + u_z\frac{\partial u_\theta}{\partial z} + \frac{u_r u_\theta}{r} \\
&\quad = -\frac{1}{\rho r}\frac{\partial p}{\partial \theta} + \nu\left(\boldsymbol{\nabla}^2 u_\theta + \frac{2}{r^2}\frac{\partial u_r}{\partial \theta} - \frac{u_\theta}{r^2}\right) + K_\theta
\end{aligned} \tag{1.93}$$

$$\begin{aligned}
&\frac{\partial u_z}{\partial t} + u_r\frac{\partial u_z}{\partial r} + \frac{u_\theta}{r}\frac{\partial u_z}{\partial \theta} + u_z\frac{\partial u_z}{\partial z} \\
&\quad = -\frac{1}{\rho}\frac{\partial p}{\partial z} + \nu\boldsymbol{\nabla}^2 u_z + K_z
\end{aligned} \tag{1.94}$$

where

$$\nabla^2 = \frac{1}{r}\frac{\partial}{\partial r}\left(r\frac{\partial}{\partial r}\right) + \frac{1}{r^2}\frac{\partial^2}{\partial \theta^2} + \frac{\partial^2}{\partial z^2} \tag{1.95}$$

The advection term of Eq. (1.88) can be transformed with the continuity equation, Eq. (1.69), as

$$u_j\frac{\partial u_i}{\partial x_j} = \frac{\partial u_i u_j}{\partial x_j} - u_i\frac{\partial u_j}{\partial x_j} = \frac{\partial u_i u_j}{\partial x_j}$$

Therefore, Eqs. (1.87) and (1.88) may be rewritten as

$$\frac{\partial \boldsymbol{u}}{\partial t} + \nabla\cdot(\boldsymbol{u}\otimes\boldsymbol{u}) = -\frac{1}{\rho}\nabla p + \nu\nabla^2\boldsymbol{u} + \boldsymbol{K} \tag{1.96}$$

$$\frac{\partial u_i}{\partial t} + \frac{\partial u_i u_j}{\partial x_j} = -\frac{1}{\rho}\frac{\partial p}{\partial x_i} + \nu\frac{\partial^2 u_i}{\partial x_j \partial x_j} - K_i \tag{1.97}$$

This type, conservative form, of the Navier-Stokes equation is convenient for dealing with some problems associated with turbulence. The components of Eq. (1.97) in the Cartesian coordinates are given as

$$\frac{\partial u}{\partial t} + \frac{\partial u^2}{\partial x} + \frac{\partial uv}{\partial y} + \frac{\partial uw}{\partial z} = -\frac{1}{\rho}\frac{\partial p}{\partial x} + \nu\left(\frac{\partial^2 u}{\partial x^2} + \frac{\partial^2 u}{\partial y^2} + \frac{\partial^2 u}{\partial z^2}\right) \tag{1.98}$$

$$\frac{\partial v}{\partial t} + \frac{\partial uv}{\partial x} + \frac{\partial v^2}{\partial y} + \frac{\partial vw}{\partial z} = -\frac{1}{\rho}\frac{\partial p}{\partial y} + \nu\left(\frac{\partial^2 v}{\partial x^2} + \frac{\partial^2 v}{\partial y^2} + \frac{\partial^2 v}{\partial z^2}\right) \tag{1.99}$$

$$\frac{\partial w}{\partial t} + \frac{\partial uw}{\partial x} + \frac{\partial vw}{\partial y} + \frac{\partial w^2}{\partial z} = -\frac{1}{\rho}\frac{\partial p}{\partial z} + \nu\left(\frac{\partial^2 w}{\partial x^2} + \frac{\partial^2 w}{\partial y^2} + \frac{\partial^2 w}{\partial z^2}\right) - g \tag{1.100}$$

1.3.1 Kinematic energy

The kinematic energy of fluid flow is defined by $q = (u^2 + v^2 + w^2)/2 = \boldsymbol{u}\cdot\boldsymbol{u}/2 = u_i u_i/2$. Consider a dot product of Eq. (1.86) and fluid velocity u_i:

$$u_i\frac{\partial u_i}{\partial t} + u_i u_j\frac{\partial u_i}{\partial x_j} = -\frac{1}{\rho}u_i\frac{\partial p}{\partial x_i} + \nu u_i\frac{\partial}{\partial x_j}\left(\frac{\partial u_i}{\partial x_j} + \frac{\partial u_j}{\partial x_i}\right) + u_i K_i \tag{1.101}$$

which may be transformed as

$$\begin{aligned}\frac{1}{2}\frac{\partial u_i u_i}{\partial t} + \frac{1}{2}u_j\frac{\partial u_i u_i}{\partial x_j} &= -\frac{1}{\rho}\frac{\partial p u_i}{\partial x_i}\\ &+\nu\frac{\partial}{\partial x_j}\left\{u_i\left(\frac{\partial u_i}{\partial x_j} + \frac{\partial u_j}{\partial x_i}\right)\right\} - \nu\frac{\partial u_i}{\partial x_j}\left(\frac{\partial u_i}{\partial x_j} + \frac{\partial u_j}{\partial x_i}\right) + u_i K_i\end{aligned} \tag{1.102}$$

Since the velocity gradient tensor, $\partial u_i/\partial x_j$, can be decomposed into symmetric and antisymmetric ones, as noted in Eqs. (1.49) and (1.50)

$$\frac{\partial u_i}{\partial x_j} = \frac{1}{2}\left(\frac{\partial u_i}{\partial x_j} + \frac{\partial u_j}{\partial x_i}\right) + \frac{1}{2}\left(\frac{\partial u_i}{\partial x_j} - \frac{\partial u_j}{\partial x_i}\right) \tag{1.103}$$

The third term on the right-hand side of Eq. (1.102) is thus transformed as

$$\nu \frac{\partial u_i}{\partial x_j}\left(\frac{\partial u_i}{\partial x_j}+\frac{\partial u_j}{\partial x_i}\right)=\frac{\nu}{2}\left\{\left(\frac{\partial u_i}{\partial x_j}+\frac{\partial u_j}{\partial x_i}\right)+\left(\frac{\partial u_i}{\partial x_j}-\frac{\partial u_j}{\partial x_i}\right)\right\}\left(\frac{\partial u_i}{\partial x_j}+\frac{\partial u_j}{\partial x_i}\right)$$
$$=\frac{\nu}{2}\left(\frac{\partial u_i}{\partial x_j}+\frac{\partial u_j}{\partial x_i}\right)^2+\frac{\nu}{2}\left(\frac{\partial u_i}{\partial x_j}\frac{\partial u_i}{\partial x_j}-\frac{\partial u_j}{\partial x_i}\frac{\partial u_j}{\partial x_i}\right)=\frac{\nu}{2}\left(\frac{\partial u_i}{\partial x_j}+\frac{\partial u_j}{\partial x_i}\right)^2 \tag{1.104}$$

The substitution into Eq. (1.102) gives the kinematic energy transport equation:

$$\frac{\partial q}{\partial t}+u_j\frac{\partial q}{\partial x_j}=-\frac{1}{\rho}\frac{\partial p u_i}{\partial x_i}+\nu\frac{\partial}{\partial x_j}\left\{u_i\left(\frac{\partial u_i}{\partial x_j}+\frac{\partial u_j}{\partial x_i}\right)\right\}-\frac{\nu}{2}\left(\frac{\partial u_i}{\partial x_j}+\frac{\partial u_j}{\partial x_i}\right)^2+u_iK_i \tag{1.105}$$

The left-hand side indicates the advection of kinematic energy $q(=u_iu_i/2)$. The first term on the right-hand side, pressure gradient, contributes to production of q, and the second term on the right-hand side indicates diffusion. The third term on the right-hand side, square of the strain, is always negative and contributes to reduce q, that is, this term describes the energy dissipation by viscosity:

$$\epsilon=\frac{\nu}{2}\left(\frac{\partial u_i}{\partial x_j}+\frac{\partial u_j}{\partial x_i}\right)^2 \tag{1.106}$$

1.3.2 Dimensionless numbers

Introducing the characteristic velocity U, length L, and time T of flow, the first term on the left-hand side of the Navier-Stokes equation, Eq. (1.87), the unsteady term, has the dimension of U/T. The dimension of the advection terms (the second term of the left-hand side of Eq. (1.87)) is expressed as U^2/L. The viscous and gravity terms (the second and third terms of the right-hand side of Eq. (1.87)) have $\nu U/L^2$ and g, respectively. The relative dominance of forces acting on fluid measures mechanical properties of fluid flows. A ratio of the unsteady term to the advection term is $(U/T)\,/\,(U^2/L)=L/TU$, which is called Strouhal number:

$$S_t=\frac{L}{TU} \tag{1.107}$$

S_t has been used for characterizing unsteadiness of flow, such as frequency of periodic ejections of vortices separated from boundary layers. A ratio of the advection to the viscous term, $(U^2/L)\,/\,(\nu U/L^2)=UL/\nu$, called Reynolds number:

$$Re=\frac{UL}{\nu} \tag{1.108}$$

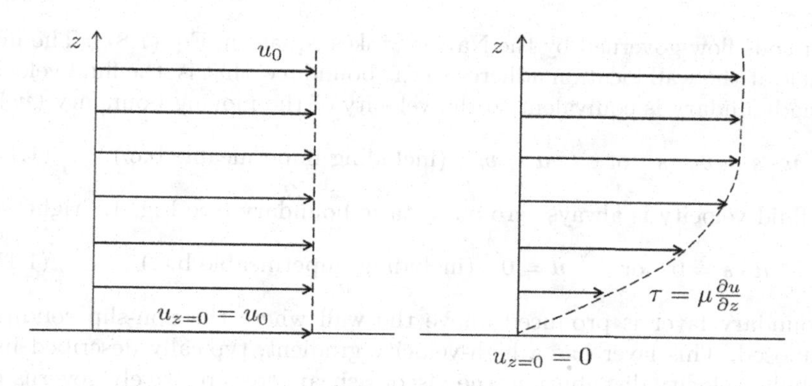

FIGURE 1.7
Flows above a slip boundary (left) and non-slip boundary (right).

Re measures the relative viscous effect in the flow and used for characterizing transition of flows, such as laminar to turbulence, fluid force acting on an object, friction on a wall. A ratio of the advection to the gravity term has $(U^2/L)/g = U^2/gL = F_r^2$.

$$F_r = \frac{U}{\sqrt{gL}} \tag{1.109}$$

Froude number, F_r, describes the relative fluid velocity with respect to long-wave speed. It is also interpreted as a rate of kinematic energy and potential energy.

1.3.3 Wall boundary conditions

When fluid flow near a solid wall is considered, the motion of equation is solved to satisfy boundary conditions on the wall. One of the wall conditions, so-called impermeable condition, imposes zero normal velocity to the wall:

$$\boldsymbol{u} \cdot \boldsymbol{n} = 0 \tag{1.110}$$

where $\boldsymbol{n}$ is the unit normal vector.

There are two kinds of the boundary conditions defining tangential velocities on a wall. 'Slip boundary condition' is used for ideal (inviscid) fluid on a solid boundary, which gives adjacent fluid velocity at the wall, that is, the ideal fluid slips on the boundary (see Fig. 1.7 left). This condition is used for flows governed by the Euler equation, Eq. (1.78), or potential flows, introduced in Section 1.6.

$$\frac{\partial}{\partial n}(\boldsymbol{u} \cdot \boldsymbol{s}) = 0 \tag{1.111}$$

where $\boldsymbol{s}$ is the unit tangential vector. 'Non-slip boundary condition' is used for viscous flow governed by the Navier-Stokes equation, Eq. (1.87). The fluid particle at the wall location adheres to the boundary; that is, the fluid velocity on the boundary is equivalent to the velocity of the moving boundary ($\boldsymbol{v}_b$).

$$\boldsymbol{u} \cdot \boldsymbol{s} = \boldsymbol{v}_b \cdot \boldsymbol{s} \quad \text{or} \quad \boldsymbol{u} = \boldsymbol{v}_b \quad \text{(including impermeable b.c.)} \tag{1.112}$$

The fluid velocity is always zero on a static boundary (see Fig. 1.7 right).

$$\boldsymbol{u} \cdot \boldsymbol{s} = 0 \quad \text{or} \quad \boldsymbol{u} = 0 \quad \text{(including impermeable b.c.)} \tag{1.113}$$

A boundary layer is produced above the wall where the non-slip condition is imposed. This layer has a high-velocity gradient, typically described by a parabolic velocity distribution; the viscous shear stress relatively governs the flow and balance with pressure gradient as inertial force is weakened owing to slower velocity near the wall.

1.3.4 Boundary layer flows

1.3.4.1 Flow between two horizontal plates (Couette flow)

Consider steady viscous flow between two horizontal plates of infinite length with spacing d under constant horizontal pressure gradient $\alpha = \partial p/\partial x$ in two-dimensional space (x, y). The upper plate horizontally moves with velocity of u_0, while the bottom one remains at rest.

In this case, as the flow is horizontally uniform, any horizontal derivatives can be ignored; $\partial/\partial x \to 0$ and $\partial^2/\partial x^2 \to 0$. Therefore, the continuity equation $\cancel{\partial u/\partial x} + \partial v/\partial y = 0$. Integrating this with respect to y, $v =$ constant. As v must be zero at the impermeable walls, Eq. (1.110), the integral constant must be zero, that is, $v = 0$ between the walls. Because of steady flow, time derivative can be also neglected; $\partial/\partial t \to 0$. The two-dimensional Navier-Stokes equation is thus reduced as

$$\frac{\cancel{\partial u}}{\cancel{\partial t}} + u\frac{\cancel{\partial u}}{\cancel{\partial x}} + \cancel{v}\frac{\partial u}{\partial y} = -\frac{1}{\rho}\frac{\overset{\alpha}{\cancel{\partial p}}}{\cancel{\partial x}} + \nu\left(\frac{\cancel{\partial^2 u}}{\cancel{\partial x^2}} + \frac{\partial^2 u}{\partial y^2}\right)$$

$$\therefore \frac{d^2 u}{dy^2} = \frac{\alpha}{\rho\nu} \tag{1.114}$$

Integrating twice, u takes the parabolic form

$$u = \frac{\alpha}{2\rho\nu}y^2 + C_1 y + C_2 \tag{1.115}$$

The non-slip condition Eq. (1.113) provides the boundary conditions $u = 0$ at $y = 0$ and $u = u_0$ at $y = d$, which determines C_1 and C_2, so that the velocity within the plates is derived as

$$u = \frac{\alpha}{2\rho\nu}\left(y^2 - dy\right) + \frac{u_0}{d}y \tag{1.116}$$

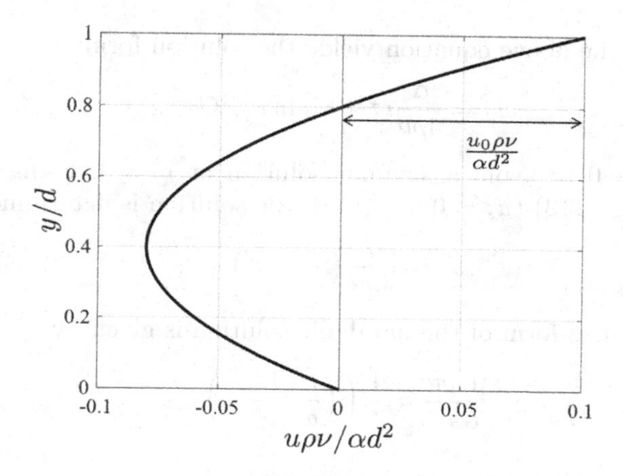

FIGURE 1.8
Dimensionless velocity profile between the horizontal plates.

Fig. 1.8 shows the distribution of the following dimensionless velocity between the plates.

$$\frac{u\rho\nu}{\alpha d^2} = \frac{1}{2}\left(\left(\frac{y}{d}\right)^2 - \left(\frac{y}{d}\right)\right) + \frac{u_0\rho\nu}{\alpha d^2}\left(\frac{y}{d}\right) \tag{1.117}$$

1.3.4.2 Hagen-Poiseuille flow

Steady flow in a horizontally located circular pipe with radius a under constant horizontal pressure gradient α is considered in a similar manner. In the cylindrical coordinate (r, θ, z), z is taken as the horizontal axis. Thus, the horizontal derivatives $(\partial/\partial z)$ can be ignored as the flow is horizontally uniform. Since the flow is axisymmetric, the azimuthal derivative does not need to consider $\partial/\partial\theta \to 0$. The continuity equation Eq. (1.70) is reduced as

$$\frac{1}{r}\frac{\partial(ru_r)}{\partial r} + \cancel{\frac{1}{r}\frac{\partial u_\theta}{\partial\theta}} + \cancel{\frac{\partial u_z}{\partial z}} = 0$$

We find $u_r = 0$ anywhere to satisfy the impermeable condition at the pipe wall, Eq. (1.110). The Navier-Stokes equation in the cylindrical coordinates, Eq. (1.94), is also reduced as

$$\cancel{\frac{\partial u_z}{\partial t}} + \cancel{u_r}\frac{\partial u_z}{\partial r} + \frac{u_\theta}{r}\cancel{\frac{\partial u_z}{\partial\theta}} + u_z\cancel{\frac{\partial u_z}{\partial z}}$$

$$= -\frac{1}{\rho}\underbrace{\frac{\partial p}{\partial z}}_{\alpha} + \nu\left(\frac{1}{r}\frac{\partial}{\partial r}\left(r\frac{\partial u_z}{\partial r}\right) + \cancel{\frac{1}{r^2}\frac{\partial^2 u_z}{\partial\theta^2}} + \cancel{\frac{\partial^2 u_z}{\partial z^2}}\right)$$

$$\therefore \frac{d^2u_z}{dr^2} + \frac{1}{r}\frac{du_z}{dr} - \frac{\alpha}{\rho\nu} = 0 \tag{1.118}$$

Integration of the above equation yields the solution form:

$$u_z = \frac{\alpha}{4\rho\nu} r^2 + C_1 \ln r + C_2 \tag{1.119}$$

We find $C_1 = 0$ to avoid a singular solution at $r = 0$. Using the non-slip condition Eq. (1.113) ($u_z = 0$ at $r = a$), the solution is determined as

$$u_z = \frac{\alpha}{4\rho\nu}\left(r^2 - a^2\right) \tag{1.120}$$

The dimensionless form of the parabolic solution is given by

$$\frac{u_z \rho\nu}{\alpha a^2} = \frac{1}{4}\left(\left(\frac{r}{a}\right)^2 - 1\right) \tag{1.121}$$

1.3.4.3 Flow above an oscillating plate

Consider viscous flow above a horizontal plate horizontally oscillating with velocity amplitude of u_0 and angular frequency σ in a x-y plane. Here fluid is at rest where far from the plate ($u \to 0$ at $y \to \infty$). This is also parallel flow to the horizontal plate and horizontally uniform, and thus there is only u component of velocity ($v = 0$) in the domain. Similarly, the Navier-Stokes equation is reduced to

$$\frac{\partial u}{\partial t} + u\cancel{\frac{\partial u}{\partial x}} + \cancel{v}\frac{\partial u}{\partial y} = -\frac{1}{\rho}\frac{\partial p}{\partial x} + \nu\left(\cancel{\frac{\partial^2 u}{\partial x^2}} + \frac{\partial^2 u}{\partial y^2}\right)$$

The pressure gradient $\partial p/\partial x$ needs to be zero to satisfy $u = 0$ at $y = \infty$ (i.e., $\partial u/\partial t = 0$ and $\partial^2 u/\partial y^2 = 0$ at $y = \infty$, and thus $\partial p/\partial x = 0$). Accordingly, the equation is reduced to

$$\frac{\partial u}{\partial t} = \nu\frac{\partial^2 u}{\partial y^2} \tag{1.122}$$

The velocity on the oscillating plate $u\mid_{y=0} = u_0 \mathrm{e}^{i\sigma t}$. As the velocity decays in the y-direction and vanish at $y \to \infty$, you may assume the solution takes the form

$$u = u_0 \mathrm{e}^{i\sigma t}\mathrm{e}^{-Cy} \tag{1.123}$$

which satisfies the velocity at $y = 0$. Substituting into Eq. (1.122), $i\sigma u_0 \mathrm{e}^{i\sigma t}\mathrm{e}^{-Cy} = \nu u_0 C^2 \mathrm{e}^{i\sigma t}\mathrm{e}^{-Cy}$, and thus the constant $C = \pm\sqrt{i\sigma/\nu}$. Eq. (1.123) is then expressed by

$$u = u_0 \mathrm{e}^{i\sigma t}\mathrm{e}^{\pm\sqrt{\frac{i\sigma}{\nu}}y} = u_0 \mathrm{e}^{i\sigma t}\mathrm{e}^{\pm\left(i\sqrt{\frac{\sigma}{2\nu}}y + \sqrt{\frac{\sigma}{2\nu}}y\right)} \tag{1.124}$$

Considering the far-field condition $u = 0$ at $y = \infty$, the solution must take the form:

$$\frac{u}{u_0} = \mathrm{e}^{i\left(\sigma t - \sqrt{\frac{\sigma}{2\nu}}y\right)}\mathrm{e}^{-\sqrt{\frac{\sigma}{2\nu}}y} \tag{1.125}$$

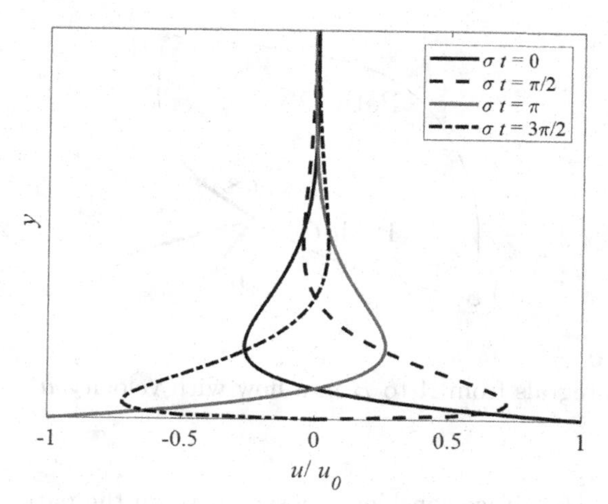

FIGURE 1.9
Dimensionless velocity profile above the oscillating horizontal plate.

The trigonometric form of the velocity (see Section 1.1.3) is then given by

$$\frac{u}{u_0} = \cos\left(\sigma t - \sqrt{\frac{\sigma}{2\nu}}y\right) e^{-\sqrt{\frac{\sigma}{2\nu}}y} \tag{1.126}$$

Eq. (1.126) describes that the velocity propagates in the y-direction with exponential decay with a length scale of $\delta = \sqrt{2\nu/\sigma}$ owing to viscosity (see Fig. 1.9). δ is interpreted as a thickness of a boundary layer over an oscillating plate.

1.4 Velocity potential

Consider a flow with velocity $\boldsymbol{u}(u, v)$ between locations A and B in the (x, y) plane (Fig. 1.10). A line integral of a tangential component of the flow on the path line C from A to B is written by

$$\phi_C(AB) = \int_A^B \boldsymbol{u} \cdot d\boldsymbol{s} = \int_A^B (udx + vdy) \tag{1.127}$$

$$= \int_A^B u_s ds \tag{1.128}$$

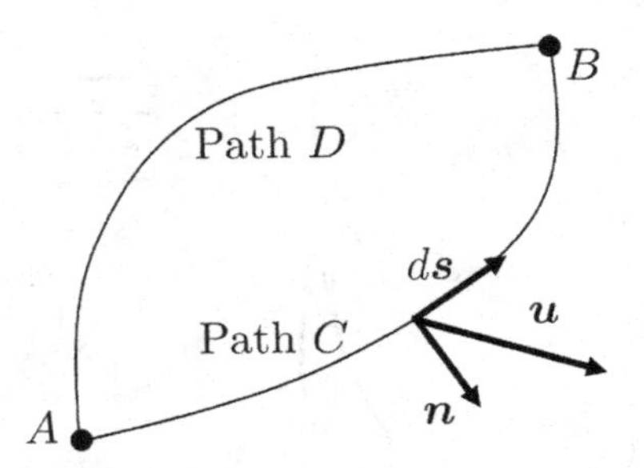

FIGURE 1.10
Paths of line integrals from A to B in a flow with velocity $\boldsymbol{u}$.

where $d\boldsymbol{s} = (dx, dy)$. If we consider a closed curve on the paths C and D, the circulation Γ is defined by

$$\Gamma_{CD} = \oint_{CD} \boldsymbol{u} \cdot d\boldsymbol{s} = \int_{A(\text{on } C)}^{B(\text{on } C)} u_s ds + \int_{B(\text{on } D)}^{A(\text{on } D)} u_s ds \qquad (1.129)$$

Considering irrotational flow, $\Gamma = 0$, the substitution of Eq. (1.128) into Eq. (1.129) gives $\Gamma_{CD} = \phi_C(AB) + \phi_D(BA) = 0$; therefore $\phi_C(AB) = \phi_D(AB)$. Accordingly, ϕ is independent of the path lines between A and B. $\phi(AB)$ can be also expressed in terms of the exact differential $d\phi = (\partial\phi/\partial x)\, dx + (\partial\phi/\partial y)\, dy$ by

$$\phi(AB) = \int_A^B d\phi = \int_A^B \left(\frac{\partial \phi}{\partial x} dx + \frac{\partial \phi}{\partial y} dy \right) \qquad (1.130)$$

$$= \int_A^B \boldsymbol{\nabla}\phi \cdot d\boldsymbol{s} \qquad (1.131)$$

Comparing Eqs. (1.127) and (1.130), we find

$$u = \frac{\partial \phi}{\partial x}, \quad v = \frac{\partial \phi}{\partial y} \qquad (1.132)$$

$$\text{or} \quad \boldsymbol{u} = \boldsymbol{\nabla}\phi \qquad (1.133)$$

The flow having the velocity potential is called potential flow. If these definitions are substituted into vorticity $\omega = \boldsymbol{\nabla} \times \boldsymbol{u}$, the irrotational condition is given

$$\frac{\partial u}{\partial y} - \frac{\partial v}{\partial x} = 0 \qquad (1.134)$$

or

$$\boldsymbol{\nabla} \times \boldsymbol{\nabla}\phi = 0 \qquad (1.135)$$

Accordingly, the potential flow must be irrotational.

1.5 Stream function

Similarly, consider the line integral of a normal component of $\boldsymbol{u}(u, v)$ on the path C from A to B (see Fig. 1.10):

$$\psi_C(AB) = \int_A^B \boldsymbol{u} \cdot \boldsymbol{n} ds = \int_A^B (u dy - v dx) \tag{1.136}$$

where $\boldsymbol{n} = (dy/ds, -dx/ds)$ is the unit normal vector and $ds = |\boldsymbol{ds}| = \sqrt{dx^2 + dy^2}$. Eq. (1.136) indicates discharge of the flow across C. If the control volume closed by C and D is considered, assuming incompressible fluid flow, the discharge coming in across the path D should come out from the path C (when flow from left to right). Accordingly, Eq. (1.136) is independent of integration paths and $\nabla\psi$ is the conserved vector field. Eq. (1.136) is therefore described with the exact differential $d\psi = (\partial\psi/\partial x)\, dx + (\partial\psi/\partial y)\, dy$ as

$$\psi(AB) = \int_A^B d\psi = \int_A^B \left(\frac{\partial\psi}{\partial x} dx + \frac{\partial\psi}{\partial y} dy \right) \tag{1.137}$$

With Eq. (1.136), the velocity is expressed in terms of ψ:

$$u = \frac{\partial\psi}{\partial y}, \quad v = -\frac{\partial\psi}{\partial x} \tag{1.138}$$

We find the continuity equation is satisfied for the flow defined by ψ:

$$\frac{\partial u}{\partial x} + \frac{\partial v}{\partial y} = \frac{\partial}{\partial x}\left(\frac{\partial\psi}{\partial y}\right) + \frac{\partial}{\partial y}\left(-\frac{\partial\psi}{\partial x}\right) = 0 \tag{1.139}$$

The contours of ϕ and ψ are mutually orthogonal, and they satisfy Cauchy-Riemann equations

$$\frac{\partial\phi}{\partial x} = \frac{\partial\psi}{\partial y} \quad \text{and} \quad \frac{\partial\phi}{\partial y} = -\frac{\partial\psi}{\partial x} \tag{1.140}$$

A line drawn in the direction of velocity vector of the flow at a certain instance is defined as a streamline. Contours of ψ (lines with constant ψ) describe the streamlines of the flow.

In a polar coordinate (r, θ), the radial velocity u_r and azimuthal velocity u_θ are expressed by

$$u_r = \frac{\partial\phi}{\partial r} = \frac{1}{r}\frac{\partial\psi}{\partial\theta} \quad \text{and} \quad u_\theta = \frac{1}{r}\frac{\partial\phi}{\partial\theta} = -\frac{\partial\psi}{\partial r} \tag{1.141}$$

1.6 Potential flows

If Eq. (1.133), $\boldsymbol{u} = \nabla\phi$, is substituted into the continuity equation of incompressible fluid, Eq. (1.69), $\nabla \cdot \boldsymbol{u} = 0$, the Laplace equation for velocity potential is given

$$\nabla^2\phi = 0 \tag{1.142}$$

Any flows with velocity potential, i.e. potential flow, are governed by the Laplace equation and solved as a boundary value problem.

As the Laplace equation is a linear differential equation, the solutions can be superposed; if ϕ_1 and ϕ_2 are the solutions, $\phi_1 + \phi_2$ is also another solution. Simple examples of steady potential flows are introduced in this section, while various wave solutions of the Laplace equation are introduced in Chapter 4.

1.6.1 Uniform flow

In the Cartesian coordinates, consider a flow expressed by

$$\phi = ax + by + cz \tag{1.143}$$

where a, b, and c are constants. As the velocity of the flow is $\boldsymbol{u} = \nabla\phi = (a, b, c)$, Eq. (1.143) describes the spatially uniform flow with constant velocity anywhere.

1.6.2 Source and sink

Consider a spherically symmetric flow with respect to the origin of the spherical coordinate (r,θ,φ) (see Section 1.1.1.3). This flow is independent of θ and φ (because of symmetry), the Laplace equation in the spherical coordinate, Eq. (1.15), is reduced to

$$\frac{1}{r^2}\frac{\partial}{\partial r}\left(r^2\frac{\partial\phi}{\partial r}\right) = 0 \tag{1.144}$$

The solution takes the form

$$\phi = -\frac{m}{r} + C \tag{1.145}$$

where m and C are the constants. The flow has only a radial component of velocity (the r-direction):

$$u_r = \frac{\partial\phi}{\partial r} = \frac{m}{r^2} \tag{1.146}$$

indicating a source or sink flow depending on a sign of m; if $m > 0$, then $u_r > 0$, indicating divergent flow at the origin (source), and if $m < 0$, convergent flow with negative u_r occurs (sink). Fig. 1.11 shows the circular (spherical) contours of velocity potential and radial streamlines of the flow of Eq. (1.145).

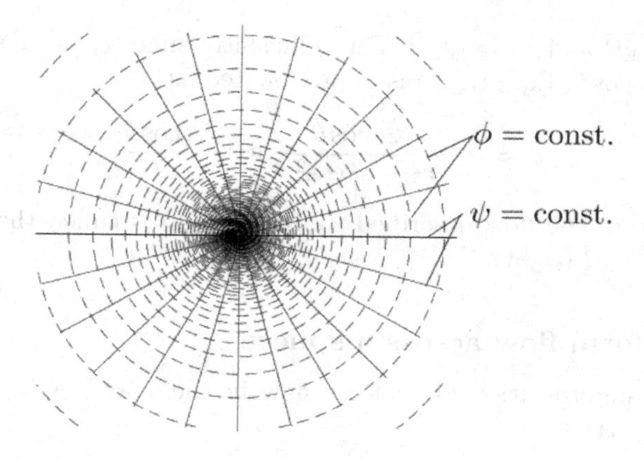

FIGURE 1.11
Contours of ϕ and ψ of source and sink flows.

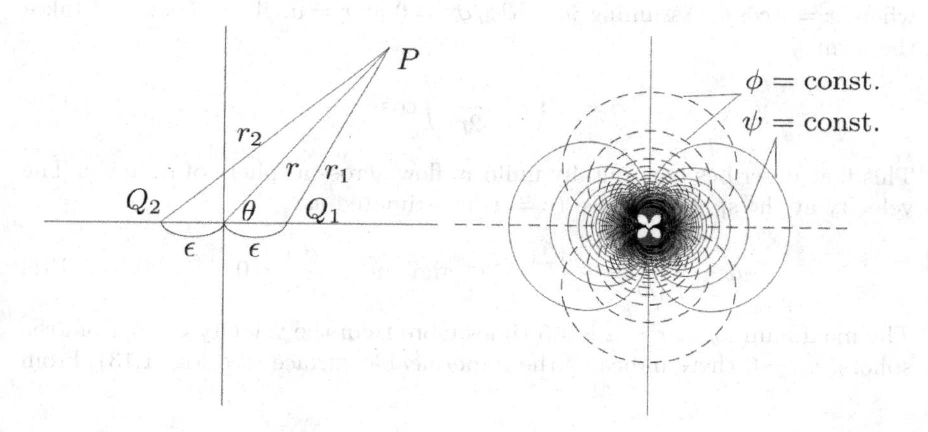

FIGURE 1.12
Locations of source Q_1, sink Q_2 and arbitrary coordinate P (left), and contours of ϕ and ψ of doublet (right).

1.6.3 Doublet

Consider the superposition of source and sink flows with the identical strength m, adjacently placed at Q_1 and Q_2 with horizontal spacing of 2ϵ (see Fig. 1.12 left):

$$\phi = -\frac{m}{r_1} + \frac{m}{r_2} \tag{1.147}$$

where $r_1 = \overline{Q_1P}$ and $r_2 = \overline{Q_2P}$. For sufficiently small ϵ, as $r_1 \approx r - \epsilon\cos\theta$ and $r_2 \approx r + \epsilon\cos\theta$, Eq. (1.147) is approximated as

$$\phi \approx -m\frac{2\epsilon\cos\theta}{r^2 - \epsilon^2\cos^2\theta} \approx -\frac{A\cos\theta}{r^2} \tag{1.148}$$

where $A = 2\epsilon m$. The flow described by Eq. (1.148) is called the doublet, as shown in Fig. 1.12 (right).

1.6.4 Uniform flow across a sphere

Consider the superposition of uniform flow in the x-axis $\phi_1 = Ux$ and the doublet $\phi_2 = -A\cos\theta/r^2$:

$$\phi = \phi_1 + \phi_2 = Ux - \frac{A\cos\theta}{r^2} = \left(Ur - \frac{A}{r^2}\right)\cos\theta \tag{1.149}$$

where $x = r\cos\theta$. Assuming $u_r = \partial\phi/\partial r = 0$ at $r = a$, $A = -Ua^3/2$, ϕ takes the form:

$$\phi = U\left(r + \frac{a^3}{2r^2}\right)\cos\theta \tag{1.150}$$

This flow describes horizontally uniform flow across a sphere of radius a. The velocity at the sphere surface ($r = a$) is estimated as

$$u_\theta = \frac{1}{r}\frac{\partial\phi}{\partial\theta} = -\frac{3}{2}U\sin\theta \quad \text{and} \quad u_r = \frac{\partial\phi}{\partial r} = 0 \tag{1.151}$$

The maximum u_θ at $r = a$ is 1.5 times more than the velocity U far from the sphere. $u_r = 0$ there indicates the impermeable surface (see Fig. 1.13). From

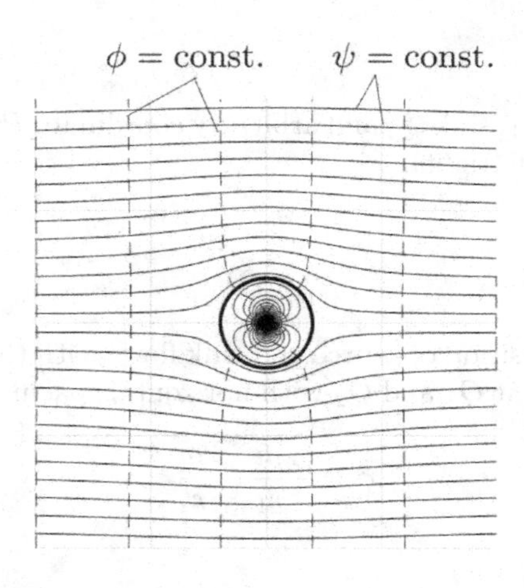

FIGURE 1.13
Contours of ϕ and ψ of the uniform flow across a sphere.

Eq. (1.141), the stream function of this flow is given by

$$\psi = -U\left(r - \frac{a^3}{r^2}\right)\sin\theta \tag{1.152}$$

1.7 Bernoulli equation

The advection terms of the Euler equation, Eqs. (1.79)–(1.81), are transformed as

$$\begin{aligned} u\frac{\partial u}{\partial x} + v\frac{\partial u}{\partial y} + w\frac{\partial u}{\partial z} &= \frac{1}{2}\left(\frac{\partial u^2}{\partial x} + \frac{\partial v^2}{\partial x} + \frac{\partial w^2}{\partial x}\right) - v\left(\frac{\partial v}{\partial x} - \frac{\partial u}{\partial y}\right) + w\left(\frac{\partial u}{\partial z} - \frac{\partial w}{\partial x}\right) \\ &= \frac{1}{2}\left(\frac{\partial u^2}{\partial x} + \frac{\partial v^2}{\partial x} + \frac{\partial w^2}{\partial x}\right) - v\omega_z + w\omega_y \end{aligned} \tag{1.153}$$

$$\begin{aligned} u\frac{\partial v}{\partial x} + v\frac{\partial v}{\partial y} + w\frac{\partial v}{\partial z} &= \frac{1}{2}\left(\frac{\partial u^2}{\partial y} + \frac{\partial v^2}{\partial y} + \frac{\partial w^2}{\partial y}\right) - w\left(\frac{\partial w}{\partial y} - \frac{\partial v}{\partial z}\right) + u\left(\frac{\partial v}{\partial x} - \frac{\partial u}{\partial y}\right) \\ &= \frac{1}{2}\left(\frac{\partial u^2}{\partial y} + \frac{\partial v^2}{\partial y} + \frac{\partial w^2}{\partial y}\right) - w\omega_x + u\omega_z \end{aligned} \tag{1.154}$$

$$\begin{aligned} u\frac{\partial w}{\partial x} + v\frac{\partial w}{\partial y} + w\frac{\partial w}{\partial z} &= \frac{1}{2}\left(\frac{\partial u^2}{\partial z} + \frac{\partial v^2}{\partial z} + \frac{\partial w^2}{\partial z}\right) - u\left(\frac{\partial u}{\partial z} - \frac{\partial w}{\partial x}\right) + v\left(\frac{\partial w}{\partial y} - \frac{\partial v}{\partial z}\right) \\ &= \frac{1}{2}\left(\frac{\partial u^2}{\partial z} + \frac{\partial v^2}{\partial z} + \frac{\partial w^2}{\partial z}\right) - u\omega_y + v\omega_x \end{aligned} \tag{1.155}$$

where vorticity $\boldsymbol{\omega} = (\omega_x, \omega_y, \omega_z) = (\partial w/\partial y - \partial v/\partial z, \partial u/\partial z - \partial w/\partial x, \partial v/\partial x - \partial u/\partial y)$. The substitutions into Eqs. (1.79)–(1.81) give

$$\frac{\partial}{\partial x}\left(\frac{1}{2}(u^2 + v^2 + w^2) + \frac{p}{\rho}\right) = v\omega_z - w\omega_y - \frac{\partial u}{\partial t} \tag{1.156}$$

$$\frac{\partial}{\partial y}\left(\frac{1}{2}(u^2 + v^2 + w^2) + \frac{p}{\rho}\right) = w\omega_x - u\omega_z - \frac{\partial v}{\partial t} \tag{1.157}$$

$$\frac{\partial}{\partial z}\left(\frac{1}{2}(u^2 + v^2 + w^2) + \frac{p}{\rho} + gz\right) = u\omega_y - v\omega_x - \frac{\partial w}{\partial t} \tag{1.158}$$

Since $\boldsymbol{u} \times \boldsymbol{\omega} = (v\omega_z - w\omega_y, w\omega_x - u\omega_z, u\omega_y - v\omega_x)$, the vector representation of these equations is given

$$\boldsymbol{\nabla}\left(\frac{1}{2}|\boldsymbol{u}|^2 + \frac{p}{\rho} + gz\right) = \boldsymbol{u} \times \boldsymbol{\omega} - \frac{\partial \boldsymbol{u}}{\partial t} \tag{1.159}$$

This is called the Bernoulli equation. Eq. (1.159) may be simplified under assumptions of flows for purposes of specific applications.

1.7.1 Bernoulli equation on a streamline

Consider the Bernoulli equation along a streamline. Since $\boldsymbol{u} \times \boldsymbol{\omega}$ is orthogonal to both fluid velocity $\boldsymbol{u}$ and vorticity $\boldsymbol{\omega}$ (i.e., $\boldsymbol{u} \times \boldsymbol{\omega} \perp \boldsymbol{u}$ and $\boldsymbol{u} \times \boldsymbol{\omega} \perp \boldsymbol{\omega}$), the

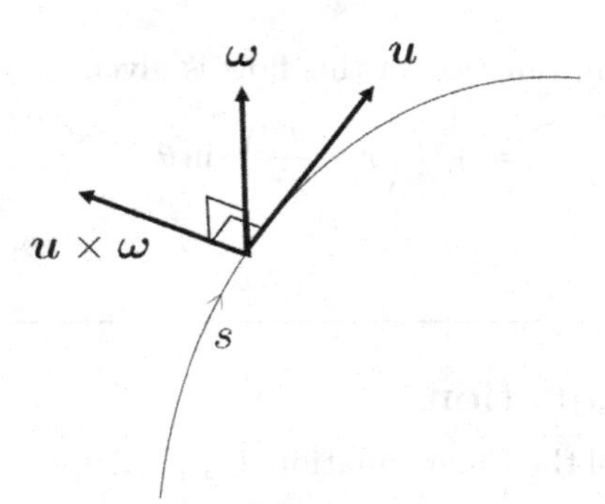

FIGURE 1.14
Relations of fluid velocity $\boldsymbol{u}$, vorticity $\boldsymbol{\omega}$, and $\boldsymbol{u} \times \boldsymbol{\omega}$ on a streamline.

tangential (s) component of $\boldsymbol{u} \times \boldsymbol{\omega}$ is always zero along the streamline (see Fig. 1.14). In this case, Eq. (1.159) is reduced to

$$\frac{\partial u_s}{\partial t} + \frac{\partial}{\partial s}\left(\frac{u_s^2}{2} + \frac{p}{\rho} + gz\right) = 0 \tag{1.160}$$

where u_s is the velocity along the streamline. Assuming steady flow and integrating over the streamline, we have

$$\frac{u_s^2}{2} + \frac{p}{\rho} + gz = C(s) \tag{1.161}$$

As the integral constant $C(s)$ depends on streamlines, this version of the Bernoulli equation cannot be used over different streamlines. Eq. (1.161) has wide applications to hydraulics and river engineering.

1.7.2 Bernoulli equation for unsteady irrotational flow

If a potential flow is assumed, the flow field is irrotational, $\boldsymbol{\omega} = 0$, and the velocity is described in terms of velocity potential, $\boldsymbol{u} = \boldsymbol{\nabla}\phi$ (see Section 1.4). In this case, Eq. (1.159) is rewritten

$$\boldsymbol{\nabla}\left(\frac{\partial \phi}{\partial t} + \frac{1}{2}|\boldsymbol{u}|^2 + \frac{p}{\rho} + gz\right) = 0 \tag{1.162}$$

Integrating the above equation, a generalized Bernoulli equation is obtained:

$$\frac{\partial \phi}{\partial t} + \frac{1}{2}|\boldsymbol{u}|^2 + \frac{p}{\rho} + gz = C(t) \tag{1.163}$$

where $C(t)$ is a function of time. This version of the Bernoulli Equation can be used for unsteady problems, which defines dynamic boundary conditions at free surfaces, introduced in Section 3.2.2.

1.8 Stokes drag

Streamwise fluid force acting on an object is called a drag force. In this section, the viscous flow around a spherical particle, known as Stokes flow, and the fluid force acting on the particle, also called Stokes law, are introduced.

1.8.1 Stokes law

Consider the horizontally uniform, viscous flow of velocity U across a sphere of radius a. The velocity $\boldsymbol{u}^*$ is assumed to be expressed as a superposition of the gradient of velocity potential $\boldsymbol{\nabla}\phi$ and the rotational velocity of viscous fluid $\boldsymbol{u}$:

$$\boldsymbol{u}^* = \boldsymbol{\nabla}\phi + \boldsymbol{u} \tag{1.164}$$

The separation of vector used for Eq. (1.164) is called a Helmholtz decomposition.[1] The boundary conditions of the flow are given

$$u^* = 0,\ v^* = 0,\ w^* = 0,\ p^* = p \quad (r = a) \tag{1.165}$$
$$u^* = U,\ v^* = 0,\ w^* = 0,\ p^* = p_\infty \quad (r \to \infty) \tag{1.166}$$

The viscous component $\boldsymbol{u}$, which should be given by the Navier-Stokes equation Eq. (1.87), contributes to achieve non-slip boundary condition on the sphere surface, Eq. (1.165), and produce a boundary layer near the surface, while the viscous effect becomes negligibly small far from the surface; $\boldsymbol{u} \to 0$ at $r \to \infty$. Eq. (1.164) gives the boundary conditions for $\boldsymbol{u}$:

$$u = -\frac{\partial \phi}{\partial x},\ v = -\frac{\partial \phi}{\partial y},\ w = -\frac{\partial \phi}{\partial z} \quad (r = a) \tag{1.167}$$
$$u = 0,\ v = 0,\ w = 0 \quad (r \to \infty) \tag{1.168}$$

We assume very slow flow around the sphere, called Stokes flow, which allows neglecting small advection terms in the Navier-Stokes equation, called Stokes approximation:

$$\frac{\partial \boldsymbol{u}}{\partial t} = -\frac{1}{\rho}\boldsymbol{\nabla}p + \nu \boldsymbol{\nabla}^2 \boldsymbol{u} \tag{1.169}$$

Taking divergence of Eq. (1.169) and considering incompressible continuity equation, Eq. (1.69), the Laplace equation for pressure is obtained:

$$\boldsymbol{\nabla}^2 p = 0 \tag{1.170}$$

[1] The Helmholtz decomposition states that arbitrary vector $\boldsymbol{f}$ is expressed by sum of the gradient of scalar valuable ϕ and the rotation of vector $\boldsymbol{A}$: $\boldsymbol{f} = \boldsymbol{\nabla}\phi + \boldsymbol{\nabla} \times \boldsymbol{A}$.

Assuming steady flow, Eq. (1.169) is reduced to

$$\nabla^2 \boldsymbol{u} = \frac{1}{\mu}\nabla p \tag{1.171}$$

The solutions for Eqs. (1.170) and (1.171) take the form:

$$p = C_1 \frac{\partial}{\partial x}\left(\frac{1}{r}\right) \tag{1.172}$$

$$\boldsymbol{u} = \frac{C_1}{2\mu}\nabla\left(\frac{\partial r}{\partial x}\right) - \frac{C_1}{\mu}\frac{\boldsymbol{e}_x}{r} \tag{1.173}$$

where C_1 is the constant and $\boldsymbol{e}_x$ is the unit vector in the x-direction. The components of the above equations are given by

$$p = -C_1 x/r^3 \tag{1.174}$$

$$u = \frac{C_1}{2\mu}\frac{\partial}{\partial x}\left(\frac{\partial r}{\partial x}\right) - \frac{C_1}{\mu}\frac{1}{r} = -\frac{C_1}{2\mu}\left(\frac{1}{r} + \frac{x^2}{r^3}\right) \tag{1.175}$$

$$v = \frac{C_1}{2\mu}\frac{\partial}{\partial y}\left(\frac{\partial r}{\partial x}\right) = -\frac{C_1}{2\mu}\frac{xy}{r^3} \tag{1.176}$$

$$w = \frac{C_1}{2\mu}\frac{\partial}{\partial z}\left(\frac{\partial r}{\partial x}\right) = -\frac{C_1}{2\mu}\frac{xz}{r^3} \tag{1.177}$$

The velocity potential in Eq. (1.164) should be expressed by the sum of the solutions of uniform flow and doublet. Considering the solution of Eq. (1.150), we may assume the form:

$$\phi = Ur\cos\theta + C_2\frac{a^3}{2r^2}\cos\theta = Ux + C_2\frac{a^3}{2r^3}x \tag{1.178}$$

where C_2 is constant, $x = r\cos\theta$ and $r^2 = x^2 + y^2 + z^2$. While $C_2 = U$ if $\boldsymbol{u} = 0$ (should agree with the potential flow solution Eq. (1.150)), in this case, C_2 should be determined to satisfy Eq. (1.167) to achieve the non-slip and impermeable boundary condition, Eq. (1.165). It should be noted that Eq. (1.178) indicates the irrotational velocity $\partial\phi/\partial x \to U$ far from the sphere ($r \to \infty$), which fulfills the far-field boundary conditions, Eqs. (1.168) and (1.168).

The superposed velocity $\boldsymbol{u}^*$ is thus given by

$$u^* = U + \frac{C_2}{2}\left(\frac{a^3}{r^3} - 3\frac{a^3x^2}{r^5}\right) - \frac{C_1}{2\mu}\left(\frac{1}{r} + \frac{x^2}{r^3}\right) \tag{1.179}$$

$$v^* = -\frac{3C_2}{2}\frac{a^3xy}{r^5} - \frac{C_1}{2\mu}\frac{xy}{r^3} \tag{1.180}$$

$$w^* = -\frac{3C_2}{2}\frac{a^3xz}{r^5} - \frac{C_1}{2\mu}\frac{xz}{r^3} \tag{1.181}$$

The boundary conditions, Eq. (1.167), determine $C_1 = 3\mu aU/2$ and $C_2 = -U/2$, which provides the solution:

$$u^* = U - \frac{aU}{4r}\left(\frac{a^2}{r^2}+3\right) + \frac{3aU}{4}\frac{x^2}{r^3}\left(\frac{a^2}{r^2}-1\right) \quad (1.182)$$

$$v^* = \frac{3aU}{4}\frac{xy}{r^3}\left(\frac{a^2}{r^2}-1\right) \quad (1.183)$$

$$w^* = \frac{3aU}{4}\frac{xz}{r^3}\left(\frac{a^2}{r^2}-1\right) \quad (1.184)$$

$$p^* = -\frac{3\mu aU}{2}\frac{x}{r^3} + p_\infty \quad (1.185)$$

The streamwise (in the x-direction) fluid force acting on the sphere is estimated by integrating the x component of the stress tensor, Eq. (1.54), over the surface of sphere:

$$F_x = \int \boldsymbol{\Sigma}_{xn} ds = \int \boldsymbol{\Sigma}_x \cdot \boldsymbol{n} ds \quad (1.186)$$

where the unit normal vector $\boldsymbol{n} = (x/r, y/r, z/r)$. The substitutions of the solutions, Eqs. (1.182)–(1.185), give

$$\begin{aligned} \boldsymbol{\Sigma}_x \cdot \boldsymbol{n}\mid_{r=a} &= \frac{x}{r}\Sigma_{xx} + \frac{y}{r}\Sigma_{xy} + \frac{z}{r}\Sigma_{xz}\mid_{r=a} \\ &= \frac{1}{a}\left(-p^*x + 2\mu x\frac{\partial u^*}{\partial x} + \mu y\left(\frac{\partial u^*}{\partial y} + \frac{\partial v^*}{\partial x}\right) + \mu z\left(\frac{\partial u^*}{\partial z} + \frac{\partial w^*}{\partial x}\right)\right) \\ &= \frac{3}{2}\frac{\mu U}{a} \end{aligned} \quad (1.187)$$

The integration of Eq. (1.187) over the sphere surface gives

$$\begin{aligned} F_x &= \int_0^\pi \left(\frac{3}{2}\frac{\mu U}{a}\right) 2a\pi \sin\theta a d\theta \\ &= 6a\pi\mu U \end{aligned} \quad (1.188)$$

The analytical solution of the drag force acting on a sphere, Eq. (1.188), is well known as the Stokes law.

1.8.2 Fall/rise velocity of a spherical particle

Consider a solid sphere of radius a with density of ρ_s, released at time $t = 0$ in quiescent fluid with density of ρ_f and dynamic viscosity of ν. Assuming the Stokes drag force and gravity are only external forces acting on the sphere, the motion of equation is given by

$$\frac{4}{3}\pi\rho_s a^3\frac{dw}{dt} = -6a\pi\rho_f\nu w - \frac{4}{3}\pi\left(\rho_s - \rho_f\right)a^3 g \quad (1.189)$$

where w is the velocity of the sphere. Eq. (1.189) is slightly rearranged to

$$\frac{dw}{dt} + \alpha w + \frac{\rho_s - \rho_f}{\rho_s} g = 0 \tag{1.190}$$

where $\alpha = (9/2)\,\rho_f \nu / \rho_s a^2$. The solution of Eq. (1.190) takes the form:

$$w = C\mathrm{e}^{-\alpha t} - \frac{\rho_s - \rho_f}{\rho_s \alpha} g$$

The initial condition $w = 0$ at $t = 0$ determines the integral constant $C = (\rho_s - \rho_f) g / \rho_s \alpha$. The velocity at arbitrary time is given by

$$w = \frac{\rho_s - \rho_f}{\rho_s \alpha} g \left(\mathrm{e}^{-\alpha t} - 1\right) \tag{1.191}$$

When $\rho_s > \rho_f$, the sphere initially accelerates downward and its fall velocity approaches the constant terminal fall velocity w_t at sufficiently long time passed, since $\mathrm{e}^{-\alpha t} \to 0$ at $t \to \infty$:

$$w_t = -\frac{2}{9} \frac{a^2}{\nu} g \left(\frac{\rho_s}{\rho_f} - 1\right) \tag{1.192}$$

If $\rho_s < \rho_f$, as the case of a bubble in liquid, the rise velocity of the sphere approaches the terminal rise velocity, which is also estimated by Eq. (1.192). The dimensionless form of Eq. (1.192) is given

$$\frac{w_t a}{\nu} = -\frac{2}{9} Ga \left(\frac{\rho_s}{\rho_f} - 1\right) \tag{1.193}$$

where the so-called Galilei number $Ga = ga^3/\nu^2$. As the left-hand side may be defined as the Reynolds number $Re = w_t a/\nu$, Re is in proportion to Ga in this problem.

The general form of a drag force acting on an arbitrary object in a flow, with velocity U, is defined in terms of a drag coefficient C_d by

$$F_d = \frac{C_d}{2} \rho_f S U^2 \tag{1.194}$$

where S is the cross-sectional area of the object. In the case of a spherical solid particle in slow flow (or a particle slowly moving in quiescent fluid), we may use the Stokes drag, Eq. (1.188), and thus the Stokes' drag coefficient can be given as

$$C_{ds} = \frac{3 d \pi \mu U}{\rho_f U^2 \pi d^2 / 8} = \frac{24 \nu}{Ud} = \frac{24}{Re} \tag{1.195}$$

where d is the diameter and Reynolds number $Re = Ud/\nu$. When $Re < 1$ (in a regime Stokes approximation is available), C_d is inversely proportional to Re, as found from Eq. (1.195), while vortices and turbulence formed behind the object in higher Reynolds number modify C_d (Fig. 2.2), which is explained in the following chapters.

In a general case, the terminal fall velocity of the sphere, assuming $\rho_s > \rho_f$, can be given by equating drag and gravity forces:

$$\frac{C_d}{2}\rho_f \frac{\pi d^2}{4} w_t^2 = (\rho_s - \rho_f)\, g \frac{\pi d^3}{6}$$
$$\therefore w_t^2 = \frac{4}{3}\frac{\rho_s - \rho_f}{\rho_f}\frac{gd}{C_d} \tag{1.196}$$

In case of $\rho_s < \rho_f$, the terminal rise velocity

$$w_t^2 = \frac{4}{3}\frac{\rho_f - \rho_s}{\rho_f}\frac{gd}{C_d} \tag{1.197}$$

2

Turbulence and Diffusion

Exact solutions of the Navier-Stokes equation have been derived for limited number of simple flows, while general cases including nonlinear effects and complex forms of boundaries, observed in natural environment, are difficult to deal with only in the analytical framework using the Navier-Stokes equation. As already noted, properties of viscous flows are well characterized by the Reynolds number $Re = UL/\nu$, where U and L are the characteristic velocity and length. Turbulent behaviors in flows, involving velocity fluctuations over a wide ranging scales, are observed in high Reynolds number. Instead of fully analytical approaches, properties of the fluctuations are statistically modeled through correlations to physical experiments. Fundamental properties of the turbulent statistics, turbulence model, turbulent boundary layer flow, which will be applied in the following chapters, are introduced in this chapter.

The transport of matter in fluid occurs through processes of advection and diffusion. In the former process, the matter is carried by a flow, mathematically expressed by the advection equation, as noted in Chapter 1. The diffusion process is important near fluid boundaries where a high gradient of concentration exists, such as an air–water interface. In particular, diffusion of gas and heat across the interface estimates ocean environment through the flux exchanges between atmosphere and ocean, which will be introduced in Chapter 8. In this chapter, the transfer of matter during the diffusion processes in laminar and turbulent flows is introduced.

2.1 Turbulence

Consider that a flow field with characteristic fluid velocity $\overline{\boldsymbol{u}(\boldsymbol{x},t)}$ is disturbed at $t = t_0$ and the velocity then becomes $\boldsymbol{u} = \overline{\boldsymbol{u}} + \boldsymbol{u}'$. If the flow is stable, the disturbance $\boldsymbol{u}'$ decreases and approaches zero at $t \to \infty$, and thus $\boldsymbol{u}$ goes back to $\overline{\boldsymbol{u}}$. This flow state may be achieved when Re is less than the critical Reynolds number, Re_c, ($Re < Re_c$). In $Re > Re_c$ where the flow becomes unstable, $\boldsymbol{u}'$ may exponentially increase and the characteristic flow field changes to another flow state that can be stable at the given Reynolds number.

Fig. 2.1 illustrates properties of the horizontal flow across a circular cylinder, depending on Re. In very low Reynolds number $Re < 1$, as the inertial

DOI: 10.1201/9781003140160-2

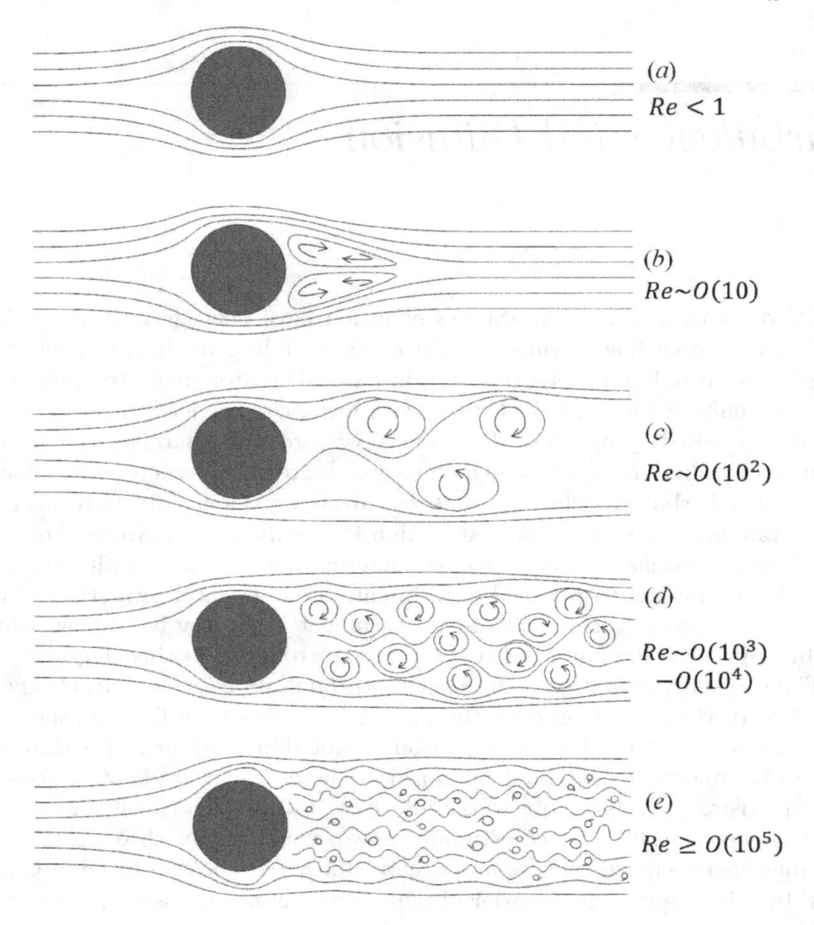

FIGURE 2.1
Illustration of flow patterns, depending on Reynolds number, behind a circular cylinder placed in a horizontal uniform flow.

effect is negligibly smaller than the viscous stress, laminar streamlines are formed around the sphere (Fig. 2.1*a*), which is well approximated by the Stokes flow, noted in Section 1.8; the analytical solution of the flow in this regime is given in Eqs. (1.182)–(1.185). With Reynolds number, the inertial effect increasingly governs the flow. When the order of Reynolds number $Re \sim O(10)$, as the viscous stress cannot maintain the boundary layer adhering on the sphere wall, the boundary layer is separated from the sphere by the inertial forces to form a symmetric pair of lee vortex behind the sphere (Fig. 2.1*b*). For higher Reynolds number, the lee vortices shed from the boundary layer to be transported down stream. Because of the vortex interaction, the flow field

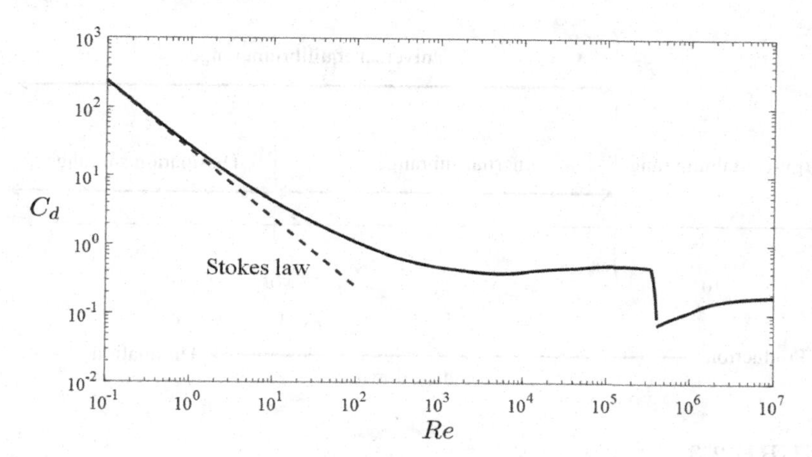

FIGURE 2.2
Drag coefficient C_d as a function of Reynolds number Re.

becomes asymmetric in a well-known Kármán's vortex street (Fig. 2.1*c*). With further increase in Reynolds number, as a large vortex cannot manage its form owing to local shear and stretch, it is decomposed into smaller vortices, evolving into turbulence characterized by small-scale velocity fluctuations (Fig. 2.1 *d* and *e*).

Fig. 2.2 shows a fitted curve of experimental drag coefficient C_d of a sphere, defined by Eq. (1.194), as a function of Re. In $Re < 1$, we find the Stokes drag, Eq. (1.195), well describing the experimental C_d. With Re, because of pressure depression, caused by the lee vortices formed behind the sphere (Fig. 2.1*b* and *c*), C_d deviates from the Stokes drag and the exponential slope of C_d become milder. In this regime, C_d may be estimated from the empirical Schiller and Naumann formula for a solid sphere[88;14]:

$$C_d = \frac{24}{Re}\left(1 + 0.15Re^{0.687}\right) \tag{2.1}$$

which is available in $Re < 800$. In the range $10^3 < Re < 10^5$, C_d is insensitive to Re in a similar flow state (see Fig. 2.1*d*). We find the depression of C_d in $Re \approx 2 \times 10^5$ where the pressure distribution changes in the transition to fully developed turbulence (Fig. 2.1*e*).

2.1.1 Turbulent statistics

Kolmogorov (1941)[44;3] theoretically explained the presence of universal statistic law defining small-scale turbulence far from boundaries in high Reynolds number. Here the small-scale turbulence means the length scale l and time scale τ of the turbulence are much smaller than the length- and

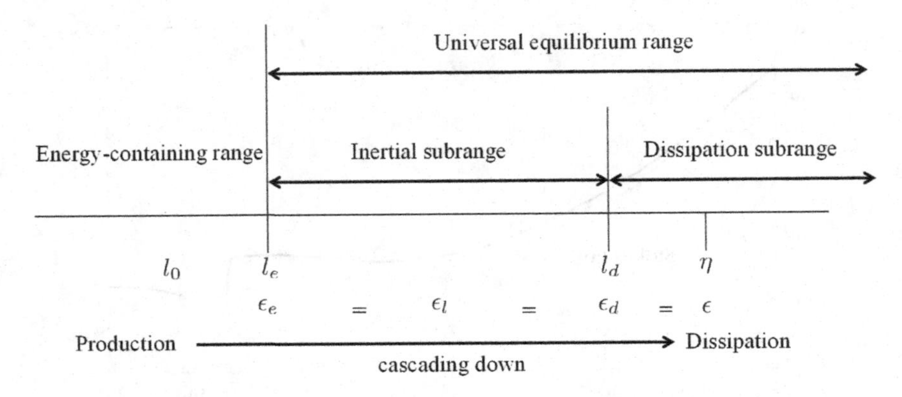

FIGURE 2.3
Scale diagram through the energy cascade.

time-scales, l_0 and $\tau_0 = l_0/u_0$, of eddy characterizing the turbulent flow, where u_0 is the velocity scale of the turbulence (see Fig. 2.3). As already noted, an eddy of the size l may be consecutively decomposed into smaller ones owing to local shear, stretch and its instability induced by nonlinear effects (see Fig. 2.1). During this process, the input of kinematic energy of the flow with the length l_0 at a rate of

$$\epsilon_0 = \frac{1}{2}\frac{d|\boldsymbol{u}^2|}{dt} \approx \frac{u_0^3}{l_0} \tag{2.2}$$

is transferred to smaller scales and finally transformed to heat at a rate of viscous dissipation, Eq. (1.106):

$$\epsilon = \frac{\nu}{2}\left(\frac{\partial u_i}{\partial x_j} + \frac{\partial u_j}{\partial x_i}\right)^2 \tag{2.3}$$

This process is called energy cascade to smaller scales (see Fig. 2.3). The universally equilibrium state, having statistically uniform and isotropic features regardless of macroscopic characteristics of turbulence, may be achieved, which is explained by Kolmogorov, based on the following hypotheses.

2.1.1.1 Kolmogorov's hypothesis of local isotropy

'At sufficiently high Reynolds number, the small scale turbulent motions ($l \ll l_0$) are statistically isotropic.' The length scale of the boundary between anisotropic eddy and isotropic smaller eddy is defined as l_e (see Fig. 2.3).

2.1.1.2 Kolmogorov's first similarity hypothesis

'In every turbulent flow, the statistics of the small-scale motions ($l < l_e$) have a universal form that is uniquely determined by ν and ϵ.' In the range of

$l < l_e$, called universal equilibrium range, through the energy cascading down, the energy transfers from large eddy to smaller ones at a rate of $\epsilon_e \sim \epsilon_l \sim \epsilon$, and finally dissipates owing to the viscous effect (Fig. 2.3). ν and ϵ measures scales of quantities: Kolmogorov's length scale, $\eta \sim (\nu^3/\epsilon)^{1/4}$, velocity scale, $u_\eta \sim (\epsilon\nu)^{1/4}$, and time scale, $\tau_\eta \sim (\nu/\epsilon)^{1/2}$. ϵ can be expressed in terms of these scales as $\epsilon = \nu(u_\eta/\eta)^2 = \nu/\tau_\eta^2$.

2.1.1.3 Kolmogorov's second similarity hypothesis

'In every turbulent flow, the statistics of the motions of scale l in the range $l_0 \gg l \gg \eta$ have a universal form that is uniquely determined by ϵ, independent on ν.' If the lower limit of this range as l_d is defined, the range $l_e > l > l_d$ is called inertial subrange (see Fig. 2.3).

In this range, as isotropic turbulent flows with various length scales are superposed, we may use a Fourier representation of velocity given

$$\boldsymbol{u}(\boldsymbol{x}) = \sum_{\boldsymbol{k}} \mathcal{U}(\boldsymbol{k}) \mathrm{e}^{i\boldsymbol{k}\cdot\boldsymbol{x}} \tag{2.4}$$

where $\boldsymbol{k}$ is the wave number, and the Fourier coefficient $\mathcal{U}(\boldsymbol{k})$ measures the velocity of eddy of $2\pi/\boldsymbol{k}$. Using Perseval's theorem (see Section 7.4.2), the kinematic energy, the mean square of Eq. (2.4) over spatial volume $V \sim L^3$, is given by

$$\frac{1}{2}\overline{\boldsymbol{u}^2(\boldsymbol{x})} = \frac{1}{2V}\int \boldsymbol{u}^2(\boldsymbol{x})d\boldsymbol{x} = \frac{1}{2}\sum_{\boldsymbol{k}} |\mathcal{U}(\boldsymbol{k})|^2 \tag{2.5}$$

If the sum in wave number space is replaced with integral by the formula

$$\sum_{\boldsymbol{k}} = \left(\frac{L}{2\pi}\right)^3 \int d\boldsymbol{k}$$

Eq. (2.5) is transformed as

$$\frac{1}{2}\overline{\boldsymbol{u}^2(\boldsymbol{x})} = \frac{1}{2}\frac{V}{(2\pi)^3}\int |\mathcal{U}(\boldsymbol{k})|^2 d\boldsymbol{k} = \int E(k)dk \tag{2.6}$$

where the energy spectrum is defined by

$$E(k) = \frac{V}{4\pi^2}|\mathcal{U}(\boldsymbol{k})|^2 k^2 \tag{2.7}$$

The second similarity hypothesis states that the energy spectrum $E(k)$ is determined by only k and ϵ. Since the dimensions of the length scale $l_k \sim k^{-1}$, time scale $\tau_k \sim \epsilon^{-1/3}k^{-2/3}$ and velocity scale $u_k \sim \epsilon^{1/3}k^{-1/3}$ at the wave number k

$$E \sim lu^2 \sim k^{-1}\left(\epsilon^{1/3}k^{-1/3}\right)^2 \sim \epsilon^{2/3}k^{-5/3} \tag{2.8}$$

$$\therefore E(k) = C\epsilon^{2/3}k^{-5/3} \tag{2.9}$$

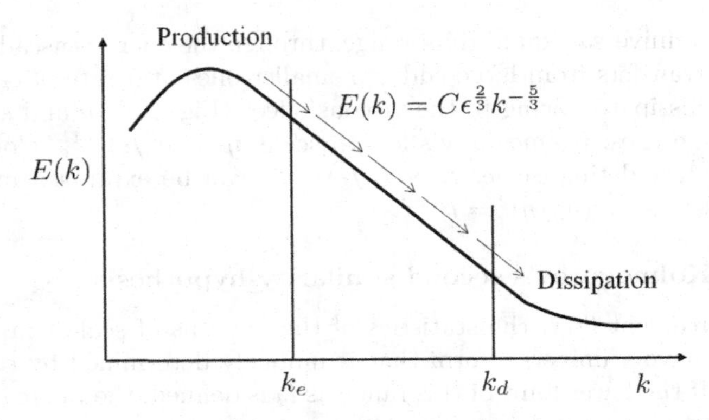

FIGURE 2.4
Turbulent energy spectrum and transfer.

This is the well-known Kolmogorov's $-5/3$ spectrum. Fig. 2.4 illustrates the spectrum form defined by Eq. (2.9). τ_k and u_k may be interpreted as the lifetime and its velocity of eddy at k. We find that a smaller eddy has shorter lifetime and slower velocity.

2.1.2 Reynolds decomposition

While instantaneous variations of a physical quantity in arbitrary turbulence cannot be deterministically identified in general, a statistical approach has been considered for defining characteristics of turbulence. Accordingly the quantities are decomposed into mean and fluctuation components (known as the Reynolds decomposition); e.g. fluid velocity $\boldsymbol{u}(\boldsymbol{x},t) = \overline{\boldsymbol{u}(\boldsymbol{x},t)} + \boldsymbol{u}'(\boldsymbol{x},t)$ where $\overline{\boldsymbol{u}}$ is the ensemble mean velocity and $\boldsymbol{u}'$ is the velocity fluctuation.

We consider mechanics contributions of the turbulent fluctuations to the mean flows in this section. The substitution of $u = \overline{u}+u'$, $v = \overline{v}+v'$, $w = \overline{w}+w'$ and $p = \overline{p}+p'$ into the x-component of the conservative form of Navier-Stokes equation, Eq. (1.98), gives

$$\frac{\partial}{\partial t}(\overline{u}+u') + \frac{\partial}{\partial x}\left(\overline{u}^2 + 2\overline{u}u' + u'^2\right) + \frac{\partial}{\partial y}(\overline{u}\,\overline{v} + \overline{u}v' + u'\overline{v} + u'v') + \frac{\partial}{\partial z}(\overline{u}\,\overline{w} + \overline{u}w' + u'\overline{w} + u'w') = -\frac{1}{\rho}\frac{\partial}{\partial x}(\overline{p}+p') + \nu\boldsymbol{\nabla}^2\overline{u} + \nu\boldsymbol{\nabla}^2u' \tag{2.10}$$

Assuming random fluctuations, double average of quantity f is equivalent to the average of f, $\overline{\overline{f}} = \overline{f}$, and the average of a fluctuation component is zero, $\overline{f'} = 0$. When all the terms of Eq. (2.10) are averaged, the equation is reduced to

$$\frac{\partial \overline{u}}{\partial t} + \frac{\partial \overline{u}^2}{\partial x} + \frac{\partial \overline{u}\,\overline{v}}{\partial y} + \frac{\partial \overline{u}\,\overline{w}}{\partial z} = -\left(\frac{\partial \overline{u'^2}}{\partial x} + \frac{\partial \overline{u'v'}}{\partial y} + \frac{\partial \overline{u'w'}}{\partial z}\right) - \frac{1}{\rho}\frac{\partial \overline{p}}{\partial x} + \nu \nabla^2 \overline{u} \tag{2.11}$$

The same operations to y- and z-components of Navier-Stoles equation, Eqs. (1.99) and (1.100), yield

$$\frac{\partial \overline{v}}{\partial t} + \frac{\partial \overline{u}\,\overline{v}}{\partial x} + \frac{\partial \overline{v}^2}{\partial y} + \frac{\partial \overline{v}\,\overline{w}}{\partial z} = -\left(\frac{\partial \overline{u'v'}}{\partial x} + \frac{\partial \overline{v'^2}}{\partial y} + \frac{\partial \overline{v'w'}}{\partial z}\right) - \frac{1}{\rho}\frac{\partial \overline{p}}{\partial y} + \nu \nabla^2 \overline{v} \tag{2.12}$$

$$\frac{\partial \overline{w}}{\partial t} + \frac{\partial \overline{u}\,\overline{w}}{\partial x} + \frac{\partial \overline{v}\,\overline{w}}{\partial y} + \frac{\partial \overline{w}^2}{\partial z} = -\left(\frac{\partial \overline{u'w'}}{\partial x} + \frac{\partial \overline{v'w'}}{\partial y} + \frac{\partial \overline{w'^2}}{\partial z}\right) - \frac{1}{\rho}\frac{\partial \overline{p}}{\partial z} + \nu \nabla^2 \overline{w} \tag{2.13}$$

Eqs. (2.11)–(2.13) are known as the Reynolds equation. The terms in the bracket of the right-hand side of Eqs. (2.11)–(2.13) describe contributions of turbulence to modify the mean velocity, which takes the form of divergence of the Reynolds stress defined as

$$\overline{\tau_{ij}} = -\rho\overline{u_i' u_j'} = -\rho \begin{pmatrix} \overline{u'^2} & \overline{u'v'} & \overline{u'w'} \\ \overline{u'v'} & \overline{v'^2} & \overline{v'w'} \\ \overline{u'w'} & \overline{v'w'} & \overline{w'^2} \end{pmatrix} \tag{2.14}$$

Accordingly the Reynolds equation can also be expressed by

$$\frac{\partial \overline{u_i}}{\partial t} + \frac{\partial \overline{u_i}\,\overline{u_j}}{\partial x_j} = -\frac{1}{\rho}\frac{\partial \overline{p}}{\partial x_i} + \frac{1}{\rho}\frac{\partial \overline{\tau_{ij}}}{\partial x_j} + \nu \frac{\partial^2 \overline{u_i}}{\partial x_j \partial x_j} \tag{2.15}$$

As found from Eq. (2.14), because of the additional six variables of the Reynolds stress, number of variables is larger than that of equations, that is, the equation system is unclosed. While numbers of turbulence models to close the system have been considered, the simplest model of Reynolds stress, assuming analogy of turbulent diffusion to molecular one and introducing eddy viscosity ν_T (Boussinesq approximation) in two-dimensional simple shear flow, is given by

$$-\overline{u'w'} = \nu_T \frac{\partial \overline{u}}{\partial z} \tag{2.16}$$

Introducing a well-known Prandtl's mixing length model:

$$-\overline{u'w'} = l^2 \left|\frac{\partial \overline{u}}{\partial z}\right| \frac{\partial \overline{u}}{\partial z} \tag{2.17}$$

the eddy viscosity may be given as

$$\nu_T = l^2 \left| \frac{\partial \overline{u}}{\partial z} \right| \tag{2.18}$$

The mixing length l for some trivial flows has been experimentally determined. For instance, in a wall region of a turbulent boundary layer, introduced in Section 2.1.4 (see Fig. 2.5), the mixing length is given as

$$l = \kappa z \tag{2.19}$$

where the Kármán constant $\kappa = 0.41$.

The general expression of Eq. (2.16) in the three-dimensional space is given by

$$\overline{\tau_{ij}} = \nu_T \left(\frac{\partial \overline{u_i}}{\partial x_j} + \frac{\partial \overline{u_j}}{\partial x_i} \right) - \frac{2}{3} \delta_{ij} q_k \tag{2.20}$$

where the turbulent kinematic energy $q_k = \frac{1}{2}\overline{u_i' u_i'} = \frac{1}{2}\left(\overline{u'^2} + \overline{v'^2} + \overline{w'^2}\right)$ is to be determined by an additional equation.

2.1.3 Turbulent energy and dissipation

Subtracting Eq. (2.15) from the Navier-Stokes equation Eq. (1.97), the equation for the velocity fluctuation is given as

$$\frac{\partial u_i'}{\partial t} + \frac{\partial \overline{u_i} u_j'}{\partial x_j} + \frac{\partial u_i' \overline{u_j}}{\partial x_j} = -\frac{1}{\rho}\frac{\partial p'}{\partial x_i} - \frac{1}{\rho}\frac{\partial \overline{\tau_{ij}}}{\partial x_j} + \frac{1}{\rho}\frac{\partial \tau_{ij}}{\partial x_j} + \nu \frac{\partial}{\partial x_j}\left(\frac{\partial u_i'}{\partial x_j} + \frac{\partial u_j'}{\partial x_i} \right) \tag{2.21}$$

Taking dot product with u_i' and averaging all terms, the turbulent energy equation can be obtained

$$\frac{\partial q_k}{\partial t} + \overline{u_j}\frac{\partial q_k}{\partial x_j} = \overline{\tau_{ij}}\frac{\partial \overline{u_i}}{\partial x_j} - \nu \overline{\frac{\partial u_i'}{\partial x_j}\frac{\partial u_i'}{\partial x_j}} + \nu \frac{\partial^2 q_k}{\partial x_j \partial x_j} - \frac{\partial}{\partial x_j}\left(\frac{1}{2}\overline{u_i' u_i' u_j'} + \frac{1}{\rho}\overline{p' u_j'} \right) \tag{2.22}$$

The second term on the right-hand side is always negative as

$$\overline{\frac{\partial u_i'}{\partial x_j}\frac{\partial u_i'}{\partial x_j}} = \overline{\left(\frac{\partial u'}{\partial x}\right)^2} + \overline{\left(\frac{\partial u'}{\partial y}\right)^2} + \overline{\left(\frac{\partial u'}{\partial z}\right)^2} + \overline{\left(\frac{\partial v'}{\partial x}\right)^2} + \ldots,$$

which is defined as the energy dissipation by viscosity, denoted by ϵ. The diffusion term, the fourth term of the right-hand side of Eq. (2.22), is modeled by a gradient hypothesis that states the nonlinear correlation is proportional to the gradient of q_k

$$\frac{1}{2}\overline{u_i' u_i' u_j'} + \frac{1}{\rho}\overline{p' u_j'} = -\frac{\nu_T}{C_k}\frac{\partial q_k}{\partial x_j} \tag{2.23}$$

where C_k is the constant estimated from turbulent statistics.

Eq. (2.22) is finally transformed as

$$\frac{\partial q_k}{\partial t} + \overline{u_j}\frac{\partial q_k}{\partial x_j} = \overline{\tau_{ij}}\frac{\partial \overline{u_j}}{\partial x_j} - \epsilon + \frac{\partial}{\partial x_j}\left\{\left(\nu + \frac{\nu_T}{C_k}\right)\frac{\partial q_k}{\partial x_j}\right\} \tag{2.24}$$

While $\overline{\tau_{ij}}$ can be given by Eq. (2.20), ϵ and ν_T need to be determined by another model.

The so-called One-equation model uses Eq. (2.24) to close Eq. (2.15), while unknown ϵ and ν_T are determined in a semi-empirical way

$$\epsilon = C q_k^{3/2}/l \tag{2.25}$$

$$\nu_T = q_k^{1/2} l \tag{2.26}$$

where the length scale l is empirically determined. While this model is very simple and easy to use, the turbulence scale l should be known in advance. Therefore the application may be limited to some specific trivial flows.

When Eq. (2.21) is differentiated by x_k, multiplied by $2\nu\partial u_i/\partial x_k$ and averaged, the equation defining the turbulent energy dissipation is derived

$$\begin{aligned}\frac{\partial \epsilon}{\partial t} + \overline{u_j}\frac{\partial \epsilon}{\partial x_j} &= -2\nu\left(\overline{\frac{\partial u_i'}{\partial x_k}\frac{\partial u_j'}{\partial x_k}}\frac{\partial \overline{u_i}}{\partial x_j} + \overline{\frac{\partial u_i'}{\partial x_k}\frac{\partial u_i'}{\partial x_j}}\frac{\partial \overline{u_j}}{\partial x_k}\right)\\ &- 2\nu\overline{u_j'\frac{\partial u_i'}{\partial x_k}}\frac{\partial^2 \overline{u_i}}{\partial x_j \partial x_k} - 2\nu\overline{\frac{\partial u_i'}{\partial x_k}\frac{\partial u_i'}{\partial x_j}\frac{\partial u_j'}{\partial x_k}} - 2\nu^2\overline{\left(\frac{\partial^2 u_i'}{\partial x_k \partial x_j}\right)^2}\\ &- \frac{2\nu}{\rho}\frac{\partial}{\partial x_j}\overline{\frac{\partial u_j'}{\partial x_k}\frac{\partial p'}{\partial x_k}} - \nu\frac{\partial}{\partial x_j}\overline{u_j'\left(\frac{\partial u_i'}{\partial x_k}\right)^2} + \nu\frac{\partial^2 \epsilon}{\partial x_j \partial x_j}\end{aligned} \tag{2.27}$$

In the two-equation model, ϵ is calculated by the modeled Eq. (2.27)

$$\frac{\partial \epsilon}{\partial t} + \overline{u_j}\frac{\partial \epsilon}{\partial x_j} = C_{\epsilon 1}\frac{\epsilon}{q_k}\overline{\tau_{ij}}\frac{\partial \overline{u_j}}{\partial x_j} - C_{\epsilon 2}\frac{\epsilon^2}{q_k} + \frac{\partial}{\partial x_j}\left\{\left(\nu + \frac{\nu_T}{C_\epsilon}\right)\frac{\partial \epsilon}{\partial x_j}\right\} \tag{2.28}$$

where

$$\nu_T = C_\mu q_k^2/\epsilon \tag{2.29}$$

The constants are theoretically determined with empirical adjustments. The system of equations, Eqs. (2.15), (2.20), (2.24), and (2.28), with (2.28) is known as $k-\epsilon$ model, has been commonly used for computations of fully developed turbulence at high Reynolds number.

2.1.4 Turbulent boundary layer

A laminar boundary layer with parabolic velocity profile is formed on a solid wall in low Reynolds number, as noted in Section 1.3.4, while high Reynolds number flow develops a turbulent boundary layer on the wall. The turbulent boundary layer has a structure of multiple layers with different velocity profiles depending on the relative Reynolds stress to the viscous one.

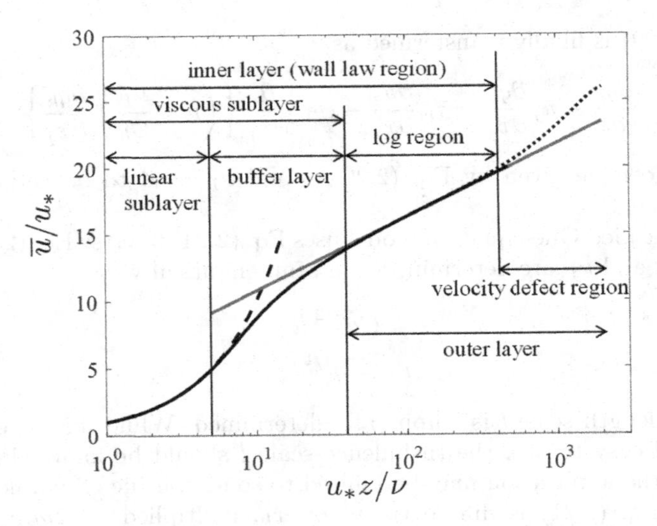

FIGURE 2.5
Layer structure of a turbulent boundary layer.

Fig. 2.5 illustrates the mean velocity distribution in the turbulent boundary layer. The boundary layer flow is scaled with the characteristic velocity u_*, termed friction velocity, defined by

$$u_* = \sqrt{\frac{\tau_w}{\rho}} \tag{2.30}$$

where the wall shear

$$\tau_w = \mu \left.\frac{\partial \overline{u}}{\partial z}\right|_{\text{at wall}} \tag{2.31}$$

The value of u_* is comparable to the fluctuation magnitude near the wall. The dimensionless distance from the wall with u_* , termed wall distance,

$$z^+ = \frac{u_* z}{\nu} \tag{2.32}$$

defines characteristic layers:
(1) linear sublayer ($z^+ \leq 5$) where the viscous stress dominates rather than negligibly small Reynolds stress suppressed by the wall
(2) buffer layer ($5 < z^+ \leq 40$) where the viscous stress and Reynolds stress are comparable
(3) log region ($z^+ > 40$ and $z < 0.2\delta$) where achieves logarithmic velocity profile (δ is thickness of boundary layer)
(4) velocity defect region ($z^+ > 40$ and $y \leq \delta$)

As shown in Fig. 2.5, the layer involving (1) and (2) is called a viscous sublayer. The region over (1), (2), and (3) is also called an inner layer, while an outer layer comprises (3) and (4).

If we consider statistically steady turbulent flow above an infinite horizontal wall in two-dimensional space (x, z), the Reynolds equation, Eq. (2.11), is expressed by

$$\frac{\partial \overline{u}^2}{\partial x} + \frac{\partial \overline{u}\overline{w}}{\partial z} = -\left(\frac{\partial \overline{u'^2}}{\partial x} + \frac{\partial \overline{u'w'}}{\partial z}\right) - \frac{1}{\rho}\frac{\partial \overline{p}}{\partial x} + \nu\left(\frac{\partial^2 \overline{u}}{\partial x^2} + \frac{\partial^2 \overline{u}}{\partial z^2}\right) \quad (2.33)$$

Assuming the velocity normal to the wall $\overline{w} \approx 0$ close to the wall, there is no horizontal variation of statistical quantities and also no horizontal gradient of pressure; Eq. (2.33) is thus reduced to

$$-\frac{\partial \overline{u'w'}}{\partial z} + \nu \frac{\partial^2 \overline{u}}{\partial z^2} \sim 0 \quad (2.34)$$

$$\therefore \nu \frac{\partial \overline{u}}{\partial z} \sim \overline{u'w'} \quad (2.35)$$

Eq. (2.35) indicates the viscous shear stress balances with the constant $\overline{u'w'}$. The viscous shear can also be expressed in terms of the wall shear, Eq. (2.31), and the friction velocity, Eq. (2.30), as

$$\nu \frac{\partial \overline{u}}{\partial z} = \tau_w / \rho \quad (2.36)$$

$$= u_*^2 \quad (2.37)$$

We find from Eqs. (2.35) and (2.37) that u_* measures the turbulent intensity near the wall; $u_*^2 \sim \overline{u'w'}$. Integrating Eq. (2.37) and using the nonslip boundary condition Eq. (1.113) at the wall, the mean velocity in the linear sublayer is given by

$$\overline{u} = \frac{u_*^2}{\nu} z \quad (2.38)$$

$$\therefore \frac{\overline{u}}{u_*} = u^+ = z^+ \quad (2.39)$$

where the dimensionless velocity is u^+. The relative effect of the Reynolds stress increases with distance from the wall. Assuming that the Reynolds stress $-\overline{u'w'} = u_*^2$ is constant in this layer, the mixing length model, Eqs. (2.17) and (2.19), provides the relation

$$u_*^2 = l^2 \left(\frac{d\overline{u}}{dz}\right)^2 = \kappa^2 z^2 \left(\frac{d\overline{u}}{dz}\right)^2 \quad (2.40)$$

and thus

$$\frac{d\overline{u}}{dz} = \frac{1}{\kappa}\frac{u_*}{z} \tag{2.41}$$

where κ is the Kármán constant. The integration of this equation gives the logarithmic velocity

$$\frac{\overline{u}}{u_*} = \frac{1}{\kappa}\ln z + C \tag{2.42}$$

where C is the constant. $\overline{u}/u_* (= u^+)$ of Eqs. (2.39) and (2.42) should connect at a certain height from the wall, $z^+ = d$ (see Fig. 2.5):

$$\frac{1}{\kappa}\ln\frac{d\nu}{u_*} + C = d$$

$$\therefore C = d - \frac{1}{\kappa}\ln\frac{d\nu}{u_*}$$

Eq. (2.42) is rewritten as

$$\frac{\overline{u}}{u_*} = \frac{1}{\kappa}\ln\frac{zu_*}{\nu} + C' \tag{2.43}$$

$$\text{or} \quad u^+ = \frac{1}{\kappa}\ln z^+ + C' \tag{2.44}$$

where $C' = d - \frac{1}{\kappa}\ln d$, and $d = 11$ is normally used. Eq. (2.44) describes the log law of the wall.

The relation between $z^+ = zu_*/\nu$ and $u^+ = \overline{u}/u_*$ over the inner layer (linear sublayer + buffer layer + log layer) is given by the Spalding formula[90] (see solid line in Fig. 2.5)

$$z^+ = u^+ + C\left(\exp\left(\kappa u^+\right) - 1 - \kappa u^+ - \frac{1}{2!}\left(\kappa u^+\right)^2 - \frac{1}{3!}\left(\kappa u^+\right)^3 - \frac{1}{4!}\left(\kappa u^+\right)^4\right) \tag{2.45}$$

where $C = 0.1108$ and the Kármán constant $\kappa = 0.4$ was used by Spalding[90].

In practice, a local skin friction coefficient is defined to describe the wall friction by

$$c_f = \frac{\tau_w}{\rho U_0^2/2} = 2\frac{u_*^2}{U_0^2} \tag{2.46}$$

$$\therefore \frac{u_*}{U_0} = \sqrt{\frac{c_f}{2}} \tag{2.47}$$

where U_0 is the velocity outside the turbulent boundary layer.

While the above turbulent boundary layer is assumed to be formed on a smooth wall surface, on rough surface, each roughness topography provides specific own flow patterns, such as flow separation, reattachment, and vortex shedding (see details in Chung et al., 2021) [16]. The effect may be characterized in terms of roughness size z_r by the roughness Reynolds number $Re_r = z_r u_* / \nu$. When the roughness is much smaller than the thickness of viscous sublayer, $Re_r \ll 1$, since the roughness is submerged in the viscous sublayer, the smooth flow along the surface may occur, and thus the velocity in the upper layer may have the identical form to Eq. (2.44). If the wall has large roughness, the flow at z higher than z_r ($z/z_r \gg 1$) should be rescaled with z_r

$$u^+ = \frac{1}{\kappa} \ln \frac{z}{z_r} + B(Re_r) \tag{2.48}$$

where the integration constant B depends on Re_r. In practice of ocean research, based on Eq. (2.48) and skin friction Eq. (2.46), a wind boundary layer flow over ocean waves and the friction at the ocean surface are characterized, which is introduced in Section 8.4.1.

2.2 Diffusion

Solute matter is transported in a diffusion process. In practice of ocean research, air-sea gas exchanges proceed via the diffusion across ocean surfaces. An analogy with turbulent diffusion of dispersive immiscible particles extends the applications in various fields, such as sediment transport in coastal region, sea spray dispersion in marine boundary layers, and bubble transport in bulk seawater. This chapter provides mathematical and physical descriptions of the diffusion processes in fluid as theoretical supports of predictions applied in coastal and ocean studies, introduced in Chapter 8.

2.2.1 Molecular and convective diffusion

The composition of the solution is characterized by concentration, c, defined as the number of particles of dissolved matter per unit volume of liquid. Molecular diffusion is the mechanism of transport of particles, causing a diffusion flux, $\boldsymbol{j}_m$, proceeding from regions of higher concentration to regions of lower one:

$$\boldsymbol{j}_m = -D \boldsymbol{\nabla} c \tag{2.49}$$

where D is the diffusion coefficient. Solute particles are also entrained and transported by the liquid flow through convective diffusion with the flux

$$\boldsymbol{j}_c = c \boldsymbol{u} \tag{2.50}$$

Total flux is thus given

$$\boldsymbol{j} = c \boldsymbol{u} - D \boldsymbol{\nabla} c \tag{2.51}$$

When a control volume V surrounded by surface S, fixed in a flow, is considered, the number of particles entering the volume is

$$Q = -\int_S \boldsymbol{j} \cdot \boldsymbol{n} ds \tag{2.52}$$

where $\boldsymbol{n}$ is the unit normal vector. Q must be equal to the temporal change in the number of particles contained in V, using the Gauss' divergence theorem Eq. (A.29):

$$\int_V \frac{\partial c}{\partial t} dv = -\int_S \boldsymbol{j} \cdot \boldsymbol{n} ds \tag{2.53}$$

$$= -\int_V \boldsymbol{\nabla} \cdot \boldsymbol{j} dv \tag{2.54}$$

The condition to satisfy Eq. (2.54) for arbitrary V is

$$\frac{\partial c}{\partial t} = -\boldsymbol{\nabla} \cdot \boldsymbol{j} \tag{2.55}$$

$$= \boldsymbol{\nabla} \cdot (D \boldsymbol{\nabla} c) - \boldsymbol{\nabla} \cdot (c \boldsymbol{u}) \tag{2.56}$$

Assuming D is independent of c and fluid is incompressible, $\boldsymbol{\nabla} \cdot \boldsymbol{u} = 0$, the advection-diffusion equation for c is given as

$$\frac{\partial c}{\partial t} + (\boldsymbol{u} \cdot \boldsymbol{\nabla}) c = D \boldsymbol{\nabla}^2 c \tag{2.57}$$

This is specifically written in the Cartesian coordinate by

$$\frac{\partial c}{\partial t} + u \frac{\partial c}{\partial x} + v \frac{\partial c}{\partial y} + w \frac{\partial c}{\partial z} = D \left(\frac{\partial^2 c}{\partial x^2} + \frac{\partial^2 c}{\partial y^2} + \frac{\partial^2 c}{\partial z^2} \right) \tag{2.58}$$

The diffusion coefficient may be described in terms of the kinematic viscosity as

$$D = \frac{\nu}{Sc} \tag{2.59}$$

where Sc is the Schmidt number. Sc is not a function of flow and is determined by the physical properties relating the momentum transfer to the transport of matter. When $Sc = 1 (D = \nu)$, there is a similarity between momentum and mass transfer. In gas flow, D and ν are of the same order; $Sc \approx 0.57$ for NH_3, 0.56 for H_2O, 1.14 for CO_2, 0.84 for O_2 (values at 25°C). In liquid phase, however, Sc of the gases takes values of the order of 100–1000, that is, the molecular diffusion is about 1000 times slower than the diffusion of momentum; $Sc \approx 360$ NH_3, 410 for CO_2, 441 for O_2 (values at 25°C), resulting in much thinner mass boundary layer than the viscous boundary layer formed on a liquid boundary.

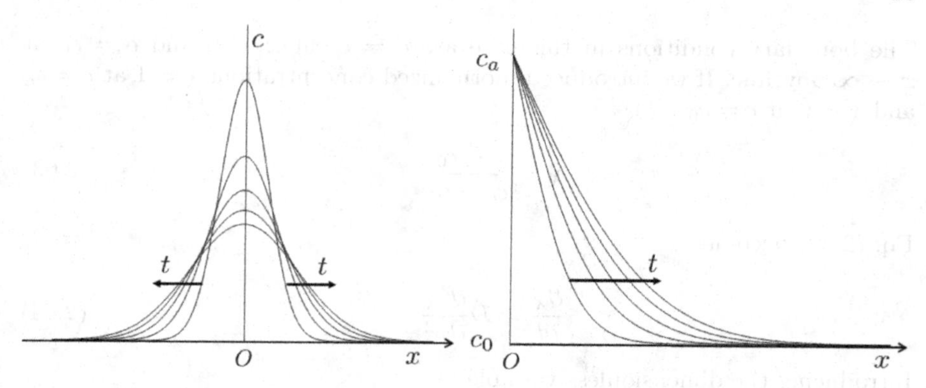

FIGURE 2.6
Variations of c following a normal distribution by Eq. (2.62) (left) and c following an error function by Eq. (2.72) (right).

2.2.2 Solution of diffusion equation

Assuming the mass flow resulting from pressure gradient (advective transport) is negligibly small in one-dimensional diffusion process, Eq. (2.58) can be reduced to

$$\frac{\partial c}{\partial t} = D\frac{\partial^2 c}{\partial x^2} \tag{2.60}$$

If we consider the initial conditions of $c = c_0$ at $x = 0$ and $c = 0$ elsewhere, the total amount of solute remains constant and equal to the amount given initially

$$\int_{-\infty}^{\infty} c(x,t)dx = c_0 \tag{2.61}$$

The solution of Eq. (2.60), satisfying the boundary condition $c = 0$ at $x = \pm\infty$ and the conservation of matter Eq. (2.61), is given by

$$c = \frac{c_0}{\sqrt{4\pi Dt}}\mathrm{e}^{-x^2/4Dt} \tag{2.62}$$

It should be noted that, at $t = 0$, c is concentrated at the origin where the value of c is infinite, while the total quantity is finite; Eq. (2.62) behaves like a delta function (see Section A.4 in Appendix). The temporal change in c is shown in Fig. 2.6 (left).

Consider another diffusion process that one fluid with c_a contacts with another fluid with c_0 at the origin at $t = 0$. This case corresponds to the simplest example of gas transfer from air with constant bulk concentration c_a to quiescent liquid of uniform concentration c_0 across the interface at $x = 0$.

The boundary conditions in this case are $c = c_a$ at $x = 0$ and $c = c_0$ at $x \to \infty$ anytime. If we introduce a normalized concentration; $\chi = 1$ at $c = c_a$ and $\chi = 0$ at $c = c_0$

$$\chi = \frac{c - c_0}{c_a - c_0} \tag{2.63}$$

Eq. (2.60) becomes

$$\frac{\partial \chi}{\partial t} = D\frac{\partial^2 \chi}{\partial x^2} \tag{2.64}$$

Introducing the dimensionless variable of

$$\eta = \frac{x}{\sqrt{4Dt}} \tag{2.65}$$

Eq. (2.64) can be transformed to

$$\frac{d^2\chi}{d\eta^2} + 2\eta\frac{d\chi}{d\eta} = 0 \tag{2.66}$$

since

$$\frac{\partial \chi}{\partial t} = \frac{\partial \chi}{\partial \eta}\frac{\partial \eta}{\partial t} = \frac{\partial \chi}{\partial \eta}\left(-\frac{\eta}{2t}\right) \quad \text{and} \quad D\frac{\partial^2 \chi}{\partial x^2} = \frac{\partial^2 \chi}{\partial \eta^2}\left(\frac{1}{4t}\right).$$

The boundary conditions are also transformed as $\chi = 1$ at $\eta = 0$ and $\chi = 0$ at $\eta \to \infty$. Introducing the transformation $\zeta = d\chi/d\eta$, Eq. (2.66) can be rewritten as

$$\frac{d\zeta}{d\eta} + 2\eta\zeta = 0 \tag{2.67}$$

The solution of Eq. (2.67) thus takes the form

$$\zeta = A_1 e^{-\eta^2} \tag{2.68}$$

where the constant A_1. Integrating Eq. (2.68), the solution of χ is given by

$$\chi = A_1 \int_0^{\eta} e^{-\eta^2} d\eta + A_2 \tag{2.69}$$

We readily find the constant $A_2 = 1$ from the boundary condition $\chi = 1$ at $\eta = 0$. Since another boundary condition $\chi = 0$ at $\eta \to \infty$,

$$A_1 \int_0^{\infty} e^{-\eta^2} d\eta = -1$$

Using Gaussian integral[1], $A_1 = -2/\sqrt{\pi}$. The solution is thus derived as

$$\chi = 1 - \frac{2}{\sqrt{\pi}} \int_0^{\eta} e^{-\eta^2} d\eta \tag{2.70}$$

$$= 1 - \mathrm{erf}(\eta) \tag{2.71}$$

where $\mathrm{erf}(\eta)$ is the error function[2]. From Eqs. (2.63) and (2.65), the concentration is determined by

$$\frac{c - c_0}{c_a - c_0} = 1 - \mathrm{erf}\left(\frac{x}{\sqrt{4Dt}}\right) \tag{2.72}$$

Fig. 2.6 (right) illustrates the variations of c.

The diffusion flux, Eq. (2.49), of Eq. (2.72) at the origin is given

$$j\mid_{x=0} = -D \left.\frac{dc}{dx}\right|_{x=0} \tag{2.73}$$

$$= \sqrt{\frac{D}{\pi t}} (c_a - c_0) \tag{2.74}$$

$$= k_D (c_a - c_0) \tag{2.75}$$

where $k_D (= \sqrt{D/\pi t})$ is called a mass transfer velocity.

2.2.2.1 Mass transfer across a spherical bubble

Consider a spherical air bubble of radius a in a liquid. The concentration c_a is fixed by local equilibrium at the interface, and the concentration far from the interface is identical to the one in the bulk c_0. As the simplest case, assuming negligibly small convection, the diffusion process is spherically symmetric and determined as a function of only r, regardless of polar and azimuthal angles, in spherical coordinates (see Eq. (1.15)). In this case, Eq. (2.57) is reduced to

$$\frac{\partial c}{\partial t} = D\left(\frac{\partial^2 c}{\partial r^2} + \frac{2}{r}\frac{\partial c}{\partial r}\right) \tag{2.76}$$

If we consider steady-state solution satisfying the boundary conditions $c = c_a$ at $r = a$ and $c = c_0$ at $r \to \infty$, the solution is given by

$$c = c_0 - \frac{a}{r}(c_0 - c_a) \tag{2.77}$$

The diffusion flux at the interface ($r = a$) is thus given

$$j\mid_{r=a} = -D \left.\frac{\partial c}{\partial r}\right|_{r=a} \tag{2.78}$$

$$= \frac{D}{a}(c_a - c_0) \tag{2.79}$$

---

[1] The Gauss integral states $\int_{-\infty}^{\infty} e^{-x^2} dx = \sqrt{\pi}$.

[2] The error function is defined as $\mathrm{erf}(x) = \frac{2}{\sqrt{\pi}} \int_0^x e^{-t^2} dt$.

The mass transfer velocity in this case is found to be $k_D = D/a$.

2.2.2.2 Higbie penetration model

A Higbie penetration model is known as a surface renewal model in which any fluid element will only come into contact with a surface for a limited time period, T; that is, the limited diffusion across the surface is allowed during the exposure time T.

The integration of the diffusion flux at the surface Eq. (2.74) over T gives the estimate of the mean flux

$$\bar{j} = \frac{1}{T}\int_0^T j dt = 2\sqrt{\frac{D}{\pi T}}\left(c_a - c_0\right) \tag{2.80}$$

The transfer velocity in this case is thus given

$$k_D = 2\sqrt{\frac{D}{\pi T}} \tag{2.81}$$

If a spherical droplet of diameter d falling with terminal velocity w_t in a fluid or a spherical bubble of diameter d rising at w_t is considered, appropriate time exposed by the surface in either case is $T = d/w_t$. Substituting into Eq. (2.81), the transfer velocity is expressed in terms of $Sc = \nu/D$ and $Re = w_t d/\nu$ as

$$k_D = \frac{2}{\sqrt{\pi}}\sqrt{\frac{Dw_t}{d}} \tag{2.82}$$

$$= \frac{2}{\sqrt{\pi}}\sqrt{ScRe}\frac{D}{d} \tag{2.83}$$

2.2.3 Effects of turbulence

The momentum diffusion by the effects of Reynolds stress, interpreted by Eq. (2.20), induces turbulent diffusion of concentration. Following the Reynolds decomposition noted in Section 2.1.2, the fluid velocity and concentration are decomposed into the ensemble mean and fluctuation, $\boldsymbol{u} = \overline{\boldsymbol{u}} + \boldsymbol{u}'$, $c = \overline{c} + c'$, and substituted to the conservation form of Eq. (2.58)

$$\frac{\partial c}{\partial t} + \frac{\partial uc}{\partial x} + \frac{\partial vc}{\partial y} + \frac{\partial wc}{\partial z} = D\left(\frac{\partial^2 c}{\partial x^2} + \frac{\partial^2 c}{\partial y^2} + \frac{\partial^2 c}{\partial z^2}\right) \tag{2.84}$$

$$\frac{\partial}{\partial t}\left(\overline{c} + c'\right) + \frac{\partial}{\partial x}\left(\overline{u}\,\overline{c} + u'\overline{c} + c'\overline{u} + c'u'\right) + \frac{\partial}{\partial y}\left(\overline{v}\,\overline{c} + v'\overline{c} + c'\overline{v} + c'v'\right)$$
$$+ \frac{\partial}{\partial z}\left(\overline{w}\,\overline{c} + w'\overline{c} + c'\overline{w} + c'w'\right) = D\boldsymbol{\nabla}^2\left(\overline{c} + c'\right) \tag{2.85}$$

Averaging all terms of Eq. (2.85) and considering $\overline{f'} = 0$ and $\overline{\overline{f}} = \overline{f}$, Eq. (2.85) is reduced to

$$\frac{\partial \overline{c}}{\partial t} + \frac{\partial \overline{u}\,\overline{c}}{\partial x} + \frac{\partial \overline{v}\,\overline{c}}{\partial y} + \frac{\partial \overline{w}\,\overline{c}}{\partial z} = D\nabla^2 \overline{c} - \left(\frac{\partial \overline{u'c'}}{\partial x} + \frac{\partial \overline{v'c'}}{\partial y} + \frac{\partial \overline{w'c'}}{\partial z} \right) \tag{2.86}$$

or the index representation

$$\frac{\partial \overline{c}}{\partial t} + \frac{\partial \overline{u}_j \overline{c}}{\partial x_j} = D \frac{\partial^2 \overline{c}}{\partial x_j \partial x_j} - \frac{\partial \overline{u'_j c'}}{\partial x_j} \tag{2.87}$$

The turbulence component, last term on the right-hand side of Eq. (2.87), takes the form of divergence of the flux $\overline{u'_j c'}$. Assuming this turbulent diffusion flux proceeds from regions of higher concentration to lower regions, like as molecular diffusion Eq. (2.49), it is often modeled by gradient of $\overline{c}$ (gradient hypothesis, Eq. (2.23))

$$\overline{u'_j c'} = -D_T \frac{\partial \overline{c}}{\partial x_j} \tag{2.88}$$

where the turbulent diffusion coefficient D_T may be described by the ratio of the eddy viscosity ν_T to turbulent Schmidt number Sc_T

$$D_T = \frac{\nu_T}{Sc_T} \tag{2.89}$$

While Sc is determined by properties of the solution and solute matter, and Sc_T represents a characteristic feature of the turbulent flow. Any universal value of Sc_T may not be able to be established. A first approximation of unity ($Sc_T \approx 1$) is often used, while empirical values in $0.1 \leq Sc_T \leq 1$ have been used in various problems.

Substituting Eqs. (2.88) and (2.89) into Eq. (2.87), the equation for mean concentration in turbulence is given as the form

$$\frac{\partial \overline{c}}{\partial t} + \frac{\partial \overline{u}_j \overline{c}}{\partial x_j} = D \frac{\partial^2 \overline{c}}{\partial x_j \partial x_j} + \frac{\partial}{\partial x_j} \left(\frac{\nu_T}{Sc_T} \frac{\partial \overline{c}}{\partial x_j} \right) \tag{2.90}$$

2.2.4 Turbulent mass boundary layer

A concentration profile in a turbulent boundary layer formed on a solid wall (Section 2.1.4) is considered in this section.

We assume the horizontally uniform flow near a flat surface where concentration c_a is fixed by local equilibrium, and the constant bulk concentration c_0, due to turbulent mixing, outside a boundary layer of thickness δ (in $z \geq \delta$). In the viscous sublayer ($z < \delta_0$), the flow is governed by viscosity as the Reynolds stress is suppressed by the wall. Therefore the viscous diffusion dominates in this layer. The integration of the molecular diffusion flux $j = D\partial c/\partial z$ gives

$$\overline{c} = \frac{j}{D} z + A \tag{2.91}$$

Considering the concentration at the surface, $\overline{c} = c_a$ at $z = 0$, $A = c_a$. Eq. (2.91) is thus

$$\overline{c} = \frac{j}{D}z + c_a \tag{2.92}$$

The turbulent diffusion flux, Eq. (2.88), is considered in the outer layer ($z > \delta_0$)

$$j = \frac{\nu_T}{Sc_T}\frac{\partial \overline{c}}{\partial z} \tag{2.93}$$

The eddy viscosity, defined by Eq. (2.16), is assumed to be expressed in terms of the Prandtl's mixing length l (see Eqs. (2.17) and (2.19)) by

$$\nu_T = l^2\frac{\partial \overline{u}}{\partial z} = \kappa^2 z^2 \frac{\partial \overline{u}}{\partial z} \tag{2.94}$$

As the velocity gradient in this region is given by Eq. (2.41), ν_T is determined as

$$\nu_T = \kappa z u_* \tag{2.95}$$

where κ is the Kármán constant and u_* is the friction velocity. The substitution into Eq. (2.93) gives

$$j = \frac{\kappa u_*}{Sc_T} z \frac{\partial \overline{c}}{\partial z} \tag{2.96}$$

The integration of Eq. (2.96) provides a logarithmic distribution of $\overline{c}$

$$\overline{c} = j\frac{Sc_T}{\kappa u_*}\ln z + A \tag{2.97}$$

The boundary condition far from the surface, $\overline{c} = c_0$ at $z = \delta$ determines the integral constant

$$A = c_0 - j\frac{Sc_T}{\kappa u_*}\ln \delta \tag{2.98}$$

$$\therefore \overline{c} = j\frac{Sc_T}{\kappa u_*}\ln \frac{z}{\delta} + c_0 \tag{2.99}$$

As $\overline{c}$ by Eq. (2.92) should agree with the one by Eq. (2.99) at $z = \delta_0$, j is confined

$$\frac{j}{D}\delta_0 + c_a = j\frac{Sc_T}{\kappa u_*}\ln \frac{\delta_0}{\delta} + c_0 \tag{2.100}$$

$$\therefore j = \frac{(c_0 - c_a)\,\nu/Sc}{\delta_0 - \frac{Sc_T}{Sc}\frac{\nu}{\kappa u_*}\ln \frac{\delta_0}{\delta}} \tag{2.101}$$

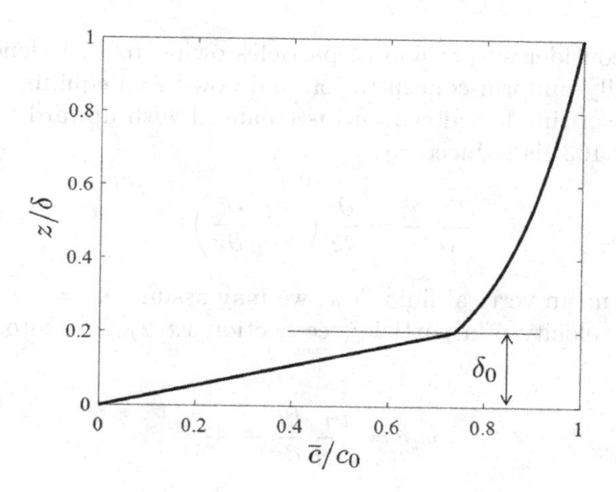

FIGURE 2.7
Concentration profile in a turbulent boundary layer by Eq. (2.102) ($c_a = 0, Sc = 1, Sc_T = 1, d = 11, \delta_0/\delta = 0.2$).

In summary, the mean concentration is given

$$\overline{c} = \begin{cases} \dfrac{c_0 - c_a}{1 - Sc_T \ln(\delta_0/\delta)/Sc\kappa d} \dfrac{z}{\delta_0} + c_a & (z < \delta_0) \\ \dfrac{c_0 - c_a}{\kappa d Sc/Sc_T - \ln(\delta_0/\delta)} \ln \dfrac{z}{\delta} + c_0 & (\delta_0 < z < \delta) \end{cases} \tag{2.102}$$

where $d = u_* \delta_0/\nu$ (δ_0 in wall unit). Fig. 2.7 illustrates a vertical profile of the turbulent mass boundary layer described by Eq. (2.102).

2.2.5 Immiscible particles

Similar diffusion processes to the soluble matter in fluid have been observed in turbulence involving immiscible particles in natural environment, such as the sediment dispersion above river and sea beds disturbed by turbulence, and sea spray, resulting from wave breaking, dispersed by wind. In such flows, molecular diffusion, associated with chemical effects, may be unimportant unless particles own chemical and electrical properties reacting to fluid. In this case, only turbulent diffusion may be considered. The concentration is redefined as the volume of immiscible particles per unit volume of fluid, and Eq. (2.90) is replaced for such flows by

$$\frac{\partial \overline{c}}{\partial t} + \frac{\partial \overline{u_{pj}}\, \overline{c}}{\partial x_j} = \frac{\partial}{\partial x_j} \left(\frac{\nu_T}{Sc_T} \frac{\partial \overline{c}}{\partial x_j} \right) \tag{2.103}$$

where $\overline{u_{pj}} = (u_p, v_p, w_p)$ is the mean velocity of particles.

Here we consider suspension of particles owing to turbulence. Assuming the horizontally uniform concentration and flow in an equilibrium state that gravitational settling flux downward is balanced with upward turbulence diffusion, Eq. (2.103) is reduced to

$$\frac{\partial \overline{w_p}\, \overline{c}}{\partial z} = \frac{\partial}{\partial z}\left(\frac{\nu_T}{Sc_T}\frac{\partial \overline{c}}{\partial z}\right) \tag{2.104}$$

If there is no mean vertical fluid flow, we may assume $\overline{w_p} = -w_t$, where w_t is the terminal velocity of a particle (see Section 1.8.2). The integration of Eq. (2.104) gives

$$-w_t \overline{c} = \frac{\nu_T}{Sc_T}\frac{\partial \overline{c}}{\partial z} + A_1 \tag{2.105}$$

where A_1 is the constant. The solution of Eq. (2.105) depends on a turbulence property defining a function form of ν_T.

2.2.5.1 Homogeneous turbulence

Consider homogeneous turbulence that the eddy viscosity is uniform in space, as the simplest example. The solution of Eq. (2.105) takes the form

$$\overline{c} = A_2 \exp\left[-\frac{Sc_T w_T}{\nu_T} z\right] - \frac{A_1}{w_t} \tag{2.106}$$

When the reference concentration, c_0, is known at the reference level, ζ_0, i.e. the boundary condition $\overline{c} = c_0$ at $z = \zeta_0$,

$$c_0 = A_2 \exp\left[-\frac{Sc_T w_T}{\nu_T}\zeta_0\right] - \frac{A_1}{w_t}$$

Assuming no suspension occurs far from ζ_0, i.e. $\overline{c} \to 0$ at $z \to \infty$, $A_1 = 0$. The solution is thus identified as

$$\frac{\overline{c}}{c_0} = \exp\left[\frac{Sc_T w_t}{\nu_T}(\zeta_0 - z)\right] \tag{2.107}$$

2.2.5.2 Turbulent boundary layer

Consider particle suspension in a turbulent boundary layer. As ν_T is given by Eq. (2.95), Eq. (2.105) can be rewritten as

$$-w_t \overline{c} = \frac{\kappa u_*}{Sc_T} z \frac{\partial \overline{c}}{\partial z} + A_1 \tag{2.108}$$

The solution takes the form

$$\overline{c} = \frac{A_2}{z^R} - \frac{A_1}{w_t} \tag{2.109}$$

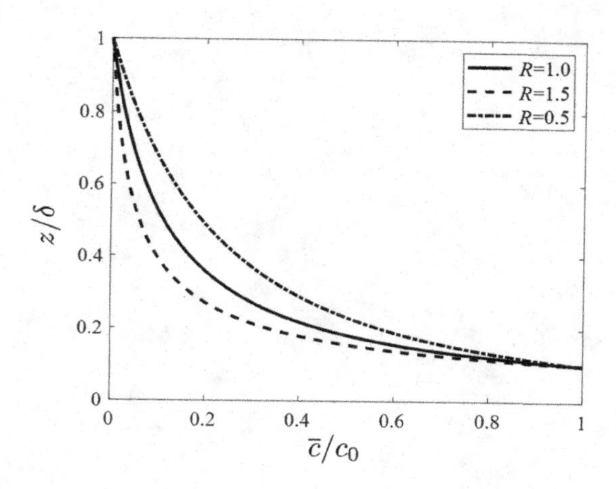

FIGURE 2.8
Distribution of the sediment concentration in a turbulent boundary layer by Eq. (2.111) ($\zeta_0 = 0.1$).

where the Rause number $R = Sc_T w_t/\kappa u_*$. Considering the boundary condition $\overline{c} = c_0$ at $z = \zeta_0$

$$c_0 = \frac{A_2}{\zeta_0^R} - \frac{A_1}{w_t} \tag{2.110}$$

Assuming no suspension occurs outside the boundary layer of thickness δ; $\overline{c} = 0$ at $z = \delta$

$$\frac{A_1}{w_t} = \frac{A_2}{\delta^R}$$

$$\therefore A_2 = \frac{c_0 \delta^R \zeta_0^R}{\delta^R - \zeta_0^R} \quad \text{and} \quad \frac{A_1}{w_t} = \frac{c_0 \zeta_0^R}{\delta^R - \zeta_0^R}$$

The concentration at arbitrary z is determined by

$$\frac{\overline{c}}{c_0} = \frac{\zeta_0^R}{z^R} \frac{\delta^R - z^R}{\delta^R - \zeta_0^R} \tag{2.111}$$

Fig. 2.8 illustrates the concentration of suspended sediment estimated by Eq. (2.111).

3

Surface and Vorticity Dynamics

Any flow has anisotropic features at a boundary (interface) which is a jump of phases with different physical properties. The flow near the interface may be constrained by the distinct mechanical response of the interface. The interaction of flow and surface may induce surface waves, roughness, currents, instability, and breakup of interfaces.

On the one hand, a vortex has anisotropic nature depending on orientation of rotation. When multiple vortices are closely located, the vortices are displaced, depending on relative distance and strengths, and induce flows between them through a vortex–vortex interactions. When they approach free-surface or interface, a surface–vortex interaction also contributes to local deformation, replacement of the surface and energy dissipation of the flow.

In this chapter, starting from the mechanical and kinematic features of free-surfaces/interfaces, dynamics of vorticity are addressed for understanding the vortex–surface interaction resulting in surface scars and vortex rings. Motions of air bubbles, drops, and foams, governed by capillary effects, are also introduced in this chapter.

3.1 Mathematical descriptions of a surface form

As the simplest example, we consider a surface of circle with radius R between the two points P at the position $\boldsymbol{r}(s) = (x(s), y(s), z(s))$, and Q at $\boldsymbol{r}(s+\Delta s)$, where s is the arc length of the curved surface (see Fig. 3.1). When Q approaches P, the relative position of P and Q, $\Delta\boldsymbol{r} = \boldsymbol{r}(s+\Delta s) - \boldsymbol{r}(s)$, describes the tangent vector $\boldsymbol{t}(s)$:

$$\boldsymbol{t}(s) = \lim_{\Delta s \to 0} \frac{\Delta \boldsymbol{r}}{\Delta s} = \frac{d\boldsymbol{r}}{ds} \tag{3.1}$$

As $\Delta\boldsymbol{r} \equiv \Delta s$ when $\Delta s \to 0$, Eq. (3.1) has the unit length, that is, $\boldsymbol{t}$ is the unit tangent vector. As the center angle of the sector of circle OPQ, $\Delta\theta$, has a relation $\Delta s = R\Delta\theta$, $\Delta\theta/\Delta s$ at the limit $\Delta s \to 0$ is given as

$$\frac{d\theta}{ds} = \lim_{\Delta s \to 0} \frac{\Delta \theta}{\Delta s} = \frac{1}{R} \tag{3.2}$$

DOI: 10.1201/9781003140160-3

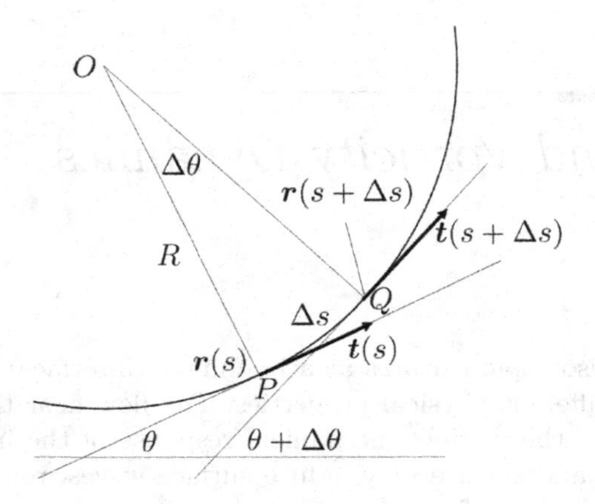

FIGURE 3.1
Illustration of a circular surface form.

where R is the radius of curvature. As $|\Delta \boldsymbol{t}| = |\boldsymbol{t}(s+\Delta s) - \boldsymbol{t}(s)| = \Delta\theta$,

$$\left|\frac{\Delta \boldsymbol{t}}{\Delta s}\right| = \frac{\Delta\theta}{R\Delta\theta} = \frac{1}{R} \tag{3.3}$$

$$\therefore \left|\frac{d\boldsymbol{t}}{ds}\right| = \lim_{\Delta s \to 0}\left|\frac{\Delta \boldsymbol{t}}{\Delta s}\right| = \frac{1}{R} \tag{3.4}$$

The differentiation of the trivial relation $\boldsymbol{t}\cdot\boldsymbol{t} = 1$ with respect to s gives $\boldsymbol{t}\cdot(d\boldsymbol{t}/ds) = 0$; that is, $d\boldsymbol{t}/ds$ is perpendicular to $\boldsymbol{t}$. Therefore the unit normal vector $\boldsymbol{n}$ can be defined by

$$\boldsymbol{n} = \frac{d\boldsymbol{t}/ds}{|d\boldsymbol{t}/ds|} = R\frac{d\boldsymbol{t}}{ds} \tag{3.5}$$

If we consider the surface conveniently described by $y = \eta(x)$ in the Cartesian coordinate (x, y), $\boldsymbol{r} = (x, \eta(x))$ and $ds = \sqrt{dx^2 + dy^2} = \sqrt{1 + (d\eta/dx)^2}dx$. The unit tangent vector Eq. (3.1) is also expressed by

$$\boldsymbol{t} = \frac{d\boldsymbol{r}}{ds} = \frac{dx}{ds}\frac{d\boldsymbol{r}}{dx} = \frac{(1, d\eta/dx)}{\sqrt{1 + (d\eta/dx)^2}} \tag{3.6}$$

The unit normal vector, perpendicular to Eq. (3.6) ($\boldsymbol{t}\cdot\boldsymbol{n} = 0$), is given by

$$\boldsymbol{n} = \frac{(-d\eta/dx, 1)}{\sqrt{1 + (d\eta/dx)^2}} \tag{3.7}$$

The gradient of the tangential line is expressed in terms of θ (see Fig. 3.1):

$$\frac{d\eta}{dx} = \tan\theta$$

$$\therefore \theta = \tan^{-1}\frac{d\eta}{dx} \tag{3.8}$$

The differentiation of Eq. (3.8) gives

$$\frac{d\theta}{dx} = \frac{1}{1+(d\eta/dx)^2}\frac{d^2\eta}{dx^2} \tag{3.9}$$

Considering Eqs. (3.2) and (3.9), the curvature $\kappa = 1/R$ is derived as

$$\frac{d\theta}{dx} = \frac{d\theta}{ds}\frac{ds}{dx} = \frac{d\theta}{ds}\sqrt{1+\left(\frac{d\eta}{dx}\right)^2} \tag{3.10}$$

$$\therefore \frac{d\theta}{ds} = \frac{1}{\left(1+(d\eta/dx)^2\right)^{\frac{3}{2}}}\frac{d^2\eta}{dx^2} = \frac{1}{R} \tag{3.11}$$

It should be noted that, while the radius of curvature R is defined above as a positive number, the sign of the curvature is ambiguous as long as the physical aspect is uncertain. Regarding our application to surface tension, consistent definition of the direction of normal vector is required to determine positive or negative surface tension depending the curvature on convex or concave surfaces, which is addressed in the next section.

In three-dimensional space, the normal vector $\boldsymbol{n}$ at the point of interest on the surface and a pair of orthogonal planes through it (Fig. 3.2*a*). The curvature of the surface is defined as the sum of the curvatures of two cross section curves on these planes:

$$\kappa = \pm\left(\frac{1}{R_1}+\frac{1}{R_2}\right) = \pm\frac{1}{R_m} \tag{3.12}$$

where R_m is the mean radius of curvature. When R_1 is the maximum and R_2 is the minimum, they are principal radii of curvature.

In the special case of a sphere, any normal vector passes through the center of the sphere, and any cross section containing $\boldsymbol{n}$ is a circle of radius R (Fig. 3.2*b*). The curvature of the sphere surface is thus:

$$\kappa = \pm\left(\frac{1}{R}+\frac{1}{R}\right) = \pm\frac{2}{R} \tag{3.13}$$

For a cylindrical surface (Fig. 3.2*c*), we may choose the circular cross section on the plane which is perpendicular to the axis of the cylinder. In this case, another cross section perpendicular to the circle forms a line parallel to the axis of the cylinder, that is, the radius of the curvature is infinity:

$$\kappa = \pm\left(\frac{1}{R}+\frac{1}{\infty}\right) = \pm\frac{1}{R} \tag{3.14}$$

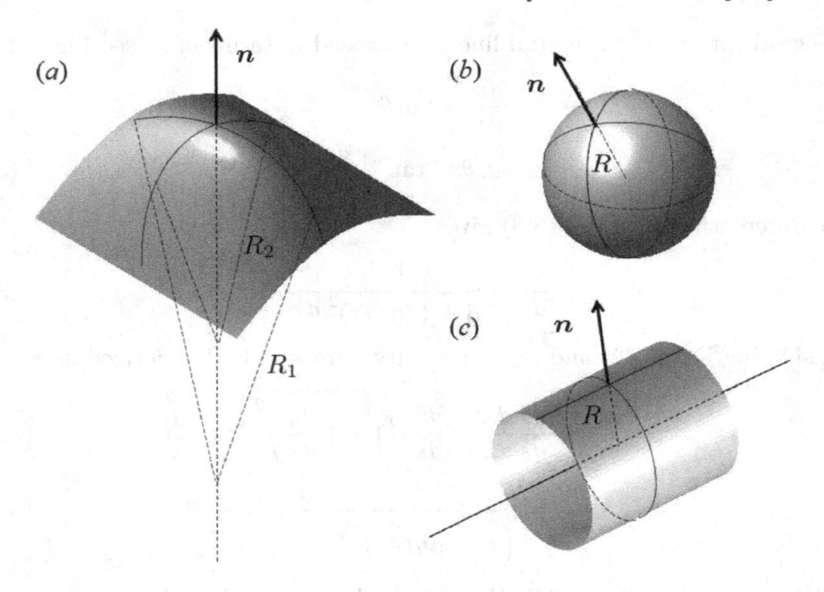

FIGURE 3.2
Curvatures on (a) arbitrary surface, (b) sphere, and (c) cylinder.

3.1.1 Arbitrary surface forms

We introduce a surface function $F(x, y) = y - \eta(x)$ where the surface location is defined at $F(x, y) = 0$. The gradient of F is:

$$\nabla F = \left(\frac{\partial F}{\partial x}, \frac{\partial F}{\partial y}\right) = \left(-\frac{\partial \eta}{\partial x}, 1\right) \tag{3.15}$$

Comparing with Eq. (3.16), we find ∇F is the normal vector. The unit normal vector is thus given

$$\boldsymbol{n} = \frac{\nabla F}{|\nabla F|} = \frac{(-d\eta/dx, 1)}{\sqrt{1 + (d\eta/dx)^2}} \tag{3.16}$$

The divergence of $\boldsymbol{n}$ gives

$$\nabla \cdot \boldsymbol{n} = -\frac{1}{\left(1 + (d\eta/dx)^2\right)^{\frac{3}{2}}} \frac{d^2\eta}{dx^2} \tag{3.17}$$

We find the absolute value of Eq. (3.17) is identical with Eq. (3.11), indicating $\nabla \cdot \boldsymbol{n}$ expresses the curvature with the negative sign. To describe the positive and negative surface tension on convex and concave surfaces, we define a signed curvature:

$$\kappa = \nabla \cdot \boldsymbol{n} \tag{3.18}$$

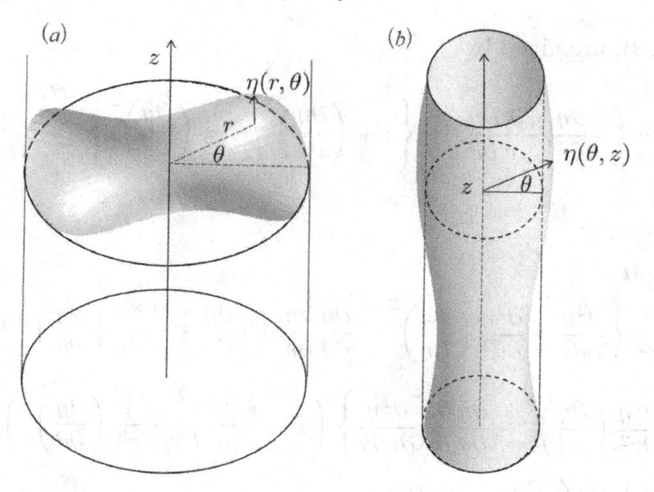

FIGURE 3.3
Definitions of the surfaces of liquid contained in a cylindrical container (*a*) and the surface of a cylindrical liquid jet (*b*).

For small $d\eta/dx$, as $(d\eta/dx)^2 \ll 1$ in Eq. (3.17), Eq. (3.18) is approximated as

$$\kappa \approx -\frac{d^2\eta}{dx^2} \tag{3.19}$$

In a three-dimensional space, since the surface function is defined as $F(x,y,z) = z - \eta(x,y)$, the normal unit vector can be expressed by

$$\boldsymbol{n} = \frac{\boldsymbol{\nabla} F}{|\boldsymbol{\nabla} F|} = \frac{(-\partial\eta/\partial x, -\partial\eta/\partial y, 1)}{\sqrt{1 + (\partial\eta/\partial x)^2 + (\partial\eta/\partial y)^2}} \tag{3.20}$$

The curvature is then given as

$$\begin{aligned}
\kappa &= \boldsymbol{\nabla}\cdot\boldsymbol{n} \\
&= -\left\{\frac{\partial^2\eta}{\partial x^2} + \frac{\partial^2\eta}{\partial y^2} + \frac{\partial^2\eta}{\partial x^2}\left(\frac{\partial\eta}{\partial y}\right)^2 + \frac{\partial^2\eta}{\partial y^2}\left(\frac{\partial\eta}{\partial x}\right)^2\right. \\
&\quad \left. -2\frac{\partial^2\eta}{\partial x\partial y}\frac{\partial\eta}{\partial x}\frac{\partial\eta}{\partial y}\right\}\left(1 + \left(\frac{\partial\eta}{\partial x}\right)^2 + \left(\frac{\partial\eta}{\partial y}\right)^2\right)^{-3/2} \\
&\approx -\left(\frac{\partial^2\eta}{\partial x^2} + \frac{\partial^2\eta}{\partial y^2}\right)
\end{aligned} \tag{3.21}$$

The normal vector and curvature for liquid filled in a circular container in a cylindrical coordinate (r, θ, z) (see Fig. 3.3*a*), where the surface is located at $z = \eta(r, \theta)$, are given by

$$\boldsymbol{n} = \left(-\frac{\partial \eta}{\partial r}, -\frac{1}{r}\frac{\partial \eta}{\partial \theta}, 1\right)\left\{1 + \left(\frac{\partial \eta}{\partial r}\right)^2 + \frac{1}{r^2}\left(\frac{\partial \eta}{\partial \theta}\right)^2\right\}^{-1/2} \tag{3.22}$$

and

$$\begin{aligned}\kappa &= \nabla \cdot \boldsymbol{n} \\ &= -\frac{1}{r^2}\left\{\frac{\partial^2 \eta}{\partial \theta^2} + \frac{\partial^2 \eta}{\partial \theta^2}\left(\frac{\partial \eta}{\partial r}\right)^2 - \frac{\partial \eta}{\partial \theta}\frac{\partial \eta}{\partial r} + r\frac{\partial \eta}{\partial r} + \frac{2}{r}\frac{\partial \eta}{\partial r}\left(\frac{\partial \eta}{\partial \theta}\right)^2 + r^2\frac{\partial^2 \eta}{\partial r^2}\right. \\ &\quad \left. + \frac{\partial^2 \eta}{\partial r^2}\left(\frac{\partial \eta}{\partial \theta}\right)^2 - \frac{\partial \eta}{\partial \theta}\frac{\partial \eta}{\partial r}\frac{\partial^2 \eta}{\partial r \partial \theta}\right\}\left(1 + \left(\frac{\partial \eta}{\partial r}\right)^2 + \frac{1}{r^2}\left(\frac{\partial \eta}{\partial \theta}\right)^2\right)^{-3/2} \\ &\approx -\left(\frac{1}{r}\frac{\partial}{\partial r}\left(r\frac{\partial \eta}{\partial r}\right) + \frac{1}{r^2}\frac{\partial^2 \eta}{\partial \theta^2}\right)\end{aligned} \tag{3.23}$$

Similarly for a surface of cylindrical jet extending in the z-direction, where the surface located at $r = \eta(\theta, z)$ (Fig. 3.3*b*), $\boldsymbol{n}$ and κ are given

$$\boldsymbol{n} = \left(1, -\frac{1}{r}\frac{\partial \eta}{\partial \theta}, -\frac{\partial \eta}{\partial z}\right)\left\{1 + \frac{1}{r^2}\left(\frac{\partial \eta}{\partial \theta}\right)^2 + \left(\frac{\partial \eta}{\partial z}\right)^2\right\}^{-1/2} \tag{3.24}$$

$$\therefore \kappa = \nabla \cdot \boldsymbol{n} \approx \frac{1}{\eta} - \frac{1}{\eta^2}\frac{\partial^2 \eta}{\partial \theta^2} - \frac{\partial^2 \eta}{\partial z^2} \tag{3.25}$$

3.1.2 Surface tension

Consider a closed system of a liquid sphere in gas phase as shown in Fig. 3.4. Assuming isothermal condition and no material exchange between the phases, and ignoring gravity, Helmholtz free energy is given by

$$F = F_g + F_l + 4\pi R^2 \gamma \tag{3.26}$$

where γ is surface tension of the liquid and gas, and the subscripts g and l represent gas and liquid phases, respectively. The change of F (differential of the free energy) is thus

$$dF = -p_g dV_g - p_l dV_l + d\left(4\pi R^2 \gamma\right) \tag{3.27}$$

As the total volume in the system is unchanged, $dV_g = -dV_l$, and the spherical volume $V_l = (4/3)\pi R^3$, Eq. (3.27) becomes

$$\begin{aligned}dF &= (p_g - p_l)\, dV_l + d\left(4\pi R^2 \gamma\right) \\ &= (p_g - p_l)\, 4\pi R^2 dR + 8\pi \gamma R dR\end{aligned} \tag{3.28}$$

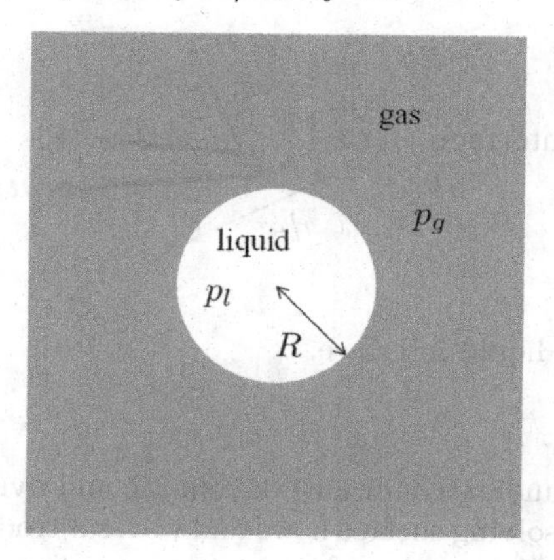

FIGURE 3.4
Laplace pressure of a spherical droplet surrounded by the gas phase; pressure in a liquid droplet (p_l) is higher that of outside (p_g).

In the equilibrium state (F is unchanged), the well-known Young-Laplace equation for a sphere surface (Eq. (3.13)) is given as

$$p_l - p_g = \frac{2\gamma}{R} \tag{3.29}$$

In a general case, according to Eqs. (3.18) and (3.12), the pressure jump (Laplace pressure) is given in terms of the surface tension:

$$p_l - p_g = \gamma\left(\frac{1}{R_1} + \frac{1}{R_2}\right) \tag{3.30}$$

$$= \gamma\kappa = \gamma\boldsymbol{\nabla}\cdot\boldsymbol{n} \tag{3.31}$$

3.2 Boundary conditions at free-surface/interface

An interface locates between two fluid phases with different densities and viscosities, typically liquid and gas. If the flow of the primary phase (liquid) is mechanically unaffected by another phase (air), the interface of the primary phase is free from any mechanical effects of the exterior fluid, which defines free surface. The mechanical effects of free surface and interface to fluid motions are mathematically given as boundary conditions for differential equations governing the interior fluid motion. While the velocity to fulfill at the fixed fluid-solid boundary has been given in Section 1.3.3, the general forms of boundary conditions at deformable free surface and interfaces are introduced in this section.

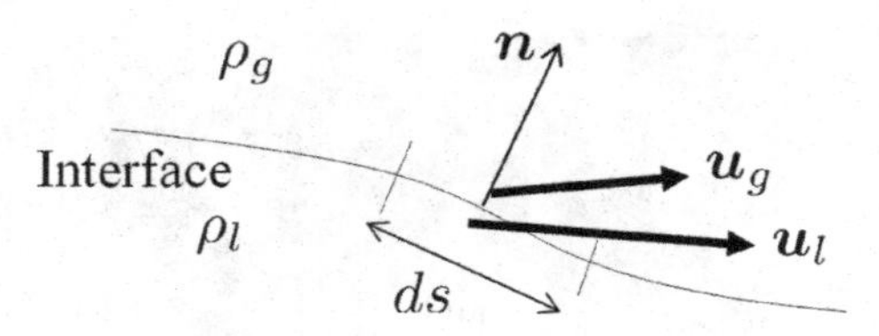

FIGURE 3.5
Illustration of a gas–liquid interface.

Two kinds of boundary conditions, kinematic and dynamic boundary conditions, are used for solving surface flows and waves introduced in the following chapters. The former defines the velocities to be satisfied at the surface boundary, while the mechanical balance at the boundary is imposed by the latter condition.

3.2.1 Kinematic boundary condition

Consider a two fluid system of liquid with density ρ_l and gas of density ρ_g separated by the interface (see Fig. 3.5). If the fluid velocity just on the liquid side of the interface $\boldsymbol{u}_l$ and the air side $\boldsymbol{u}_g$ are defined, the mass conservation requires that mass fluxes of liquid and gas, passing through unit area ds of the interface per unit time, are identical:

$$\rho_l \boldsymbol{u}_l \cdot \boldsymbol{n} = \rho_g \boldsymbol{u}_g \cdot \boldsymbol{n} \tag{3.32}$$

where $\boldsymbol{n}$ is the unit normal vector of the interface. This is a general form of the kinematic boundary condition. Assuming the immiscible fluids, Eq. (3.32) is replaced by

$$\boldsymbol{u}_l \cdot \boldsymbol{n} - \boldsymbol{u}_g \cdot \boldsymbol{n} = (\boldsymbol{u}_l - \boldsymbol{u}_g) \cdot \boldsymbol{n} = 0 \tag{3.33}$$

Accordingly, the normal component of the relative fluid particle velocity on the interface is zero. This condition states that, once a fluid particle exists on a surface at a certain instant, the particle also exits on the surface at the following time. We consider the mathematical expression of this statement.

Assuming an arbitrary shape of a curved surface is described in terms of the surface function $F(x, y, z, t)$ which defines the surface boundary located at $F(x, y, z, t) = 0$ (see also Section 3.1.1). The fluid particle, located at $\boldsymbol{x} = (x, y, z)$ where $F(x, y, z, t) = 0$ at time t, is displaced to $\boldsymbol{x} + \Delta\boldsymbol{x} = \boldsymbol{x} + \boldsymbol{u}\Delta t$ on the surface where $F(x + u\Delta t, y + v\Delta t, z + w\Delta t, t + \Delta t) = 0$ at time $t + \Delta t$. The Taylor approximation of F about small-time interval Δt is:

$$\begin{aligned} &F(x + u\Delta t, y + v\Delta t, z + w\Delta t, t + \Delta t) \\ &\quad = F(x, y, z, t) + u\Delta t \frac{\partial F}{\partial x} + v\Delta t \frac{\partial F}{\partial y} + w\Delta t \frac{\partial F}{\partial z} + \Delta t \frac{\partial F}{\partial t} + O\left(\Delta t^2\right) \end{aligned} \tag{3.34}$$

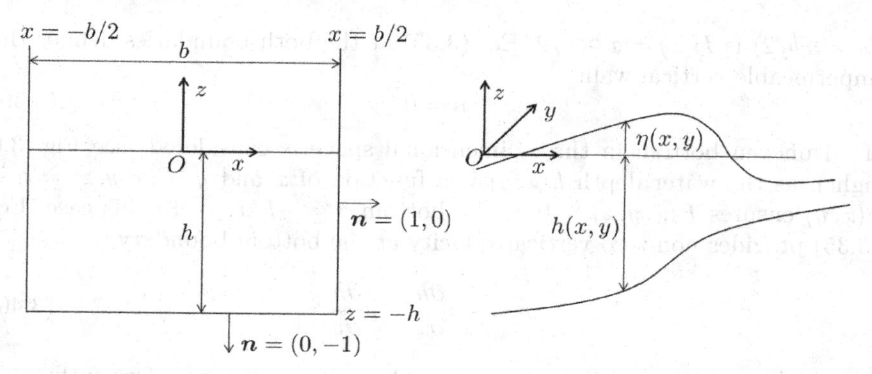

FIGURE 3.6
Coordinates of a rectangular water tank (left) and surface flow in uneven depth.

Neglecting the second and higher order terms, the substitutions of $F(x,y,z,t) = 0$ and $F(x+u\Delta t, y+v\Delta t, z+w\Delta t, t+\Delta t) = 0$ into Eq. (3.34) give the advection equation for F:

$$\frac{\partial F}{\partial t} + u\frac{\partial F}{\partial x} + v\frac{\partial F}{\partial y} + w\frac{\partial F}{\partial z} = 0 \quad \text{or} \quad \frac{DF}{Dt} = 0 \tag{3.35}$$

Since $\boldsymbol{n} = \boldsymbol{\nabla} F/|\boldsymbol{\nabla} F|$ as noted by Eq. (3.16), Eq. (3.35) may be rewritten as

$$\frac{\partial F}{\partial t} + \boldsymbol{u}\cdot\boldsymbol{\nabla} F = \frac{\partial F}{\partial t} + \boldsymbol{u}\cdot\boldsymbol{n} \mid \boldsymbol{\nabla} F \mid = 0 \tag{3.36}$$

$$\therefore \boldsymbol{u}\cdot\boldsymbol{n} = -\frac{\partial F/\partial t}{\mid \boldsymbol{\nabla} F \mid} \tag{3.37}$$

If a still water surface under quiescent air is considered, as $\partial F/\partial t{=}0$, $\boldsymbol{u}_l \cdot \boldsymbol{n} = \boldsymbol{u}_g \cdot \boldsymbol{n} = 0$, corresponding to the kinematic boundary condition Eq. (3.33).

The kinematic boundary condition for any fixed or moving boundary is given by Eq. (3.35) if appropriate F is specified. As the simplest example, consider the fixed boundaries of the rectangle tank with width b and water depth h (see Fig. 3.6 left). We define the Cartesian coordinates with the origin at the center of the tank in the horizontal x-axis, and at the still water level in the vertical z-axis. Since $F(z) = z + h$ fulfills $F(z) = 0$ at the fixed horizontal bottom boundary of $z = -h$, the substitution into Eq. (3.35) gives

$$w = 0 \tag{3.38}$$

which indicates no vertical velocity across the bottom boundary, that is, the bottom is impermeable, $\boldsymbol{u}\cdot\boldsymbol{n} = 0$ (identical to the impermeable wall condition Eq. (1.110)). As the appropriate $F(x)$ to satisfy $F(x) = 0$ at the side walls

($x = \pm b/2$) is $F(x) = x \mp b/2$, Eq. (3.35) at the both boundaries defines the impermeable vertical wall:

$$u = 0 \tag{3.39}$$

If an uneven bottom in three-dimensional space is considered (see Fig. 3.6 right), as the water depth $h(x, y)$ is a function of x and y, $F(x, y, z) = z + h(x, y)$ ensures $F(x, y, z) = 0$ at the bottom $z = -h(x, y)$. In this case, Eq. (3.35) provides non-zero vertical velocity at the bottom boundary

$$w = -u\frac{\partial h}{\partial x} - v\frac{\partial h}{\partial y} \tag{3.40}$$

The kinematic condition at a moving boundary (such as a free-surface) is also determined in the same manner. If the time-dependent surface elevation $\eta(x, y, t)$ with respect to the still water level is defined, as the surface locates at $z = \eta(x, y, t)$, the surface function $F(x, y, z, t) = z - \eta(x, y, t)$ and thus the kinematic condition at the surface is given by

$$w = \frac{\partial \eta}{\partial t} + u\frac{\partial \eta}{\partial x} + v\frac{\partial \eta}{\partial y} \tag{3.41}$$

In the cylindrical coordinate, the advection equation for F is given by

$$\frac{\partial F}{\partial t} + u_r\frac{\partial F}{\partial r} + \frac{u_\theta}{r}\frac{\partial F}{\partial \theta} + u_z\frac{\partial F}{\partial z} = 0 \tag{3.42}$$

Therefore the kinematic condition of a free surface of liquid in a circular tank (see Fig. 3.3a) with $F(r, \theta, z) = z - \eta(r, \theta)$ is given as

$$u_z = \frac{\partial \eta}{\partial t} + u_r\frac{\partial \eta}{\partial r} + \frac{u_\theta}{r}\frac{\partial \eta}{\partial \theta} \tag{3.43}$$

For the cylindrical liquid jet, extending in the z-direction, with the surface located at $r = \eta(\theta, z)$ (see Fig. 3.3b), $F(r, \theta, z) = r - \eta(\theta, z)$ leads

$$u_r = \frac{\partial \eta}{\partial t} + \frac{u_\theta}{\eta}\frac{\partial \eta}{\partial \theta} + u_z\frac{\partial \eta}{\partial z} \tag{3.44}$$

3.2.2 Dynamic boundary condition

The momentum conservation requires the identical momentum passing through the interface from each fluid phase:

$$\Sigma^l_{ij}n_j + \rho_l u^l_j n_j = \Sigma^g_{ij}n_j + \rho_l u^g_j n_j \tag{3.45}$$

where Σ_{ij} is the stress tensor, Eq. (1.54), n_j is the unit normal vector, and the superscripts l and g indicate the liquid and gas phases, respectively. The kinematic boundary condition, Eq. (3.32), reduces Eq. (3.45) to

$$\Sigma^l_{ij}n_j = \Sigma^g_{ij}n_j \tag{3.46}$$

When the surface tension γ acts on a curved interface with curvature κ, which balances with the pressure difference (Laplace pressure) given by Eq. (3.31), Eq. (3.46) is modified by

$$\Sigma_{ij}^{l} n_j - \Sigma_{ij}^{g} n_j = \gamma\kappa \tag{3.47}$$

The substitution of the stress tensor Eq. (1.54) into Eq. (3.47) gives

$$p_l \delta_{ij} n_j - p_g \delta_{ij} n_j = \gamma\kappa + \mu_l \left(\frac{\partial u_i^l}{\partial x_j} + \frac{\partial u_j^l}{\partial x_i} \right) n_j - \mu_g \left(\frac{\partial u_i^g}{\partial x_j} + \frac{\partial u_j^g}{\partial x_i} \right) n_j \tag{3.48}$$

where μ is the viscosity. The normal and tangential components of the dynamic boundary condition Eq. (3.48) are given by

$$p_l - p_g = \gamma\kappa + 2\mu_l \frac{\partial u_n^l}{\partial n} - 2\mu_g \frac{\partial u_n^g}{\partial n} \tag{3.49}$$

$$\mu_l \left(\frac{\partial u_n^l}{\partial s} + \frac{\partial u_s^l}{\partial n} \right) - \mu_a \left(\frac{\partial u_n^a}{\partial s} + \frac{\partial u_s^a}{\partial n} \right) = \frac{\partial \gamma}{\partial s} \tag{3.50}$$

where s and n are the tangential and normal axes, respectively. u_s and u_n are the tangential and normal components of fluid velocity on the interface. Eq. (3.49) describes the pressure jump, caused by the surface tension, at the interface. The right-hand side of Eq. (3.50) indicates a gradient of surface tension along the interface, called Marangoni stress, which occurs where is a variation in temperature or surfactant over the surface.

For free surfaces, where is free from any stresses in the air phase, Eqs. (3.49) and (3.50) are reduced to

$$p_l = \gamma\kappa + 2\mu_l \frac{\partial u_n^l}{\partial n} \tag{3.51}$$

$$\mu_l \left(\frac{\partial u_n^l}{\partial s} + \frac{\partial u_s^l}{\partial n} \right) = \frac{\partial \gamma}{\partial s} \tag{3.52}$$

Assuming constant γ along the surface, the condition of zero tangential shear to be fulfilled at the free surface is given

$$\mu_l \left(\frac{\partial u_n^l}{\partial s} + \frac{\partial u_s^l}{\partial n} \right) = 0 \tag{3.53}$$

When the surface tension and viscous effects are negligibly smaller than gravity, Eq. (3.51) is reduced to the simplest dynamic boundary condition at the free surface:

$$p_l = 0 \tag{3.54}$$

Assuming irrotational flow, we may use the Bernoulli equation Eq. (1.163) to specify the pressure at the free surface. Supposing atmospheric pressure

$p_g = 0$ in gauge pressure at the free surface, as the Laplace pressure $p_l = \gamma\kappa$, the dynamic boundary condition at the surface is given

$$\frac{\partial \phi}{\partial t} + \frac{1}{2}(u^2 + v^2 + w^2) + \frac{\gamma}{\rho}\kappa + gz = C(t) \quad (z = \eta) \tag{3.55}$$

The approximation of κ by Eq. (3.21) gives

$$\frac{\partial \phi}{\partial t} + \frac{1}{2}(u^2 + v^2 + w^2) - \frac{\gamma}{\rho}\left(\frac{\partial^2 \eta}{\partial x^2} + \frac{\partial^2 \eta}{\partial y^2}\right) + gz = C(t) \quad (z = \eta) \tag{3.56}$$

In the case of negligibly small surface tension, Eq. (3.55) is simplified to provide the dynamic boundary condition for gravity-dominant surface waves:

$$\frac{\partial \phi}{\partial t} + \frac{1}{2}(u^2 + v^2 + w^2) + gz = C(t) \quad (z = \eta) \tag{3.57}$$

3.3 Vorticity

As noted in Section 2.1, vortices characterize flow patterns and local pressure in flows. Vorticity is an important parameter to define strength and orientation of the vortex, which has been used in analyses of various flow field, introduced in the following chapters. In this section, the governing equation of vorticity and its solutions are given to address interactions between vortices and surfaces.

3.3.1 Vorticity equation

The equation governing variations of vorticity is derived in this section. Starting from the modified form of the Euler equation, Eq. (1.159):

$$\boldsymbol{\nabla}\left(\frac{1}{2}|\boldsymbol{u}|^2 + \frac{p}{\rho} + gz\right) = \boldsymbol{u} \times \boldsymbol{\omega} - \frac{\partial \boldsymbol{u}}{\partial t} \tag{3.58}$$

Considering vector formula Eq. (A.2) in Appendix, the rotation of Eq. (3.58) is reduced to

$$\frac{\partial \boldsymbol{\omega}}{\partial t} = \boldsymbol{\nabla} \times (\boldsymbol{u} \times \boldsymbol{\omega}) \tag{3.59}$$

where the vorticity

$$\boldsymbol{\omega} = \boldsymbol{\nabla} \times \boldsymbol{u} = (\omega_x, \omega_y, \omega_z) = \left(\frac{\partial w}{\partial y} - \frac{\partial v}{\partial z}, \frac{\partial u}{\partial z} - \frac{\partial w}{\partial x}, \frac{\partial v}{\partial x} - \frac{\partial u}{\partial y}\right) \tag{3.60}$$

From vector formula Eq. (A.7) in Appendix, the right-hand side of Eq. (3.59) is expanded as

$$\boldsymbol{\nabla} \times (\boldsymbol{u} \times \boldsymbol{\omega}) = (\boldsymbol{\nabla} \cdot \boldsymbol{\omega})\,\boldsymbol{u} - (\boldsymbol{\nabla} \cdot \boldsymbol{u})\,\boldsymbol{\omega} + (\boldsymbol{\omega} \cdot \boldsymbol{\nabla})\,\boldsymbol{u} - (\boldsymbol{u} \cdot \boldsymbol{\nabla})\,\boldsymbol{\omega}$$

Since $\boldsymbol{\nabla}\cdot\boldsymbol{\omega} = \boldsymbol{\nabla}\cdot(\boldsymbol{\nabla}\times\boldsymbol{u}) = 0$, from vector formula Eq. (A.1) in Appendix, and the incompressible continuity condition $\boldsymbol{\nabla}\cdot\boldsymbol{u} = 0$, Eq. (3.59) is transformed to

$$\frac{\partial \boldsymbol{\omega}}{\partial t} + (\boldsymbol{u}\cdot\boldsymbol{\nabla})\,\boldsymbol{\omega} = (\boldsymbol{\omega}\cdot\boldsymbol{\nabla})\,\boldsymbol{u} \tag{3.61}$$

For viscous fluid, the identical operation is performed for the Navier-Stokes equation, i.e. the rotation of the viscous term $\nu\boldsymbol{\nabla}^2\boldsymbol{u}$ is added to Eq. (3.61):

$$\frac{\partial \boldsymbol{\omega}}{\partial t} + (\boldsymbol{u}\cdot\boldsymbol{\nabla})\,\boldsymbol{\omega} = (\boldsymbol{\omega}\cdot\boldsymbol{\nabla})\,\boldsymbol{u} + \nu\boldsymbol{\nabla}^2\boldsymbol{\omega} \tag{3.62}$$

$$\frac{\partial \omega_i}{\partial t} + u_j\frac{\partial \omega_i}{\partial x_j} = \omega_j\frac{\partial u_i}{\partial x_j} + \nu\frac{\partial^2 \omega_i}{\partial x_j \partial x_j} \tag{3.63}$$

The left-hand side indicates the advection of vorticity by the flow, the first and second terms of the right-hand side indicate production of vorticity owing to fluid deformation, and the viscous diffusion of vorticity, respectively. The components of the equation are thus specified:

$$\begin{aligned}&\frac{\partial \omega_x}{\partial t} + u\frac{\partial \omega_x}{\partial x} + v\frac{\partial \omega_x}{\partial y} + w\frac{\partial \omega_x}{\partial z}\\ &\quad = \omega_x\frac{\partial u}{\partial x} + \omega_y\frac{\partial u}{\partial y} + \omega_z\frac{\partial u}{\partial z} + \nu\left(\frac{\partial^2 \omega_x}{\partial x^2} + \frac{\partial^2 \omega_x}{\partial y^2} + \frac{\partial^2 \omega_x}{\partial z^2}\right)\end{aligned} \tag{3.64}$$

$$\begin{aligned}&\frac{\partial \omega_y}{\partial t} + u\frac{\partial \omega_y}{\partial x} + v\frac{\partial \omega_y}{\partial y} + w\frac{\partial \omega_y}{\partial z}\\ &\quad = \omega_x\frac{\partial v}{\partial x} + \omega_y\frac{\partial v}{\partial y} + \omega_z\frac{\partial v}{\partial z} + \nu\left(\frac{\partial^2 \omega_y}{\partial x^2} + \frac{\partial^2 \omega_y}{\partial y^2} + \frac{\partial^2 \omega_y}{\partial z^2}\right)\end{aligned} \tag{3.65}$$

$$\begin{aligned}&\frac{\partial \omega_z}{\partial t} + u\frac{\partial \omega_z}{\partial x} + v\frac{\partial \omega_z}{\partial y} + w\frac{\partial \omega_z}{\partial z}\\ &\quad = \omega_x\frac{\partial w}{\partial x} + \omega_y\frac{\partial w}{\partial y} + \omega_z\frac{\partial w}{\partial z} + \nu\left(\frac{\partial^2 \omega_z}{\partial x^2} + \frac{\partial^2 \omega_z}{\partial y^2} + \frac{\partial^2 \omega_z}{\partial z^2}\right)\end{aligned} \tag{3.66}$$

In a two-dimensional space (x, y), as there is only $\omega_z = \partial v/\partial x - \partial u/\partial y$ vorticity component resulting from rotation of velocity (u, v), the equation is reduced to

$$\frac{\partial \omega_z}{\partial t} + u\frac{\partial \omega_z}{\partial x} + v\frac{\partial \omega_z}{\partial y} = \nu\left(\frac{\partial^2 \omega_z}{\partial x^2} + \frac{\partial^2 \omega_z}{\partial y^2}\right) \tag{3.67}$$

Introducing the stream function, ψ defined by Eq. (1.138), Eq. (3.67) is rewritten in terms of ψ:

$$\frac{\partial \omega_z}{\partial t} + \frac{\partial \psi}{\partial y}\frac{\partial \omega_z}{\partial x} - \frac{\partial \psi}{\partial x}\frac{\partial \omega_z}{\partial y} = \nu\left(\frac{\partial^2 \omega_z}{\partial x^2} + \frac{\partial^2 \omega_z}{\partial y^2}\right) \tag{3.68}$$

where

$$\omega_z = -\boldsymbol{\nabla}^2\psi \tag{3.69}$$

$$= -\left(\frac{\partial^2 \psi}{\partial x^2} + \frac{\partial^2 \psi}{\partial z^2}\right) \tag{3.70}$$

The closed system of Eqs. (3.68) and (3.70), with only two variables ω_z and ψ, can provide the flow solution, instead of the system of Navier-Stokes equation. However, the application is limited to two-dimensional flows as the stream function can be defined only on a two-dimensional space. In the cylindrical coordinate (r, θ, z), the vorticity vector is defined by

$$\begin{aligned}\boldsymbol{\omega} &= (\omega_r, \omega_\theta, \omega_z)\\ &= \left(\frac{1}{r}\left(\frac{\partial u_z}{\partial \theta} - \frac{\partial}{\partial z} r u_\theta\right), \frac{\partial u_r}{\partial z} - \frac{\partial u_z}{\partial r}, \frac{1}{r}\left(\frac{\partial}{\partial r} r u_\theta - \frac{\partial u_r}{\partial \theta}\right)\right)\end{aligned} \tag{3.71}$$

The vorticity equation in this system is given

$$\begin{aligned}&\frac{\partial \omega_r}{\partial t} + u_r \frac{\partial \omega_r}{\partial r} + \frac{u_\theta}{r}\frac{\partial \omega_r}{\partial \theta} + u_z \frac{\partial \omega_r}{\partial z}\\ &= \omega_r \frac{\partial u_r}{\partial r} + \frac{\omega_\theta}{r}\frac{\partial u_r}{\partial \theta} + \omega_z \frac{\partial u_r}{\partial z} + \nu\left(\nabla^2 \omega_r - \frac{\omega_r}{r^2} - \frac{2}{r^2}\frac{\partial \omega_\theta}{\partial \theta}\right)\end{aligned} \tag{3.72}$$

$$\begin{aligned}&\frac{\partial \omega_\theta}{\partial t} + u_r \frac{\partial \omega_\theta}{\partial r} + \frac{u_\theta}{r}\frac{\partial \omega_\theta}{\partial \theta} + u_z \frac{\partial \omega_\theta}{\partial z}\\ &= \omega_r \frac{\partial u_\theta}{\partial r} + \frac{\omega_\theta}{r}\frac{\partial u_\theta}{\partial \theta} + \omega_z \frac{\partial u_\theta}{\partial z} - \frac{u_\theta \omega_\theta}{r} + \nu\left(\nabla^2 \omega_\theta - \frac{\omega_\theta}{r^2} + \frac{2}{r^2}\frac{\partial \omega_r}{\partial \theta}\right)\end{aligned} \tag{3.73}$$

$$\begin{aligned}&\frac{\partial \omega_z}{\partial t} + u_r \frac{\partial \omega_z}{\partial r} + \frac{u_\theta}{r}\frac{\partial \omega_z}{\partial \theta} + u_z \frac{\partial \omega_z}{\partial z}\\ &= \omega_r \frac{\partial u_z}{\partial r} + \frac{\omega_\theta}{r}\frac{\partial u_z}{\partial \theta} + \omega_z \frac{\partial u_z}{\partial z} + \nu \nabla^2 \omega_z\end{aligned} \tag{3.74}$$

where

$$\nabla^2 = \frac{\partial^2}{\partial r^2} + \frac{1}{r}\frac{\partial}{\partial r} + \frac{1}{r^2}\frac{\partial^2}{\partial \theta^2} + \frac{\partial^2}{\partial z^2} \tag{3.75}$$

3.3.2 Vorticity in turbulence

In sufficiently high Reynolds number, the vorticity equation Eq. (3.62) is difficult to solve because of nonlinearity. In such case, instead of direct analysis of instantaneous vorticity, statistical quantities are studied in the same manner as turbulence (Section 2.1). Following Reynolds decomposition, introduced in Section 2.1.2, the equation for the mean vorticity is derived in this section.

Eq. (3.63) is modified using relations $\partial u_j/\partial x_j = 0$ ($\nabla \cdot \boldsymbol{u} = 0$) and $\partial \omega_j/\partial x_j = 0$ ($\nabla \cdot \boldsymbol{\omega} = 0$) to obtain the conservative form of the vorticity equation through the similar operation to derive the conservative form of the Navier-Stokes equation, Eqs. (1.96)–(1.100):

$$\frac{\partial \omega_i}{\partial t} + \frac{\partial \omega_i u_j}{\partial x_j} = \frac{\partial u_i \omega_j}{\partial x_j} + \nu \frac{\partial^2 \omega_i}{\partial x_j \partial x_j} \tag{3.76}$$

Introducing the Reynolds decomposition for velocity and vorticity, $u_i = \overline{u_i} + u_i'$ and $\omega_i = \overline{\omega_i} + \omega_i'$ are substituted into Eq. (3.76):

$$\frac{\partial}{\partial t}\left(\overline{\omega_i} + \omega_i'\right) + \frac{\partial}{\partial x_j}\left(\overline{\omega_i}\,\overline{u_j} + \omega_i'\overline{u_j} + u_j'\overline{\omega_i} + \omega_i' u_j'\right)$$

$$= \frac{\partial}{\partial x_j}\left(\overline{u_i}\,\overline{\omega_j} + u_i'\overline{\omega_j} + \omega_j'\overline{u_i} + u_i'\omega_j'\right) + \nu\frac{\partial^2}{\partial x_j \partial x_j}\left(\overline{\omega_i} + \omega_i'\right) \tag{3.77}$$

Considering $\overline{\overline{u_i}} = \overline{u_i}$, $\overline{\overline{\omega_i}} = \overline{\omega_i}$, $\overline{u_i'} = 0$ and $\overline{\omega_i'} = 0$ (see Section 2.1.2), and averaging all the terms of Eq. (3.77), the equation for the mean vorticity is given

$$\frac{\partial \overline{\omega_i}}{\partial t} + \frac{\partial}{\partial x_j}\left(\overline{\omega_i}\,\overline{u_j}\right) = \frac{\partial}{\partial x_j}\left(\overline{u_i}\,\overline{\omega_j}\right) + \frac{\partial \overline{u_i'\omega_j'}}{\partial x_j} - \frac{\partial \overline{\omega_i' u_j'}}{\partial x_j} + \nu\frac{\partial^2 \overline{\omega_i}}{\partial x_j \partial x_j} \tag{3.78}$$

The left-hand side describes transport of the mean vorticity. The first and second terms of right-hand side indicates vorticity production by mean and turbulent fluid strains, respectively. The third term takes the form of divergence of the flux $\overline{\omega_j' u_j'}$, indicating turbulent diffusion. According to the gradient hypothesis, used in Eqs. (2.23) and (2.88), this term may be modeled as

$$\overline{\omega_i' u_j'} = -\frac{\nu_t}{C_\omega}\frac{\partial \overline{\omega_i}}{\partial x_j} \tag{3.79}$$

where ν_T is the eddy viscosity and C_ω is the constant. The last term describes viscous diffusion of the mean vorticity.

3.3.3 Rankine vortex

Consider an axisymmetric rotational flow with uniform vorticity ω within a circle of radius a, surrounded by irrotational flow, in two-dimensional polar coordinate (r, θ), where the stream function is defined in Eq. (1.141) and ∇^2 is given in Eq. (1.10). Eq. (3.70) in this coordinate is given by

$$\omega = -\nabla^2\psi \tag{3.80}$$

$$= -\left(\frac{1}{r}\frac{\partial}{\partial r}\left(r\frac{\partial \psi}{\partial r}\right) + \frac{1}{r^2}\frac{\partial^2 \psi}{\partial \theta^2}\right) \tag{3.81}$$

As the flow is axisymmetric and thus independent on θ, ψ is governed by

$$\frac{1}{r}\frac{d}{dr}\left(r\frac{d\psi}{dr}\right) = \begin{cases} -\omega & (r < a) \\ 0 & (r > a) \end{cases} \tag{3.82}$$

In $r < a$, the solution of Eq. (3.82) takes the form:

$$\psi = -\frac{\omega}{4}r^2 + C_1 \ln r + C_2 \quad (r < a) \tag{3.83}$$

As ψ becomes singular at $r = 0$, the constant C_1 needs to be zero. Assuming arbitrary constant $C_2 = 0$ is given, the solution is determined by

$$\psi = -\frac{\omega}{4}r^2 \quad (r < a) \tag{3.84}$$

In $r > a$, the solution form is given

$$\psi = C_3 \ln r + C_4 \quad (r > a) \tag{3.85}$$

Since the value of ψ and the azimuthal velocity $u_\theta = -d\psi/dr$ need to match at $r = a$, two equations $-\omega a^2/4 = C_3 \ln a + C_4$ and $-\omega a/2 = C_3/a$ determine the constants; $C_3 = -\omega a^2/2$ and $C_4 = -\omega a^2/4 + (\omega a^2/2)\ln a$:

$$\psi = -\frac{\omega}{2}\ln\frac{r}{a} - \frac{\omega}{4}a^2 \quad (r > a) \tag{3.86}$$

The flow described by Eqs. (3.84) and (3.86) is the well-known Rankine vortex. The azimuthal velocity in this flow is determined as

$$u_\theta = -\frac{\partial\psi}{\partial r} = \begin{cases} \dfrac{1}{2}\omega r & (r < a) \\ \dfrac{a^2}{2}\dfrac{\omega}{r} & (r > a) \end{cases} \tag{3.87}$$

The trivial result of radial velocity $u_r = r^{-1}\partial\psi/\partial\theta = 0$ is mentioned.

Next we consider local pressure in this flow. Substituting $u_r = 0$ and $u_z = 0$ (as no z-axis in the two-dimensional polar system) into the radial component of Euler equation in the cylindrical coordinate, Eq. (1.82), the pressure gradient, balancing with the centrifugal force, is given as

$$\frac{dp}{dr} = \rho\frac{u_\theta^2}{r} \tag{3.88}$$

Substituting Eq. (3.87) into Eq. (3.88) and integrating, the pressure can be determined by matching inner and outer pressure at $r = a$ and imposing the far-field boundary condition $p = p_\infty$ at $r \to \infty$:

$$p = \begin{cases} -\dfrac{\rho}{8}\omega^2\left(2a^2 - r^2\right) + p_\infty & (r < a) \\ -\dfrac{\rho}{8}\omega^2\dfrac{a^4}{r^2} + p_\infty & (r > a) \end{cases} \tag{3.89}$$

Fig. 3.7 shows the distribution of azimuthal velocity and the pressure of the Rankine vortex. The absolute velocity is maximum at the edge of a vortex core ($r/a = 1$), while the pressure depresses in the core and achieved to be minimum at the center ($r/a = 0$). Any vortices causes the pressure depression within the cores, resulting in various mechanical effects to forces acting on objects in flows and motion of the objects such as drag force variations shown in Fig. 2.2.

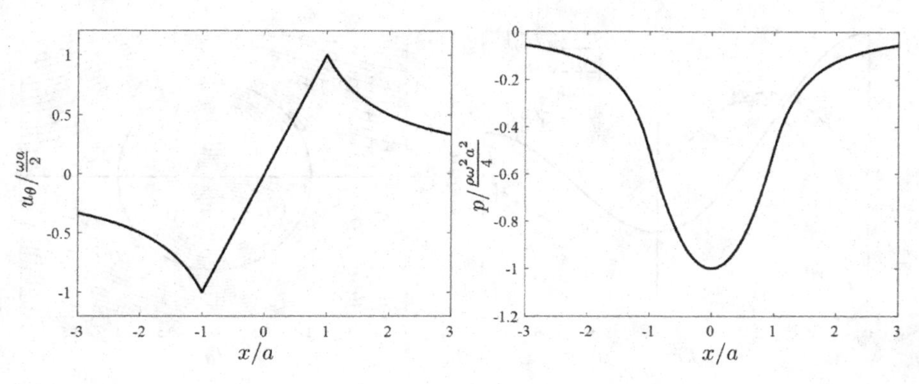

FIGURE 3.7
Radial variations of azimuthal velocity (left) and pressure (right) in a Rankine vortex.

3.3.4 Batchelor's vortex pair

Consider a flow with vorticity in proportion to the stream function:

$$\omega = k^2\psi \tag{3.90}$$

where k is the constant. The substitution into Eq. (3.81) yields Bessel's differential equation:

$$\frac{\partial^2\psi}{\partial r^2} + \frac{1}{r}\frac{\partial\psi}{\partial r} + \frac{1}{r^2}\frac{\partial^2\psi}{\partial\theta^2} + k^2\psi = 0 \tag{3.91}$$

The solution takes the form (see Section A.8.1 in Appendix):

$$\psi = C_1 J_1(kr)\sin\theta + C_2 Y_1(kr)\sin\theta \tag{3.92}$$

where J_1 and Y_1 is the Bessel function of the first kind and second kind, respectively. C_1 and C_2 are the constant. Since Y_1 is singular at $kr = 0$ (see Fig. 3.8 left), C_2 must be zero. The solution form is then:

$$\psi = C_1 J_1(kr)\sin\theta \tag{3.93}$$

According to derivative, Eq. (A.57) in Appendix, and recurrence property, Eq. (A.51), of the Bessel function $J_n(x)$, $dJ_1(x)/dx = J_0(x) - J_1(x)/x$. The velocity is given

$$u_\theta = -\frac{\partial\psi}{\partial r} = -C_1 k\left(J_0(kr) - \frac{J_1(kr)}{kr}\right)\sin\theta \tag{3.94}$$

$$u_r = \frac{1}{r}\frac{\partial\psi}{\partial\theta} = \frac{C_1}{r}J_1(kr)\cos\theta \tag{3.95}$$

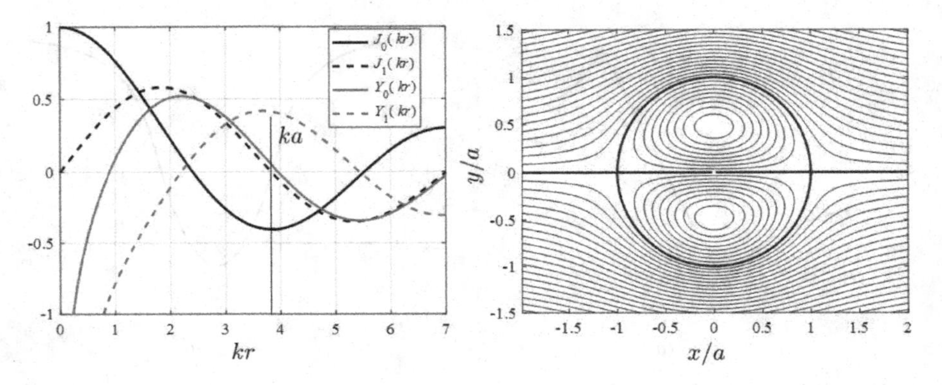

FIGURE 3.8
Bessel functions of the first kind $J_0(kr)$ and $J_1(kr)$ (left) and streamlines of the Batchelor's vortex pair (right).

We find from Eq. (3.93) that $\psi = 0$ for any θ when $J_1(kr)$=0; if ka is the first root of $J_1(kr)$, $J_1(ka) = 0$ at $r = a$ (see Fig. 3.8 left). Accordingly, the streamline of $\psi = 0$ draws a circle of radius a. The velocity on the circle $(r = a)$ is given from Eqs. (3.94) and (3.95) by

$$u_\theta = -C_1 k J_0(ka) \sin\theta \quad \text{and} \quad u_r = 0 \tag{3.96}$$

$u_r = 0$ at $r = a$ indicates the impermeable surface (see Section 1.6.4). If irrotational uniform flow across a sphere, introduced in Section 1.6.4, is considered in $r > a$, u_θ, given in Eq. (1.151), must match with Eq. (3.96):

$$-\frac{3}{2}U \sin\theta = -C_1 k J_0(ka) \sin\theta$$

$$\therefore C_1 = \frac{3}{2}\frac{U}{kJ_0(ka)} \tag{3.97}$$

$$\therefore \psi = \frac{3}{2}\frac{U}{k}\frac{J_1(kr)}{J_0(ka)} \sin\theta \tag{3.98}$$

Fig. 3.8 (right) shows the streamlines drawn by Eq. (3.98) in $r < a$ and Eq. (1.152) in $r > a$, the so-called Batchelor's vortex pair[8)], which describes a vortex pair in uniform flow or a pair of counter-rotating vortices moving with translation velocity (introduced in Section 3.4.3). As a similar flow pattern of a vortex sphere traveling with constant velocity, Hill's spherical vortex is well known[49)]. The counter-rotating flow in a sphere, derived in these studies, has been observed within a rising air bubble in liquid. The internal circulation contributes to reduce viscous drag of a bubble, resulting in faster terminal velocity than that of a solid particle, which is addressed in Section 3.6.2.

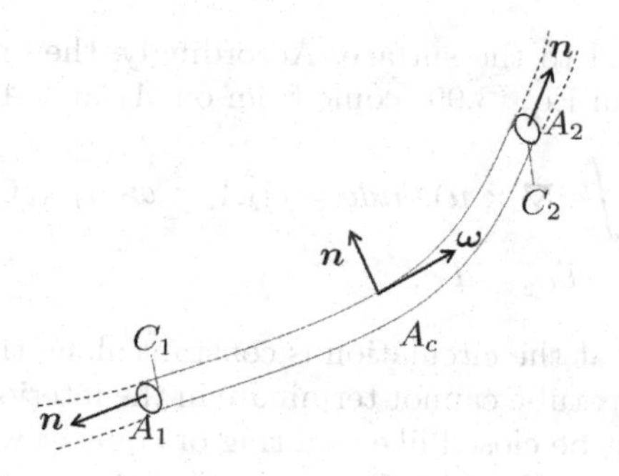

FIGURE 3.9
Illustration of a section of a vortex tube.

3.4 Vorticity dynamics

In this section, we consider properties of a vortex following the definitions: In vorticity field, a vortex line is defined as a curve along the tangential direction of local $\boldsymbol{\omega}$. A surface bounded by many vortex lines through a specific curve is called the vortex surface. A vortex tube formed along vortex lines has the closed vortex surface. A vortex filament is a thin vortex tube adjacently surrounded by irrotational flow.

According to Stokes' theorem (see Eq. (A.33) in Appendix), a circulation defined by tangential line integral of fluid velocity along a closed curve C (see Eq. (1.129)) relates to the vorticity $\boldsymbol{\omega} = \nabla \times \boldsymbol{u}$ integrated over the area A enclosed by C:

$$\Gamma_C = \oint_C \boldsymbol{u} \cdot d\boldsymbol{x} = \int_A \boldsymbol{\omega} \cdot \boldsymbol{n} da \tag{3.99}$$

where $d\boldsymbol{x}$ is differential vector along boundary curve C and $\boldsymbol{n}$ is the outward unit normal vector of the surface.

Consider a section of a vortex tube with a closed surface of the area A_c, which is bounded by two plane surfaces of the area A_1 and A_2, as shown in Fig. 3.9. The curves C_1 and C_2 enclose S_1 and S_2. The circulations at both the cross sections are thus given by Eq. (3.99) as $\Gamma_{C_1} = \boldsymbol{\omega} \cdot \boldsymbol{n} A_1 = \omega_1 A_1$ and $\Gamma_{C_2} = \boldsymbol{\omega} \cdot \boldsymbol{n} A_2 = \omega_2 A_2$. On the one hand, the last term of Eq. (3.99) is associated with the divergence theorem, Eq. (A.29) in Appendix:

$$\int_A \boldsymbol{\omega} \cdot \boldsymbol{n} da = \int_V \nabla \cdot \boldsymbol{\omega} dv \tag{3.100}$$

Here as $\nabla \cdot \boldsymbol{\omega} = \nabla \cdot (\nabla \times \boldsymbol{u}) = 0$, from the vector formula Eq. (A.1) in Appendix, $\int_A \boldsymbol{\omega} \cdot \boldsymbol{n} da = 0$. We readily find that $\boldsymbol{\omega} \cdot \boldsymbol{n}$=0 on the closed surface

A_c, as $\boldsymbol{\omega}$ is tangential to the surface. Accordingly, the only contributions to the surface integral in Eq. (3.99) come from on A_1 and A_2:

$$\int_A (\boldsymbol{\nabla} \times \boldsymbol{u}) \cdot \boldsymbol{n} da = \omega_2 A_2 - \omega_1 A_1 = 0 \tag{3.101}$$

$$\therefore \varGamma_{C_2} = \varGamma_{C_1} \tag{3.102}$$

This result states that the circulation is constant along the vortex tube, indicating that the vortex tube cannot terminate in the interior fluid. Accordingly, the vortex tube must be closed like as a ring or connect with the fluid boundary. We also find from Eq. (3.101), $\omega_2 A_2 = \omega_1 A_1 = \text{const.}$, indicating the mean vorticity is inversely proportional to the cross-sectional area, which results in vortex intensification owing to stretching of a vortex tube (a so-called stretch-and-intensification process).

3.4.1 Biot-Savart law

Following the Helmholtz decomposition, Eq. (1.164), arbitrary velocity field $\boldsymbol{u}$ is expressed by the sum of irrotational and rotational velocities:

$$\boldsymbol{u} = \boldsymbol{\nabla} \phi + \boldsymbol{u}_v \tag{3.103}$$

where ϕ is the velocity potential and the rotational velocity $\boldsymbol{u}_v$ defined by

$$\boldsymbol{u}_v = \boldsymbol{\nabla} \times \boldsymbol{A} \tag{3.104}$$

where $\boldsymbol{A}$ is the vector potential satisfying $\boldsymbol{\nabla} \cdot \boldsymbol{A} = 0$. Taking the divergence of Eq. (3.103) and considering vector formula Eq. (A.1) in Appendix,

$$\boldsymbol{\nabla} \cdot \boldsymbol{u} = \boldsymbol{\nabla} \cdot \boldsymbol{\nabla} \phi + \cancel{\boldsymbol{\nabla} \cdot (\boldsymbol{\nabla} \times \boldsymbol{A})} = \boldsymbol{\nabla}^2 \phi \tag{3.105}$$

Accordingly, Eq. (3.103) satisfies the continuity condition $\boldsymbol{\nabla} \cdot \boldsymbol{u} = 0$ if the Laplace equation for ϕ is satisfied.

The rotation of Eq. (3.103) gives (see vector formula Eqs. (A.2) and (A.6) in Appendix):

$$\begin{aligned} \boldsymbol{\nabla} \times \boldsymbol{u} &= \boldsymbol{\omega} \\ &= \cancel{\boldsymbol{\nabla} \times \boldsymbol{\nabla} \phi} + \boldsymbol{\nabla} \times (\boldsymbol{\nabla} \times \boldsymbol{A}) = \boldsymbol{\nabla}(\boldsymbol{\nabla} \cdot \boldsymbol{A}) - \boldsymbol{\nabla}^2 \boldsymbol{A} = -\boldsymbol{\nabla}^2 \boldsymbol{A} \end{aligned} \tag{3.106}$$

$$\therefore \boldsymbol{\nabla}^2 \boldsymbol{A} = -\boldsymbol{\omega} \tag{3.107}$$

In two-dimensional space, $\boldsymbol{A}$ corresponds to the stream function, ψ (Section 1.5), and Eq. (3.107) thus describes the Poisson equation for ψ, Eq. (3.70). The solution of the Poisson equation Eq. (3.107) is given by a Green function (Eq. (A.44) in Appendix):

$$\boldsymbol{A} = \frac{1}{4\pi} \int \frac{\boldsymbol{\omega}(\boldsymbol{x}')}{|\boldsymbol{r}|} d\boldsymbol{x}' \tag{3.108}$$

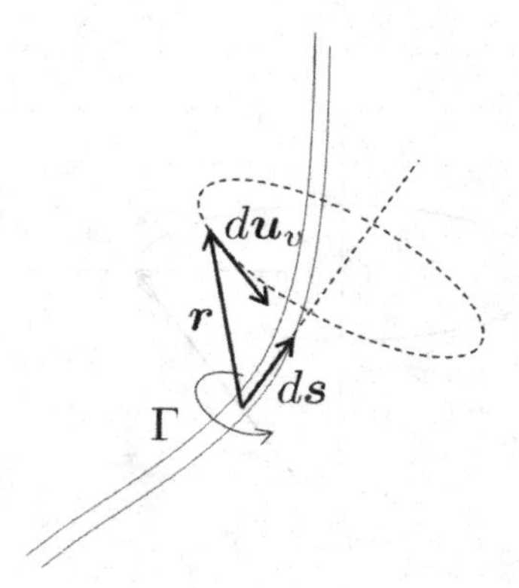

FIGURE 3.10
Illustration of a vortex filament.

where $\boldsymbol{r} = \boldsymbol{x} - \boldsymbol{x}'$. $\boldsymbol{u}_v$, Eq. (3.104), is thus given by

$$\boldsymbol{u}_v = \frac{1}{4\pi}\int \frac{\boldsymbol{\omega}(\boldsymbol{x}') \times \boldsymbol{r}}{|\boldsymbol{r}|^3} d\boldsymbol{x}' \tag{3.109}$$

Consider a small section of a vortex filament with length $ds = |d\boldsymbol{s}|$ ($\boldsymbol{s}$ is the tangent vector to the filament) and cross section area of a (see Fig. 3.10). As the definition of a vortex filament states that ω=0 outside the filament, the integration of Eq. (3.109) over the domain is reduced to the line integral along the filament. The contribution of this section, ds, of the filament to induce fluid velocity is thus given by

$$d\boldsymbol{u}_v = \frac{1}{4\pi}\frac{\boldsymbol{\omega} \times \boldsymbol{r}}{|\boldsymbol{r}|^3} ads \tag{3.110}$$

Since $\boldsymbol{\omega} ads = \omega a d\boldsymbol{s} = \Gamma d\boldsymbol{s}$, as $\boldsymbol{\omega} \parallel d\boldsymbol{s}$, Eq. (3.110) is transformed to the well-known Biot-Savart law:

$$d\boldsymbol{u}_v = \frac{\Gamma}{4\pi}\frac{d\boldsymbol{s} \times \boldsymbol{r}}{|\boldsymbol{r}|^3} \tag{3.111}$$

3.4.2 Point vortex

When a vortex filament with a straight rotating axis is considered, the vortex induces rotational flow on the plane perpendicular to the axis (Fig. 3.11). If the center of the vortex is located at the origin of the cylindrical coordinate (r, θ, z), there is only vorticity component of ω_z over z-axis; $\boldsymbol{\omega} = (0, 0, \omega_z)$. Since the fluid velocity induced by the vortex ($\boldsymbol{u}_v$) is defined by the Biot-Savart law, Eq. (3.111), the induced velocity $\boldsymbol{u}_v = (u_{vr}, u_{v\theta}, u_{vz})$ at $(r, 0, 0)$

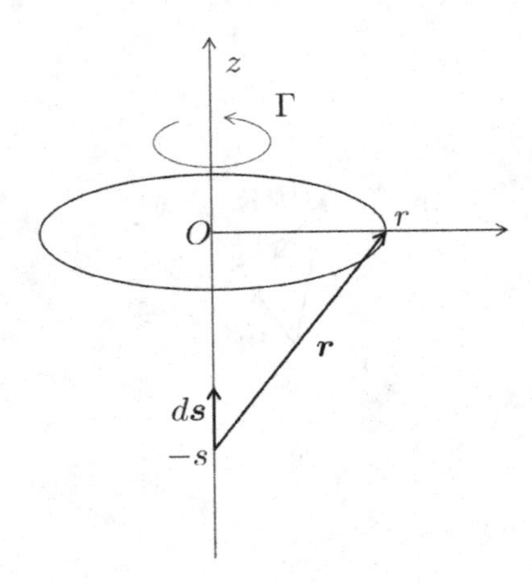

FIGURE 3.11
Illustration of a vortex filament with a straight rotating axis.

by the portion of the filament ds at $z = -s$ is given by

$$d\boldsymbol{u}_v = \frac{\Gamma}{4\pi}\frac{d\boldsymbol{s} \times \boldsymbol{r}}{|\boldsymbol{r}|^3} = \frac{\Gamma}{4\pi|\boldsymbol{r}|^3}(0, rds, 0) \tag{3.112}$$

where $\boldsymbol{r} = (r, 0, -s)$ and $d\boldsymbol{s} = (0, 0, ds)$. There is only $u_{v\theta}$ component, as anticipated. The integration of Eq. (3.112) gives

$$u_{v\theta} = \frac{\Gamma}{4\pi}\int_{-\infty}^{\infty}\frac{ads}{(r^2 + s^2)^{3/2}} = \frac{\Gamma}{2\pi r} \tag{3.113}$$

Since the vortex is axisymmetric and uniform over z, Eq. (3.113), derived at $z = 0$, is uniquely defined over z coordinate. In this case, we may regard that the vortex is placed on the two-dimensional plane perpendicular to the rotating axis. Accordingly, a point vortex can be defined to have infinite vorticity localized at a point, which may be considered as a singularity of the vorticity field:

$$\omega(\boldsymbol{x}) = \Gamma\delta(\boldsymbol{x} - \boldsymbol{x}') \tag{3.114}$$

where Γ is the circulation, δ is the Dirac delta function (see Section A.4 in Appendix), and $\boldsymbol{x}'$ is the location of the point vortex. The vorticity field formed in a system of N point vortices is described as the superposition of Eq. (3.114) (see Fig. 3.12):

$$\omega(\boldsymbol{x}) = \sum_{n=1}^{N}\Gamma_n\delta(\boldsymbol{x} - \boldsymbol{x}_n) \tag{3.115}$$

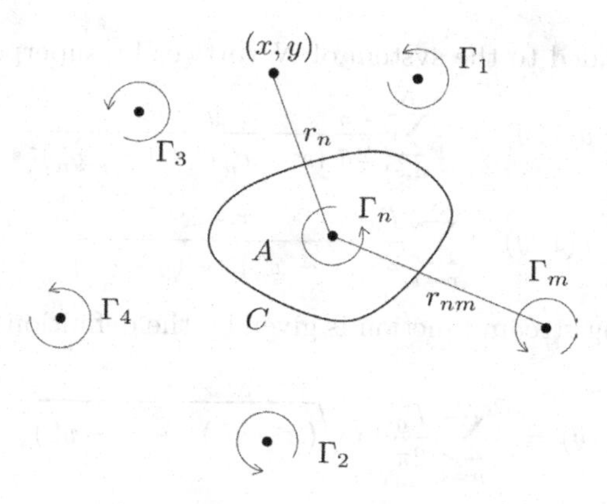

FIGURE 3.12
System of point vortices.

According to Stokes' theorem (see Section A.5 in Appendix), the circulation around C in a region bounded by C in the point vortex field (Fig. 3.12) is expressed by

$$\Gamma_C = \oint_C \boldsymbol{u} \cdot d\boldsymbol{x} = \int_A (\boldsymbol{\nabla} \times \boldsymbol{u}) \cdot \boldsymbol{n} da = \int_A \omega da \tag{3.116}$$

In the case of a single vortex within A, the substitution of Eq. (3.114) into Eq. (3.116) states consistent result:

$$\Gamma_C = \int_A \Gamma \delta(\boldsymbol{x} - \boldsymbol{x}') da = \Gamma \tag{3.117}$$

Here positive Γ indicates that anticlockwise rotation of velocity is induced by the vortex. The integration of Eq. (3.114) multiplied by $\boldsymbol{x}$ gives the vortex center location (by definition of the delta function):

$$\boldsymbol{x}' = \frac{1}{\Gamma} \int_A \boldsymbol{x} \omega da \left(= \int_A \delta (\boldsymbol{x} - \boldsymbol{x}') da \right) \tag{3.118}$$

The velocity at (x, y), induced by the nth single point vortex, can be estimated by Eq. (3.113):

$$u(x, y) = -\frac{\Gamma_n}{2\pi} \frac{y - y'_n}{(x - x'_n)^2 + (y - y'_n)^2} \tag{3.119}$$

$$v(x, y) = \frac{\Gamma_n}{2\pi} \frac{x - x'_n}{(x - x'_n)^2 + (y - y'_n)^2} \tag{3.120}$$

This can be extended to the system of N vortices by superposition:

$$u(x,y) = -\sum_{n=1}^{N} \frac{\Gamma_n}{2\pi} \frac{y - y'_n}{(x - x'_n)^2 + (y - y'_n)^2} \tag{3.121}$$

$$v(x,y) = \sum_{n=1}^{N} \frac{\Gamma_n}{2\pi} \frac{x - x'_n}{(x - x'_n)^2 + (y - y'_n)^2} \tag{3.122}$$

The corresponding stream function is given by the definition of Eq. (1.138) as

$$\psi(x,y) = -\sum_{n=1}^{N} \frac{\Gamma_n}{2\pi} \ln \sqrt{(x - x'_n)^2 + (y - y'_n)^2} \tag{3.123}$$

Considering the nth vortex is transported by the flow induced by the other $N-1$ vortices, the moving velocity of the nth vortex is given by

$$\frac{dx'_n}{dt} = -\sum_{n \neq m}^{N} \frac{\Gamma_m}{2\pi} \frac{y'_n - y_m}{(x'_n - x'_m)^2 + (y'_n - y'_m)^2} \tag{3.124}$$

$$\frac{dy'_n}{dt} = \sum_{n \neq m}^{N} \frac{\Gamma_m}{2\pi} \frac{x'_n - x'_m}{(x'_n - x'_m)^2 + (y'_n - y'_m)^2} \tag{3.125}$$

3.4.3 Pair vortices

Two different types of a vortex pair, counter-rotating and corotating vortices, yield distinct behaviors and induced flow field. The former consists two adjacent vortices with the opposite circulations; $\Gamma_1 = -\Gamma_2 = \Gamma$, where Γ_1 and Γ_2 are the circulations of the first and second point vortices, respectively. The latter one has the identical circulations $\Gamma_1 = \Gamma_2 = \Gamma$. The interactions of these pair vortices have important roles in ocean fluid dynamics, especially in breaking waves (introduced in Chapter 8). In this section, the fundamental features of the pair vortices are introduced in a framework of the point vortex analysis.

3.4.3.1 Counter-rotating pair

Consider two-point vortices horizontally aligned with spacing ϵ. The vortex 1 located at $\boldsymbol{x}_1$=$(-\epsilon/2, 0)$ has the circulation $\Gamma_1 = \Gamma$ (anticlockwise rotation is defined to be positive), and the vortex 2 located at $\boldsymbol{x}_2$=$(\epsilon/2, 0)$ with $\Gamma_2 = -\Gamma$, shown in Fig. 3.13. The fluid velocity induced at the middle of these vortices (at the origin of the coordinate) is given by Eqs. (3.121) and (3.122) with $N = 2$:

$$u\,|_{(0,0)} = 0, \quad v\,|_{(0,0)} = \frac{2\Gamma}{\pi\epsilon} \tag{3.126}$$

The vertical flow is induced along the straight streamline passing the midpoint (Fig. 3.13 right), and the maximum velocity, Eq. (3.126), is achieved at the

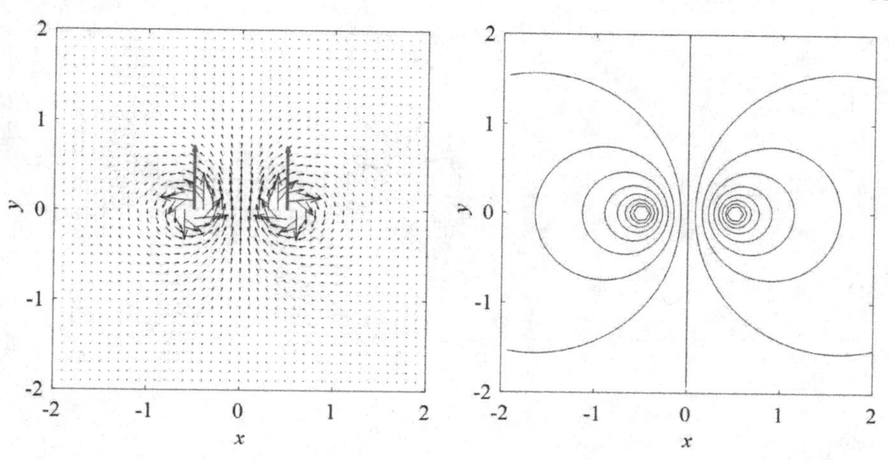

FIGURE 3.13
Fluid velocity (black vector) induced by the counter-rotating pair vortices located at $\boldsymbol{x}_1 = (-0.5, 0)$ and $\boldsymbol{x}_2 = (0.5, 0)$ and translational moving velocity of the vortices (gray vector with different arrow scale with the black one); left, and the streamline of the vortex-induced flow; right.

origin (see Fig. 3.13 left). The moving velocity of the vortices is estimated from Eqs. (3.124) and (3.125) as

$$\frac{dx'_1}{dt} = \frac{dx'_2}{dt} = 0, \quad \frac{dy'_1}{dt} = \frac{dy'_2}{dt} = \frac{\Gamma}{2\pi\epsilon} \tag{3.127}$$

Accordingly, the both vortices are in translating motion in the direction perpendicular to the vortex alignment.

In a reference frame moving with the translation velocity of the counter-rotating pair, an interesting feature of the vortex-induced flow can be observed (Fig. 3.14 left). The vortices create inner and outer regions bounded by a separatrix streamline. The boundary of the regions behaves like an impermeable surface of solid cylinder installed in a uniform flow (compare with the corresponding potential flow in Fig. 1.13). In the inner region, there are a pair of cells containing closed streamlines, similar to an internal circulation pattern of the Batchelor's vortex pair (Fig. 3.8). It should be noted that the internal cells have been experimentally observed in the flow inside an air-bubble and droplet [14].

When multiple counter-rotating pairs align horizontally, upward and downward flows induced between neighboring vortices in turn, as anticipated (Fig. 3.14). The regular fluctuations induced among multiple pairs have been observed in practice to contribute to transport process of particles, such as fluctuations of motion in a bubble swarm [110], sediment suspension in the surf zone [39]. When the counter-rotating vortices approach free surface, the surface

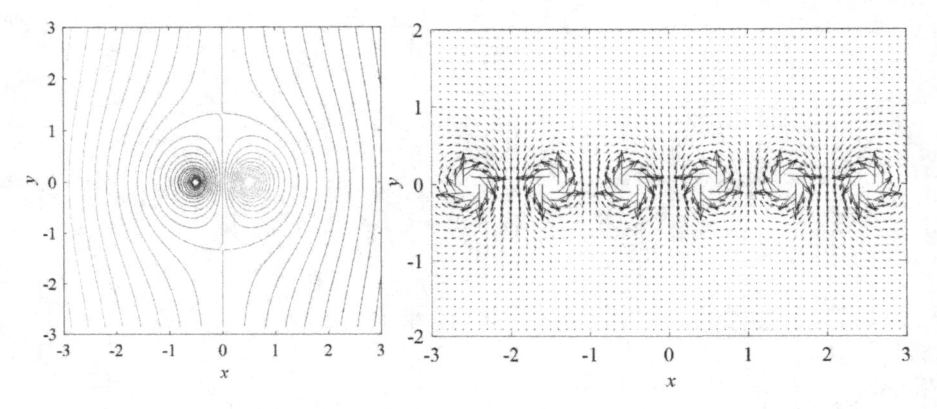

FIGURE 3.14
Streamlines of a counter-rotating pair in the moving reference with the translational moving velocity of the vortices, $\Gamma/2\pi\epsilon$, (left), and induced velocity in three pairs of counter-rotating vortices horizontally aligned (right).

is regularly deformed as the upward flow lifts the surface above the vortices and the downward one depresses it [87;120;1], which is introduced in the following sections.

3.4.3.2 Corotating pair

Similarly consider the system of vortices, horizontally placed with spacing ϵ, with the same circulation $\Gamma_1 = \Gamma_2 = \Gamma$ (Fig. 3.15). The fluid velocity at the middle point of the vortices is estimated by Eqs. (3.121) and (3.122) as $(u_{(0,0)}, v_{(0,0)}) = (0,0)$; that is, a stagnation point appears at the origin. Typically streamlines intersect at the stagnation point (see Fig. 3.15 right) where the fluid is highly stretched in the oblique direction to the vortex alignment (Fig. 3.15 left). This effect may cause instability of vortices in the direction of the vortex axis, resulting in a formation of rib-like vortex structure in Kármán vortex field [111] and in braking waves [123].

The vortices move the opposite directions with antisymmetrical velocities about the middle point, $dy'_1/dt = -\Gamma/2\pi\epsilon = -dy'_2/dt$ (Fig. 3.15 left). Accordingly, the vortices rotate around the origin with an angular velocity $\Omega = \Gamma/\pi\epsilon^2$.

3.4.3.3 Rotational behavior

In the case of $\Gamma_1 + \Gamma_2 \neq 0$, motions of the vortices exhibit rotational behavior. Fig. 3.16 shows the moving velocity of vortices together with the vortex-induced fluid velocity for the counter-rotating system of $\Gamma_1 = 2\Gamma$, $\Gamma_2 = -\Gamma$ (left) and the corotating system of $\Gamma_1 = 2\Gamma$, $\Gamma_2 = \Gamma$ (right). The both vortices of counter-rotating system (left) rotate anticlockwise on the circular

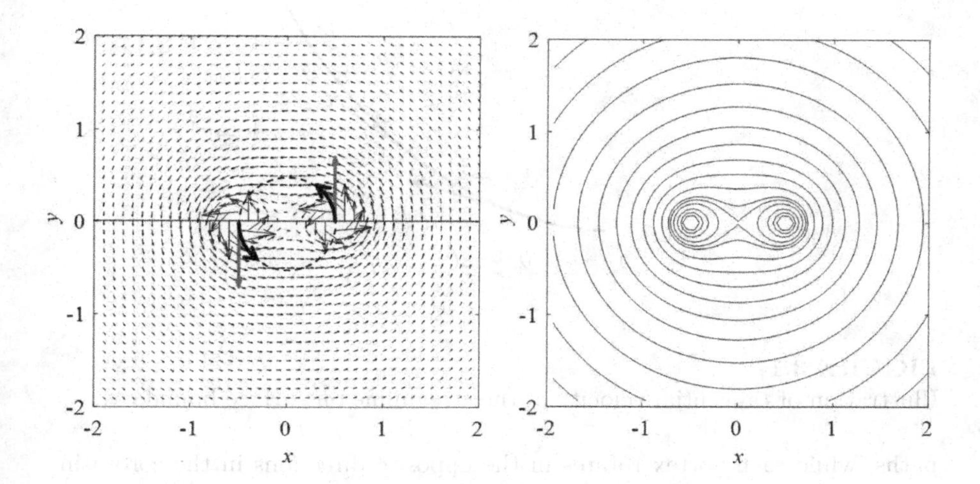

FIGURE 3.15
Fluid velocity (black vector) induced by the corotating pair vortices located at $\boldsymbol{x}_1 = (-0.5,\ 0)$ and $\boldsymbol{x}_2 = (0.5,\ 0)$ and rotational moving velocity of the vortices (gray vector with different arrow scale with the black one); left, and the streamline of the vortex-induced flow; right.

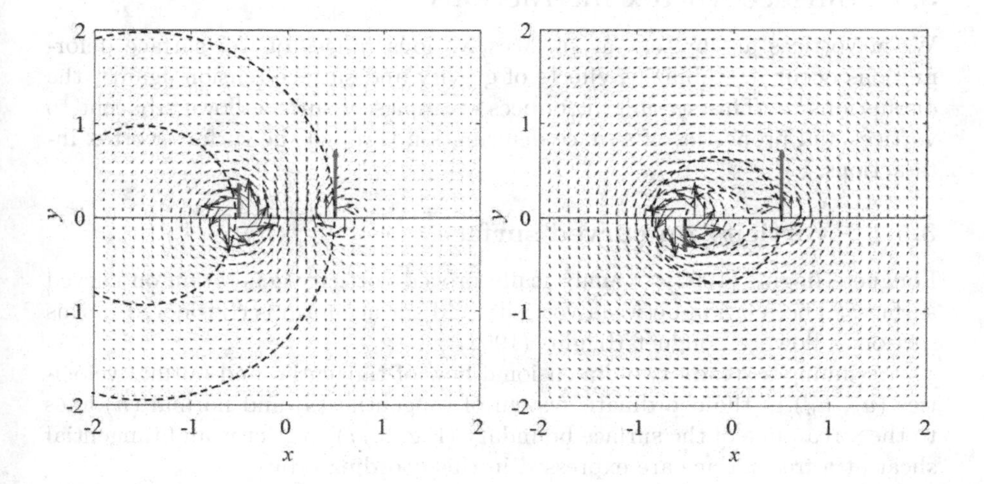

FIGURE 3.16
Fluid velocity induced by the vortex pair with different circulations; $\Gamma_1 = 2\Gamma$ and $\Gamma_2 = -\Gamma$ (left) and $\Gamma_1 = 2\Gamma$ and $\Gamma_2 = \Gamma$ (right). The broken line indicates the rotational path of each vortex.

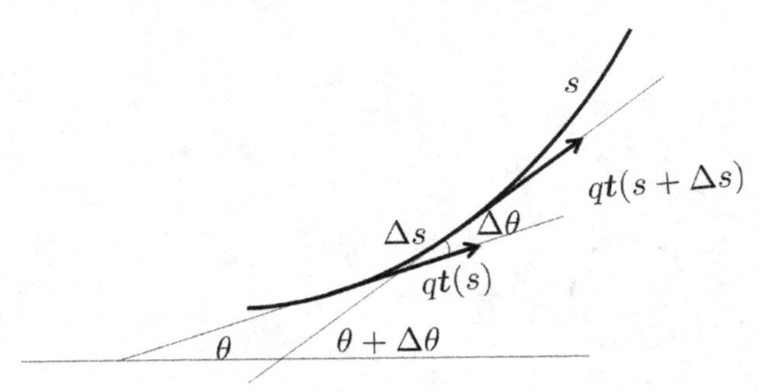

FIGURE 3.17
Illustration of tangential velocity of the streamline on surface boundary.

paths, while each vortex rotates in the opposite directions in the corotating system (anticlockwise rotation on the outer path, and clockwise on the inner path). These orbital motions occur around the vorticity center of the pair $\boldsymbol{x}_c = (\Gamma_1 \boldsymbol{x}'_1 + \Gamma_2 \boldsymbol{x}'_2) / (\Gamma_1 + \Gamma_2)$, and the angular velocity of the motion is given by $\Omega = (\Gamma_1 + \Gamma_2) / 2\pi\epsilon^2$.

3.5 Surface–vortex interactions

When vortices are present at surfaces, various ranges of free-surface deformations occur in stabilizing effects of gravity and surface tension against the disruptions[11]. This section introduces examples of surface flows adjacent to vortices and properties of surface deformation through the surface–vortex interactions.

3.5.1 Vorticity on curved surfaces

Longuet-Higgins (1992)[72] analytically proved vorticity generation on curved surfaces. The relation between vorticity and surface form is considered in this section, following Longuet-Higgins (1992)[72].

Assuming a steady two-dimensional flow of tangential and normal velocities (u_s, u_n) in the coordinate system of tangential (s) and normal (n) axes to the streamline of the surface boundary (Fig. 3.17), vorticity, and tangential shear at a free surface are expressed in this coordinate by

$$\omega = \frac{\partial u_n}{\partial s} - \frac{\partial u_s}{\partial n} \tag{3.128}$$

$$\tau_{sn} = \nu \left(\frac{\partial u_s}{\partial n} + \frac{\partial u_n}{\partial s} \right) \tag{3.129}$$

Since the tangential shear vanishes at a free surface, Eq. (3.53); $\tau_{sn} = 0$, Eqs.

(3.128) and (3.129) give the vorticity at the surface:

$$\omega = 2\frac{\partial u_n}{\partial s} \tag{3.130}$$

Introducing tangential velocity vector of the streamline on the boundary, $\boldsymbol{q} = q\boldsymbol{t}$, where $\boldsymbol{t}$ is the unit tangential vector and q is the absolute value of $\boldsymbol{q}$ (Fig. 3.17), as $\boldsymbol{t}(s+\Delta s) - \boldsymbol{t}(s) = \Delta\theta$ (see also Fig. 3.1), $\Delta u_n = q\boldsymbol{t}(s+\Delta s) - \boldsymbol{t}(s) = q\Delta\theta$:

$$\frac{\partial u_n}{\partial s} = \lim_{\Delta s \to 0} \frac{\Delta u_n}{\Delta s} = q \lim_{\Delta s \to 0} \frac{\Delta \theta}{\Delta s} = q\frac{\partial \theta}{\partial s} \tag{3.131}$$

As the curvature κ can be defined by Eq. (3.2), $d\theta/ds = \kappa$, Eq. (3.131) may give

$$\frac{\partial u_n}{\partial s} = q\kappa \tag{3.132}$$

Substituting into Eq. (3.130), the vorticity at the free surface is given by

$$\omega = 2\kappa q \tag{3.133}$$

It should be noted that the original equation derived by Longuet-Higgins (1992) is $\omega = -2\kappa q$ as the opposite sign of curvature has been defined in this book.

Longuet-Higgins (1992) also provided two interesting interpretations of Eq. (3.133). Considering a solid rotation of fluid bounded by a free surface imposing $\tau_{sn} = 0$ (see Fig. 3.18), the tangential velocity at the surface $q = \Omega r$, where Ω is angular velocity and r is the distance from the center of rotation. It is well known that the vorticity is twice the angular velocity of the rotation: $\omega = 2\Omega$. As the curvature $\kappa = 1/r$ in this case, the vorticity of the solid rotation

$$\omega = 2\Omega = 2\frac{q}{r} = 2q\kappa \tag{3.134}$$

agreeing with Eq. (3.133).

We next consider vorticity generated on waves having the surface elevation is described by

$$\eta = A\cos kx \tag{3.135}$$

where A is the amplitude, and $k = 2\pi/L$ is the wave number, L is wavelength (see Fig. 3.18). The curvature of the wave surface is given by Eq. (3.19):

$$\kappa \approx -\frac{\partial^2 \eta}{dx^2} = Ak^2 \cos kx \tag{3.136}$$

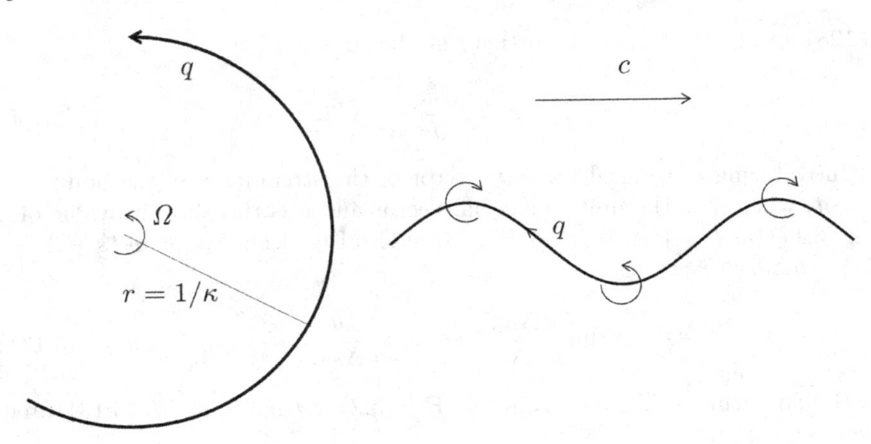

FIGURE 3.18
Illustration of solid rotation of fluid bounded by a free surface (left) and vorticity generation on surface waves (right).

As the waves propagates with speed c, $q = -c$ at the reference frame moving with the waves (see Fig. 3.18). Eq. (3.133) estimates the vorticity on the wave surface:

$$\omega = 2\kappa q = -2Ak\sigma \cos kx \tag{3.137}$$

where the angular frequency $\sigma = ck$. We find the vorticity varies in phase of wave surface elevation (Fig. 3.18).

If the steepness $Ak \to 0(A/L \to 0)$, $\omega \to 0$; that is, the curvature-induced vorticity diminishes on small amplitude waves with long wavelength. However, short, steep gravity waves, especially those with parasitic capillary waves, may contain strong vorticity near the crests (Fig. 3.19), which creates a thick boundary layer beneath the surface and contributes to diffusion and dissipation processes[72]. Fig. 3.19 shows the wave forms developed by experimental wind blowing above the surface. Irregular fluctuations of the surfaces observed at very short fetch (left) evolve into the capillary roller wave having parasitic capillary waves in front of the roller crest (right). According to Longuet-Higgins (1992)[72], the effect of capillary waves can be 50 times larger than the wind shear at the surface, the energy dissipation by micro-breaking through the vorticity generation on the capillary waves largely contributes to momentum and energy transfers in open ocean (Sutherland and Melville 2013)[94].

3.5.2 Formation of scars

Sarpkaya and Shuthon (1991)[87] experimentally demonstrated three-dimensional flow structures resulting from the interaction between an ascending counter-rotating vortex pair and free-surface. When the vortices arrive the

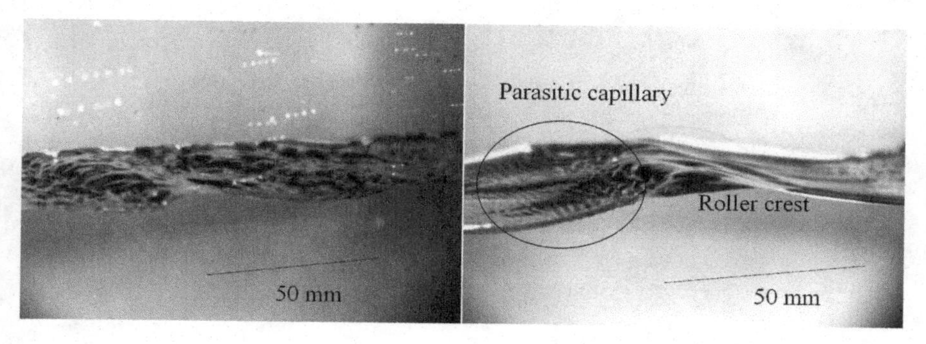

FIGURE 3.19
Water waves produced in an experimental wind tunnel; initial wave forms at 0.4 m fetch (left) and capillary roller at 1.6 m fetch (right).

free-surface, the surface is lifted to envelope the vortices and is rotated around them and entrained into bulk water at their both edges where are surface depressions (called scars). They also observed regular striations of the surfaces along the vortex axis. Tryggvason (1988) [102)] computed deformation of a free surface caused by the roll up of a vortex sheet beneath the surface. They found characteristic surface motions leading breaking wave, air entrainment, and generation of short surface waves, depending on the strength and depth of the vortex sheet. In the breaking wave type, the sub-surface vortex sheet rolls up to produce the steep wave. A scar then appears at the downstream edge. A stronger and shallower vortex sheet results in rapid growth of the scar and entrainment of the top surface fluid into the interior of flow, causing air entrainment. A weak and deep vortex sheet creates a short wave train without any breaking behavior.

The formation of scars can be confirmed by a simple computation of the flow induced by a point vortex pair, following Eqs. (3.121) and (3.122), placed beneath a free surface where imposes the kinematic boundary condition of Eq. (3.41). Fig. 3.20 shows the free surface form evolving by the flow induced by a counter-rotating vortex pair placed beneath a still water surface. The ascending flow induced between the counter-rotating pair locally lifts the free surface above the pair (Fig. 3.20 left), which is very similar to the experimental blob surface [87)]. The local depressions (scars) are observed at both sides of the blob containing the vortices. On the one hand, for the opposite orientation of the counter rotation (Fig. 3.20 right), since the vortex pair induces the surface flows converging from the both sides to the midpoint of the vortex alignment, the upper surface is entrained into the interior for forming the scar between the vortices. The identical surface behaviors owing to the subsurface counter-rotating vortices have been observed in breaking waves [123;124)]. Watanabe et al. (2005) [123)] found extensions of scars on the

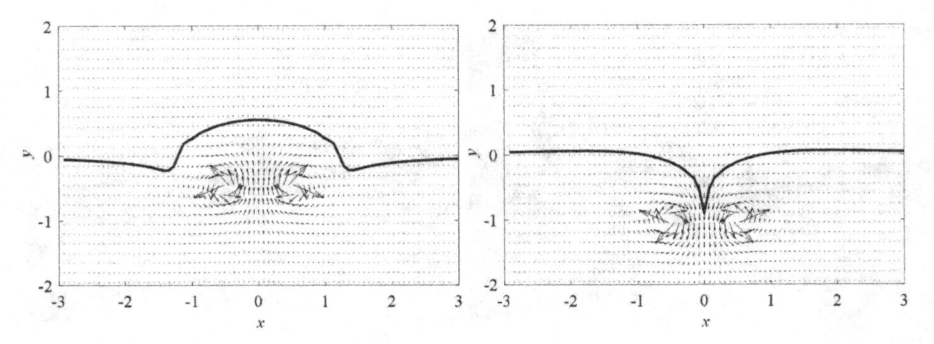

FIGURE 3.20
Computed free surface and fluid velocity induced by a counter-rotating pair placed beneath a still water level; the left and right vortices have positive and negative circulations, respectively (left). The pair having the opposite circulations (right).

rear surface of breaking waves above counter-rotating vortex array composing three-dimensional vortex structures. Watanabe and Mori (2008)[120)] observed from surface temperature measurements of breaking waves that a heated free surface on the broken wave is entrained at the scarified areas into the bulk water, and colder bulk water is ejected between the subsurface counter-rotating vortices to emerge on the surface. Further surface–vortex interactions observed in the wave breaking process are introduced in Chapter 8.

In the case of corotating pair (Fig. 3.21), the upstream surface is rolled up and the scar is produced in front of a steep wave face, which may correspond to the breaking wave scenario observed by Tryggvason (1988)[102)]. In a plunging wave breaking (barrel waves), when an overturning jet rotating around the barrel splashes onto a forward water surface, a secondary jet ejects from the plunging point to produce a preceding roller vortex. This splash-up process creates an array of multiple corotating vortices in the cross-shore direction in the surf zone. The vortex-induced surface deformation, shown in Fig. 3.21, suggests the breaking wave surface forms are determined by the own vortices.

3.5.3 Vortex ring

When a droplet splashes onto a water pool, a variety of behaviors of the receiving surfaces have been observed, depending on the Weber ($We = \rho w_t^2 d/\gamma$) and Froude ($Fr = w_t^2/gd$) numbers, where ρ is the density, w_t is the terminal velocity, the diameter d , and the surface tension γ. For sufficiently large We and Fr, say $We > 10$ and $Fr > 10$, the droplet typically creates a radial cavity crater after coalescing with the receiving surface. The cavity surface is then lifted upward for forming concentric vertical jet (Leng 2001)[53)]. A

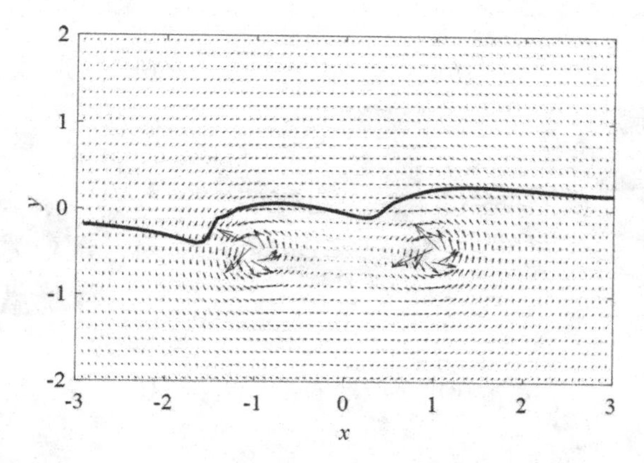

FIGURE 3.21
Computed free-surface and fluid velocity induced by a corotating pair, having positive circulation, placed beneath a still water level.

vortex ring produced at the beginning of the splash is known to evolve with interaction to the free surface (Watanabe et al. 2008) [120]. Fig. 3.22 shows the evolution of free surface and vortex cores after the droplet impact on a still water surface. The ring is initially formed on a circular edge of drop surface contacting with the receiving surface (Fig. 3.22*a*) because of vorticity generation on highly curved surface along the circular edge (Eq. (3.133)). The ring enclosing the cavity goes down when the cavity depth increases (Fig. 3.22*b* and *c*). In this phase, the free surface adjacent to the vortex is brought down by the descending flow induced by the counter-rotating vorticities along the ring axis to deepen the scar, as observed in Fig. 3.20 (right), that is, growth of the cavity is enhanced by the interaction to the enclosing vortex ring. As the counter-rotating pair moves with self-induced translation velocity, as noted in Section 3.4.3 (see Fig. 3.13), the vortex pair travels downward, as observed in Fig. 3.22*b* and *c*. Once the vortex ring moves far enough from the cavity surface (Fig. 3.22*d*), the center of the surface, released from the downward vortex-induced flow, rises up to form the vertical jet owing to the radial pressure gradients(Fig. 3.22*e*). Watanabe et al. (2008) [120] found a vorticity boundary layer produced on the cavity surface is separated by the downward induced flow and wrapped by the ring, resulting in axial counter-rotating vortices vertically extending from the cavity bottom to the ring. The induced flow by the wrapped vortices deforms the primary vortex ring along its rotational axis through an azimuthal instability, resulting in a formation of axisymmetric vortex structure involving 'petal' (counter-rotating vortices wrapped by the ring) and 'stalk' (vortex pairs connecting the free-surface and the ring). The ring keeps traveling downward with the vortex-induced translation velocity until the vorticity is sufficiently dissipated.

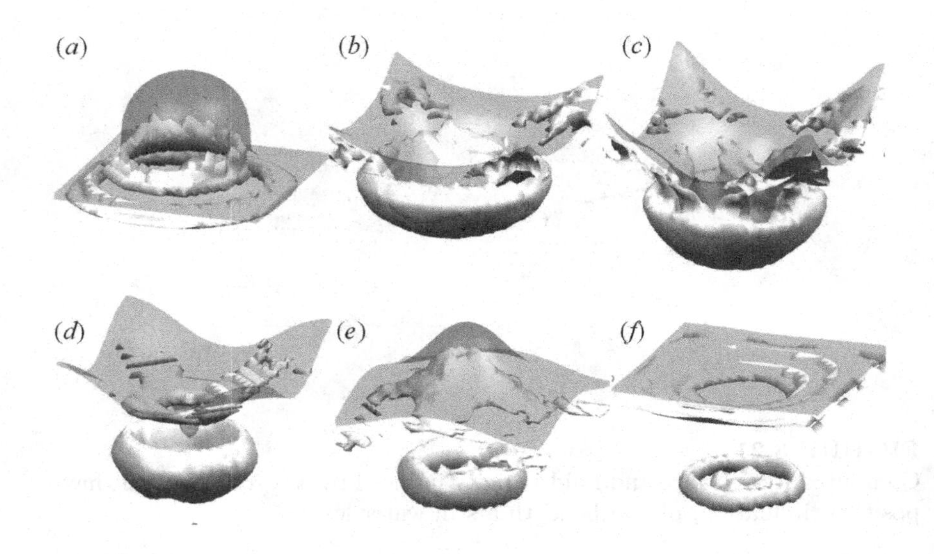

FIGURE 3.22
Evolution of cores of vortices produced by the drop impact at a still water surface.

When it rains on ocean surface, huge amounts of splashes are observed on the surface (Fig. 3.23). This event may contribute to transfer gas dissolved in raindrops and heat owned by the drops into ocean via the downward transport of vortex rings composed of rain water.

3.6 Bubbles, foams, and drops

When ocean waves break, huge amounts of air bubbles are entrained into depth. The breaking-wave-induced vortices cause complex behaviors of bubbles, while the flow is also modified by drag forces of the swarms of bubbles during active rising motions with fluctuations. When ascending bubbles arrived a sea surface, since surfactant of seawater inhibits the coalescence, a certain amount of bubbles may remain on the sea surface without the coalescence with the sea surface. The floating bubble attracts nearby bubbles, owing to bubble buoyancy on local meniscus yielding around the bubble, forms bubble clusters. The visible appearance of the bubble clusters on the sea surface is known as 'whitecap' of breaking waves. As a film of a bubble cap, i.e. the top part of the floating bubble, thins in time, it finally breaks up into sea sprays during a bursting process. Sea spray is also produced by wave crests

FIGURE 3.23
Rain fall onto ocean surface.

torn by strong wind (so-called spume process), as well as surface tension instability of overturning jets splashed on the sea surface during breaking process. Large sea sprays may fall onto the sea surface to disturb the surface flow and create another smaller bubbles and drops at the impacts, while smaller one may be evaporated during convection by wind, and suspend in atmosphere as marine aerosols for long time. A series of the processes, initially caused by wave breaking, is of huge importance in transfers of heat, moisture, gas, and momentum between atmosphere and ocean.

Fundamental theories and models describing dynamics of bubble, foam, and drop are provided in this section. The macroscopic features of the air-water two phase dynamics, whitecapping, and aeration processes in wave breaking will be introduced in Chapter 8.

3.6.1 Motion of a particle

Consider that a small discrete spherical particle of density ρ_p and diameter d moves with particle velocity $\boldsymbol{v} = (v_x, v_y, v_z)$ is contained by the element of the fluid of density ρ_f and viscosity μ moving with fluid velocity $\boldsymbol{u} = (u_x, u_y, u_z)$. The motion of equation for the particle is given by a Basset-Boussinesq-Oseen equation (so-called BBO equation):

$$\begin{aligned}\frac{\pi}{6}\rho_p d^3 \frac{d\boldsymbol{v}}{dt} &= \frac{\rho_f \pi}{8} C_d d^2 |\boldsymbol{u}-\boldsymbol{v}|\,(\boldsymbol{u}-\boldsymbol{v}) + \frac{\pi}{6}\rho_f d^3 \frac{D\boldsymbol{u}}{Dt} \\ &\quad + \frac{1}{2}\frac{\pi}{6}\rho_f d^3 \left(\frac{D\boldsymbol{u}}{Dt} - \frac{d\boldsymbol{v}}{dt}\right) + \frac{\pi}{6}\left(\rho_p - \rho_f\right) d^3 \boldsymbol{g} \\ &\quad - \frac{3}{2} d^2 \sqrt{\pi \rho_f \mu} \int_{t_0}^{t} \left(\frac{d\boldsymbol{v}}{dt'} - \frac{D\boldsymbol{u}}{Dt'}\right) (t-t')^{-1/2}\, dt' \qquad (3.138)\end{aligned}$$

where C_d is the drag coefficient, t_0 is the starting time, and the gravity vector $\boldsymbol{g} = (0, 0, -g)$. d/dt is total differentiation along the particle trajectory, which is not generally coincide with Lagrangian differentiation D/Dt along the fluid trajectory. The term on the left-hand side is the inertial force. The first term on the right-hand side is the drag force. The second one is due to the pressure gradient in the fluid surrounding the particle, caused by fluid acceleration. The third one is the added mass force of the particle relative to the ambient fluid. The fourth one indicates gravity and buoyant forces. The last term is called the Basset (or history) term that takes account of the effect of the deviation in flow pattern from steady state. The nonlinear equation, Eq. (3.138), including nonlinear parameter C_d depending on Re (see Fig. 2.2), cannot be analytically solved for arbitrary Reynolds number flow, while it is interesting to consider typical motions in some extreme cases.

For a solid particle in $Re \ll 1$, the drag force may be approximated by the Stokes law, Eq. (1.188); $3\pi\mu d\,(\boldsymbol{v} - \boldsymbol{u})$, thus $C_d = 24/Re$ (see Eq. (1.195)), where $Re = |\boldsymbol{u} - \boldsymbol{v}|d/\nu$. In this case, Eq. (3.138) may be expressed by

$$\frac{d\boldsymbol{v}}{dt} + A\boldsymbol{v} = A\boldsymbol{u} + B\frac{D\boldsymbol{u}}{Dt} + C\boldsymbol{g} + D\int_{t_0}^{t}\left(\frac{d\boldsymbol{v}}{dt'} - \frac{D\boldsymbol{u}}{Dt'}\right)\left(\sqrt{t-t'}\right)^{-1/2} dt' \quad (3.139)$$

where

$$A = \frac{36\mu}{(2\rho_p + \rho_f)\,d^2},\ B = \frac{3\rho_f}{2\rho_p + \rho_f},\ C = \frac{2\,(\rho_p - \rho_f)}{2\rho_p + \rho_f},\ D = \frac{18}{(2\rho_p + \rho_f)\,d}\sqrt{\frac{\rho_f\mu}{\pi}}$$

Consider the particle motion at the limit $\rho_p/\rho_f \to 0$; that is, ρ_p is much smaller than ρ_f, like as an air bubble in water. In this case, $B \to 3$ and $C \to -2$. If inviscid flow is assumed ($\mu \to 0$), A and D are zero, the vertical component of Eq. (3.139) is approximated by

$$\frac{dv_z}{dt} = 3\frac{Du_z}{Dt} + 2g \quad (3.140)$$

This equation interprets a bubble released in quiescent perfect fluid ($Du_z/Dt = 0$) will rise up with acceleration $2g$. If a bubble locates where is horizontal pressure gradient in water, such as a front face of water wave or bore, the horizontal component of Eq. (3.140) is:

$$\frac{dv_x}{dt} = 3\frac{Du_x}{Dt} \quad (3.141)$$

Accordingly, the bubble horizontally accelerates three times faster than the water particle. The drag force owing to the viscous stress constrains monotonic increase of the particle velocity at constant acceleration as described by Eq. (3.140) (see also Section 1.8.2).

We find a bubble is released in quiescent viscous liquid ($\boldsymbol{u} = 0$, $D\boldsymbol{u}/Dt = 0$), $A \approx 36\mu/\rho_f d^2$, $D \approx 18\sqrt{\mu/\pi\rho_f d^2}$. If the Basset term is ignored for simplicity[1], Eq. (3.139) in this case is reduced to

$$\frac{dv_z}{dt} + Av_z = 2g \tag{3.142}$$

As the initial condition $v_z = 0$ at $t = 0$, v_z is given by

$$v_z = \frac{2}{A}g\left(1 - \mathrm{e}^{-At}\right) \tag{3.143}$$

We find the acceleration $2g\mathrm{e}^{-At}$ temporally decreases and the velocity approaches terminal velocity

$$w_t = \frac{2g}{A} = \frac{gd^2}{18\nu} \tag{3.144}$$

We find the above terminal velocity is identical to Eq. (1.192) for a spherical solid particle. Accordingly, the bubble velocity Eq. (3.144) is achieved when the bubble surface is rigid and behave like the solid particle owing to surfactant effect in liquid (referred to as freezing bubble). In pure water system, the internal circulation occurring in a bubble owing to the viscous shear at the bubble surface (Fig. 3.8 right) reduces the bubble drag and thus faster terminal velocity is achieved, which is introduced in the next section.

At the inverse limit of density ratio, $\rho_f/\rho_p \to 0$, as observed for a liquid particle in air, $A \to 18\mu/d^2$, $B \to 0$, $C \to 1$, and $D \to 0$. The motion equation when a small spherical droplet is released in quiescent air field is similar to Eq. (3.142), and the identical downward $w_t = -gd^2/18\nu$ is obtained, as same as Eq. (1.192).

3.6.2 Particle drag

A drag force acting to a spherical solid particle in slow flow ($Re < 1$) is governed by the Stokes law (Section 1.8). For a fluid particle (bubble and droplet), the interfacial shear stress of the inner fluid is balanced with the outer shear caused by the rising motion, following the dynamic boundary condition Eq. (3.50). As a result, the interior fluid is driven to create the internal circulation in the spherical fluid particle (Fig. 3.8 right). Clift, Grace, and Weber (1978)[14] introduced Hadamard-Rybczynski solution to explain how the internal circulation affects the drag force and terminal velocity of a fluid particle. The solution takes the similar form of Hill's spherical vortex[49]

[1]While the Basset term has often ignored in practice, it should not be generally ignored especially for accelerated particles in turbulent flows. Readers should refer to Hinze[35] for details.

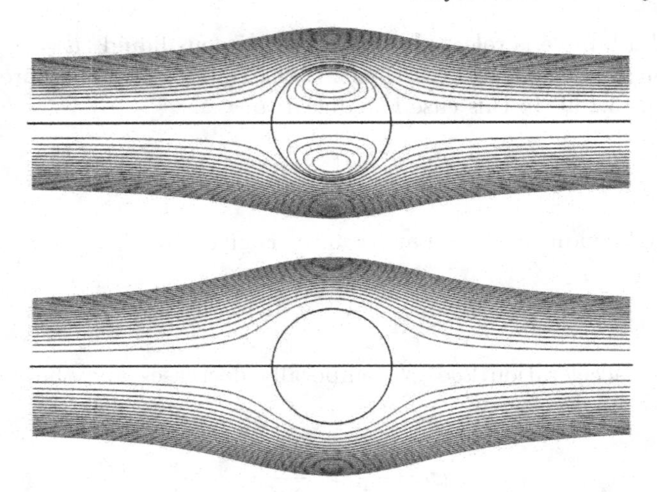

FIGURE 3.24
Streamlines of Hadamard-Rybczynski solution, Eq. (3.145), for a bubble ($\alpha = \mu_a/\mu_w$=0.017) and drop ($\alpha = \mu_w/\mu_a$=57.978) where viscosity of air $\mu_a = 1.83 \times 10^{-5}$ Pa and $\mu_w = 106.1 \times 10^{-5}$ Pa for water.

but the dynamic boundary condition Eq. (3.50) is considered at the interface:

$$\psi(r,\theta) = \begin{cases} \dfrac{Ur^2 \sin^2\theta}{4(1+\alpha)}\left(1 - \dfrac{r^2}{a^2}\right) & (r < a) \\ -\dfrac{Ur^2 \sin^2\theta}{2}\left(1 - \dfrac{a(2+3\alpha)}{2r(1+\alpha)} + \dfrac{\alpha a^3}{2r^3(1+\alpha)}\right) & (r > a) \end{cases} \tag{3.145}$$

where a is the sphere radius, U is the traveling velocity of the particle and the viscosity ratio $\alpha = \mu_p/\mu_o$. μ_p and μ_o are the viscosity of the inner fluid of the particle and that of outer one, respectively. Fig. 3.24 shows the streamlines of Eq. (3.145) for the cases of a bubble in water (top) and a droplet in air (bottom). We find the pair of circulation in a bubble but in a droplet.

Clift, Grace, and Weber (1978)[14] provide the drag coefficient for Eq. (3.145):

$$C_d = \frac{8}{Re}\left(\frac{2+3\alpha}{1+\alpha}\right) \tag{3.146}$$

We readily find that Eq. (3.146) asymptotically agrees with Stokes drag $C_{ds} = 24/Re$, Eq. (1.195), when $\alpha \to \infty$ (infinite viscosity in the particle behaves like a solid). In case of a droplet falling in air, the viscosity ratio $\alpha \approx 58$ is high enough to well approximate the drop drag by the Stokes drag C_{ds}. Accordingly, the internal circulation in a droplet does not contribute to the drag, and thus the drag coefficient of a solid particle, Eq. (2.1) (see Fig. 2.2), may be used for predicting the motion of a droplet.

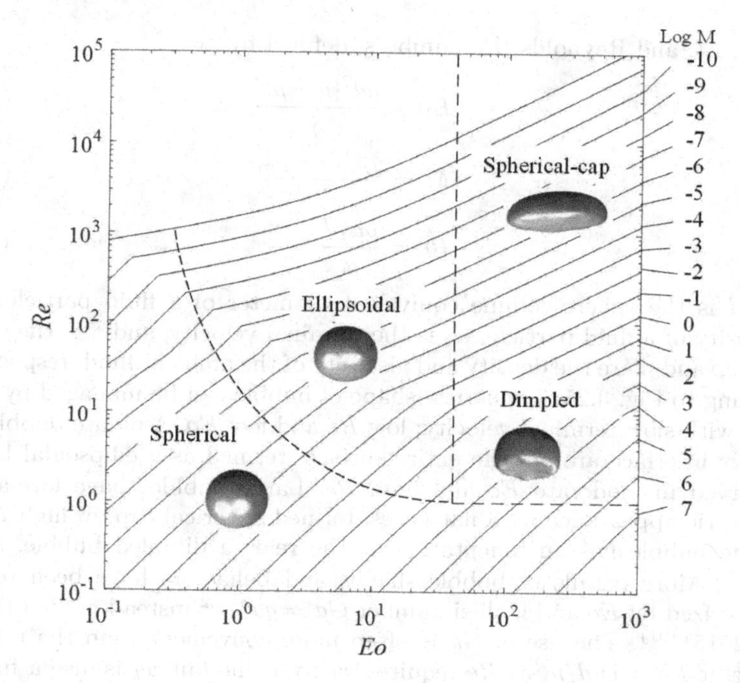

FIGURE 3.25
Grace diagram.

In another extreme limit $\alpha \rightarrow 0$, corresponding to a case of a bubble rise in water, Eq. (3.146) gives a lower drag than the Stokes' one:

$$C_d = \frac{16}{Re} \tag{3.147}$$

Accordingly, the internal circulation in a bubble reduces the viscous drag and thus increases the terminal velocity. It should be noted that surfactant freezes a bubble interface and inhibits the internal circulation. Thus, in a contaminated system, the bubble behaves like a solid particle, following the solid particle drag, Eq. (2.1).

While Eq. (3.147) is valid in the Stokes flow regime of pure water system, $Re < 1$, Levich[54] derived C_d in higher Re:

$$C_d = \frac{48}{Re} \tag{3.148}$$

It is known that C_d also depends on a bubble shape depending on local pressure, viscous stress, and surface tension. Grace (1976)[30] proposed a shape regime map for fluid particles (see Fig. 3.25) parameterized by Eötvös, Eo,

Morton, M, and Reynolds, Re numbers, defined by

$$Eo = \frac{gd^2|\rho - \rho_p|}{\gamma} \tag{3.149}$$

$$M = \frac{g\mu^4|\rho - \rho_p|}{\rho^2\gamma^3} \tag{3.150}$$

$$Re = \frac{\rho w_t d}{\mu} \tag{3.151}$$

where d is the sphere-volume equivalent diameter of a fluid particle, ρ_p is the density of a fluid particle, w_t is the terminal velocity, and γ is the surface tension. ρ and μ are the density and viscosity of the ambient fluid, respectively. According to Fig. 3.25, a spherical shape of bubble can be managed by a tiny bubble with slow terminal velocity; low Re and low Eo. A oblate bubble with a convex interface around the entire surface, termed as a ellipsoidal bubble, is observed in moderate Eo and high Re. Large bubbles have fore-and-aft asymmetric appearance with flat bases, termed spherical cap, in high Re and Eo. The bubble has an indentation at the rear, a dimpled bubble, also in high Eo. More details of bubble shapes and behaviors have been recently characterized by Eo and Galilei number $Ga = ga^3/\nu^2$ instead of Re (Tripathi et al. 2015)[101]. The use of Ga is often more convenient than the recursive parameter $Re = w_t d/\nu$, as Re requires w_t to define but w_t is also a function of Re. As Ga has a unique relation with Re, Eq. (1.193), Re in Fig. 3.25 may be replaced by Ga.

While the drag forces of spherical and ellipsoidal bubbles may be governed by viscosity and thus defined by Re, non-spherical one observed in high Eo (Fig. 3.25) produces lee vortices behind the base, causing variations in C_d owing to pressure depression in the vortices, as noted in Section 3.3.3 (see also Figs. 2.1 and 2.2). In this regime, Eo should be a parameter to determine C_d.

While many experimental investigations have been performed to identify C_d in specific ranges[14], Tomiyama et al. (1998)[99] integrated the previous models of bubble drag over different parameter ranges, and proposed a universal empirical drag coefficient, which includes the effects of bubble deformation and surfactant and covers the ranges: $10^{-2} < Eo < 10^3$, $10^{-14} < M < 10^7$ and $10^{-3} < Re < 10^5$. Tomiyama's drag for a pure system:

$$C_d = \max\left\{\min\left[\frac{16}{Re}\left(1 + 0.15Re^{0.687}\right), \frac{48}{Re}\right], \frac{8}{3}\frac{Eo}{Eo+4}\right\} \tag{3.152}$$

for a slightly contaminated system

$$C_d = \max\left\{\min\left[\frac{24}{Re}\left(1 + 0.15Re^{0.687}\right), \frac{72}{Re}\right], \frac{8}{3}\frac{Eo}{Eo+4}\right\} \tag{3.153}$$

and for a contaminated system

$$C_d = \max\left\{\frac{24}{Re}\left(1 + 0.15Re^{0.687}\right), \frac{8}{3}\frac{Eo}{Eo+4}\right\} \tag{3.154}$$

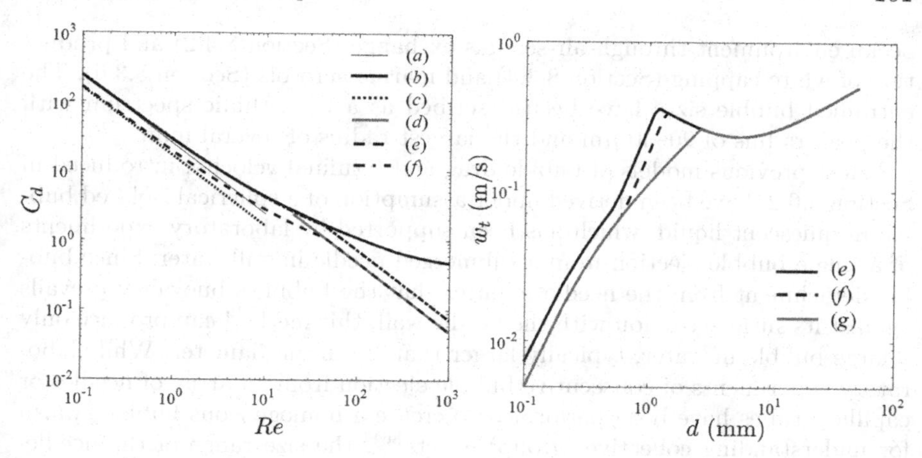

FIGURE 3.26
Drag coefficient of a spherical bubble as a function of Re (left): (a) Stokes, Eq. (1.195), (b) solid particle (Schiller and Naumann), Eq. (2.1), (c) Hadamard-Rybczynski, Eq. (3.147), (d) Levich, Eq. (3.148), (e) Tomiyama (pure system), Eq. (3.152), (f) Tomiyama (slightly contaminate system), Eq. (3.153); and terminal velocity of Tomiyama model, depending on surfactant (right): (e) Eq. (3.152), (f) Eq. (3.153), (g) Eq. (3.154).

Fig. 3.26 (left) shows the bubble drag coefficients of Eqs. (3.147) and (3.148), and Tomiyama's models Eqs. (3.152) and (3.153) (assuming small E_o), together with the Stokes drag Eq. (1.195) and empirical drag for a rigid particle Eq. (2.1) (see Fig. 2.2). The Tomiyama's drag for a pure system, Eq. (3.152), reasonably connects a gap in C_d of Eqs. (3.147) and (3.148) at $Re \sim 100$.

The freezing effect in a contaminated system, i.e. identical C_d to a rigid particle, Eq. (2.1), has been modeled by Eqs. (3.153) and (3.154). The effects of surfactant to the terminal velocity, estimated by the Tomiyama's drag, are found from Fig. 3.26 (right). In a pure system, while a local maxima of w_t appears at $d \sim 1$mm, w_t is modified for a larger bubble whose C_d is governed by Eo.

Motions of bubbles of arbitrary size can be computed by simultaneously solving Navier-Stokes equation and BBO equation, Eq. (3.138), with the Tomiyama's drag. Niida and Watanabe (2018) [118] include the turbulent effects to the system of BBO and Navier-Stokes equations to compute turbulent bubble plumes.

3.6.3 Bubble swarms

In breaking waves, numbers of air bubbles are entrained into bulk and involved by vortices and turbulence, providing a variety of contributions to the ocean environment through air-sea gas exchange (Section 8.4.2) and production of white capping (Section 8.3.4) and marine aerosols (Section 8.3.5). The entrained bubble sizes have been described as a logarithmic spectrum with the peak radius of 20–30 μm and the largest radius of several mm.

Most previous models of bubble drag and terminal velocity, introduced in Section 3.6.2, have been derived under assumption of a spherical isolated bubble in quiescent liquid, which has been supported by laboratory experiments of a single bubble ejection from a submerged needle in still water. Since bubble detachment from the needle requires that the bubble's buoyancy prevails against its surface tension with the needle wall, this method can produce only a large bubble in water, typically larger than 2 mm in diameter. While laboratory experiments of consecutive bubble ejection from an array of needles or capillary tubes have been performed to create a homogeneous bubble swarm for understanding collective (group) effects[86)], the size range of the needle-generated bubbles, say larger than 2 mm in diameter, deviates from sub-mm scale bubbles observed in aerated ocean surfaces.

In order to understand the collective effects of bubble drag in oceanic microbubble swarms, Watanabe et al. (2021)[110)] used electrolytically generated microbubble swarms in a diameter range of 10–1000 μm, which covers the dominant oceanic bubble size. In this method, macroscopic density difference between the liquid and bubble layers on the electrode induces regular variations in bubble concentration and ascending velocity through Rayleigh-Taylor instability (introduced in Section 6.1). The undulation of bubble accumulation amplifies and vertically ejects bubbles at regular intervals (see Fig. 6.2), causing buoyancy-driven upwelling liquid flows, which induces formation of counter-rotating vortices above the ejections (Fig. 3.27). The bubbles are accumulated in a convergent flow and accelerate in the vertical flow induced between the counter-rotations (see Fig. 3.13), resulting in higher terminal velocity with dispersion (Fig. 3.28). The empirical formula of the optimal terminal velocity of the bubble swarm in the equilibrium state has been proposed:

$$w_{ts} = Cd^n \tag{3.155}$$

where d is the diameter, $C = 231.59$ and $n = 1.04$. The corresponding drag coefficient is given by

$$C_d = \frac{4}{3}\frac{\rho_l - \rho_g}{\rho_l}\frac{gd}{w_{ts}^2} = \alpha Re^{\beta} \tag{3.156}$$

where $\alpha = 4.35$ and $\beta = (1 - 3n)/(n + 1)$.

The bubble size distributions of the oceanic bubbles and the specific effects of bubble plumes in breaking waves are also introduced in Section 8.3.

3.6.4 Foam

When a rising air bubble reaches an air–water interface, responses of the bubble depend on surfactant of liquid; that is, in pure water, the bubble surface

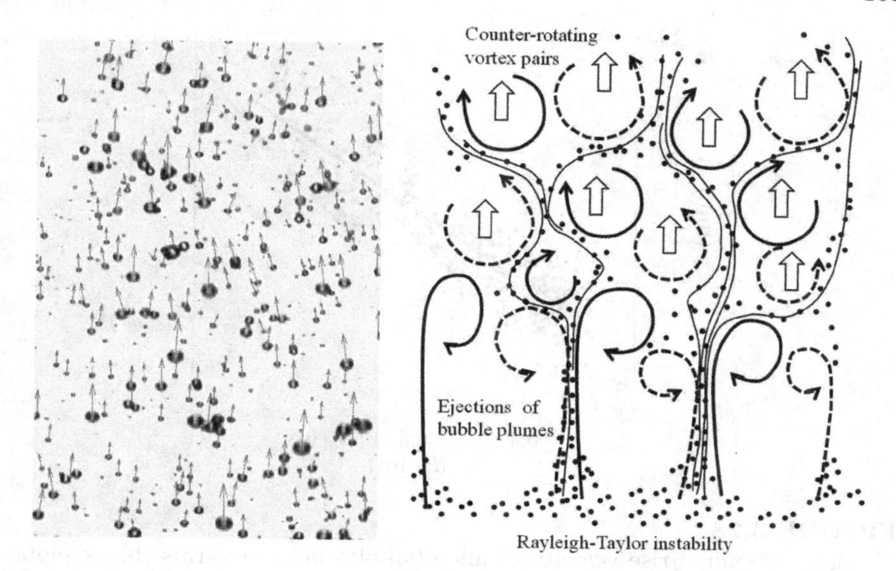

FIGURE 3.27
Measured velocity vectors of electrolytically generated microbubbles (left), and schematic illustration of bubbles aligned on the paths in the flow induced between counter-rotating vortices originated by vertical ejects of bubble plume via Rayleigh-Taylor instability (right).

may coalesce with the interface and disappear, while the bubble may survive and float on the surfactant interface since surfactant prevents coalescence of the interfaces. The appearances of floating bubbles on surfactant water surfaces are shown in Fig. 3.29. The floating bubbles are attracted each other and forms clusters. While adjacent bubbles may coalesce to create a larger bubble and the clusters are composed by various sizes of bubbles for lower concentration of surfactant (left), the clusters have larger numbers of floating bubbles with the identical sizes on higher concentration of surfactant surface where coalescence is suppressed.

The mechanism of the bubble attraction, resulting from meniscus formation, to create the bubble clusters is analytically explained as follows.

3.6.4.1 Meniscus

The meniscus is the curve caused by surface tension. The shape of the static meniscus is governed by two factors (see Fig. 3.30 left); the contact angle θ between the surface and solid wall, and the mechanical balance of hydrostatic pressure and capillary pressure, defined by Young-Laplace equation Eq. (3.31). While the former is material dependent, and determined by surface tensions of the used liquid and solid, the latter is given from Eq. (3.31) with $p_l - p_g = -\rho g \eta$ by

$$\gamma \kappa = -\rho g \eta \tag{3.157}$$

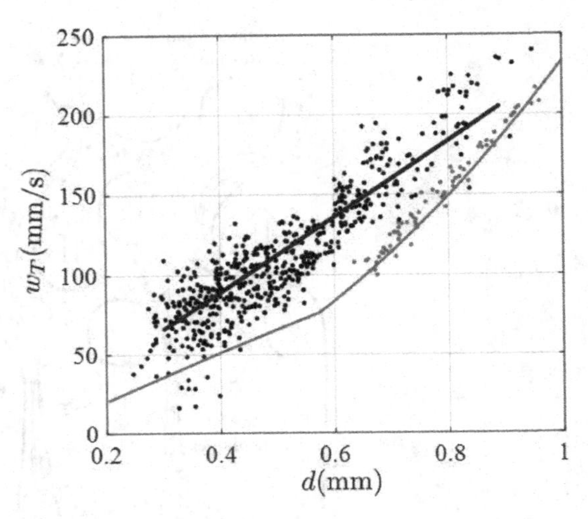

FIGURE 3.28
Measured terminal rise velocity of microbubbles in the swarms (black plots) and the empirical formula, Eq. (3.155) (black line). Gray plots indicate the rise velocity of initially generated bubbles without any bubble–bubble interaction, which follows Tomiyama's model, Eq. (3.153), denoted by the gray line.

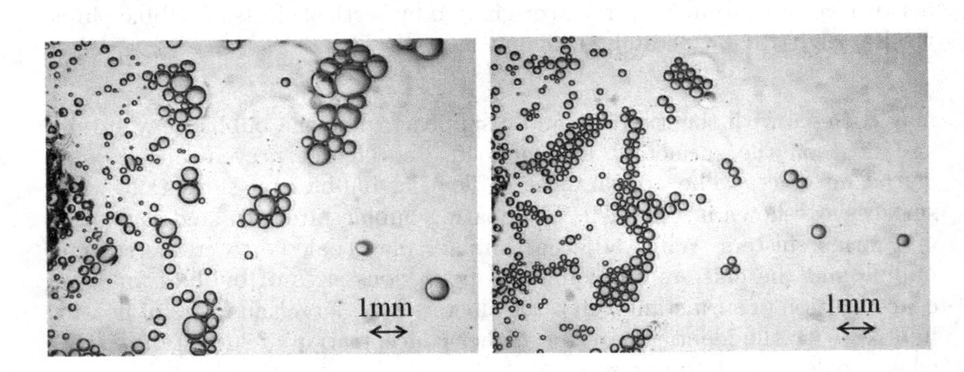

FIGURE 3.29
Backlit images of floating bubbles clustered on free surfaces; 400 μg/L (left) and 800 μg/L (right) of Triton X-100 solution. Bubbles are generated by a porous stone placed on the tank floor.

where $\eta(x)$ is the surface elevation. Eq. (3.157) can be also provided by the dynamic boundary condition Eq. (3.55) at static state:

$$\cancel{\frac{\partial \phi}{\partial t}} + \cancel{\frac{1}{2}(u^2 + v^2 + w^2)} + \frac{\gamma}{\rho}\kappa + gz = \cancelto{0}{C(t)} \quad (z = \eta)$$
$$\therefore \frac{\gamma}{\rho}\kappa + g\eta = 0 \tag{3.158}$$

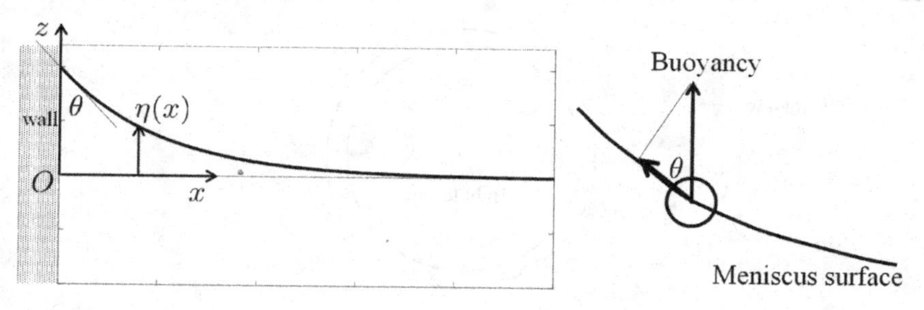

FIGURE 3.30
Surface form at the contact on a vertical wall (left) and a buoyant bubble pulled along the sloping surface (right).

In a two-dimensional space (x, z), as $\kappa \approx -\partial^2\eta/\partial x^2$ from Eq. (3.19), Eq. (3.157) is expressed by

$$\gamma\frac{\partial^2\eta}{\partial x^2} = \rho g \eta \tag{3.159}$$

This differential equation is solved with the boundary conditions (see Fig. 3.30 left):

$$\frac{\partial\eta}{\partial x} = -\cot\theta \quad (x = 0) \tag{3.160}$$

$$\eta \to 0 \quad (x \to \infty) \tag{3.161}$$

The solution is given by

$$\eta = \sqrt{\frac{\gamma}{\rho g}}\cot\theta\exp\left[-\sqrt{\frac{\rho g}{\gamma}}x\right]$$
$$= l_c\cot\theta\exp\left[-\frac{x}{l_c}\right] \tag{3.162}$$

where the capillary length $l_c = \sqrt{\gamma/\rho g}$.

When a bubble on the sloped meniscus surface (see Fig. 3.30 right), while the bubble cannot go vertically across the surface, the bubble must move to a higher place along the surface because of the $\cos\theta$ component of the buoyancy pulls the bubble along the surface. The attraction of the bubble toward the wall occurs in this mechanism, which may be confirmed by beer foams aggregated on inner surface of a beer glass.

Since a floating bubble yields meniscus around it, the identical mechanism attract bubbles on the meniscus surface, which the solution was derived by Nicolson (1949)[75].

3.6.4.2 Surface form around the floating bubble

Following Nicolson (1949)[75], we consider a floating spherical bubble of radius R, which is a cylindrical symmetric about the vertical axis through the center

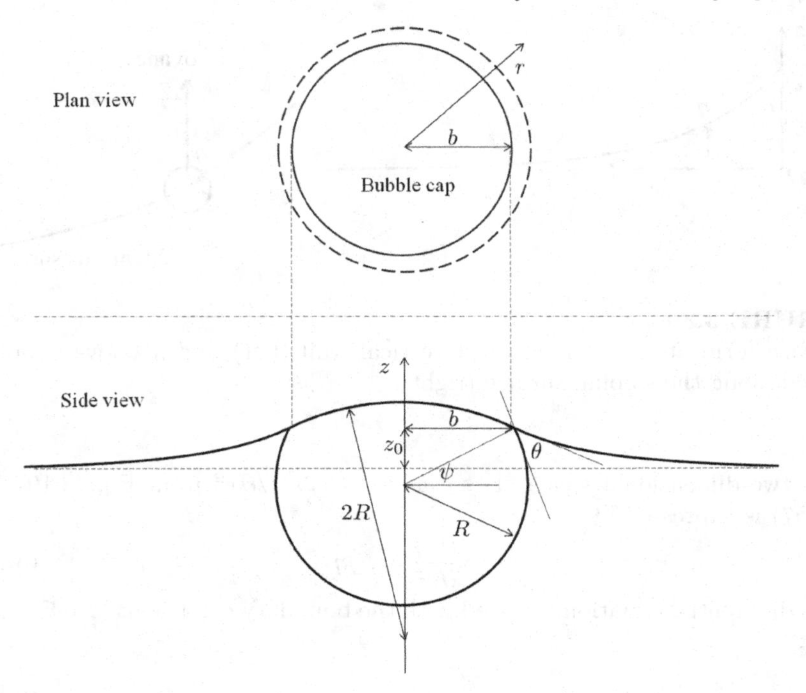

FIGURE 3.31
Illustration of a floating bubble form.

of the bubble (see Fig. 3.31). The edge of the bubble cap configures a ring of radius b where the cap of the bubble, having radius $2R$, contacts with the free surface around the bubble. The surface form and vertical position of the bubble in the equilibrium state is governed by the mechanical balance among the bubble buoyancy, hydrostatic pressure of the meniscus and the capillary pressure at the contact ring. The curvature in Eq. (3.157) can be approximated in cylindrical coordinate (r, z), from Eq. (3.23), as

$$\gamma \frac{1}{r}\frac{\partial}{\partial r}\left(\frac{\partial \eta}{\partial r}\right) = \rho g \eta \tag{3.163}$$

Introducing dimensionless variables $\chi = r/R$, $\zeta = \eta/R$, Eq. (3.163) gives the modified Bessel's differential equation (see Section A.8.2 in Appendix):

$$\frac{d^2\zeta}{d\chi^2} + \frac{1}{\chi}\frac{d\zeta}{d\chi} - \alpha^2\zeta = 0 \tag{3.164}$$

where α^2 indicates the Eötvös number Eo:

$$Eo = \alpha^2 = R^2/l_c^2 = R^2\rho g/\gamma \tag{3.165}$$

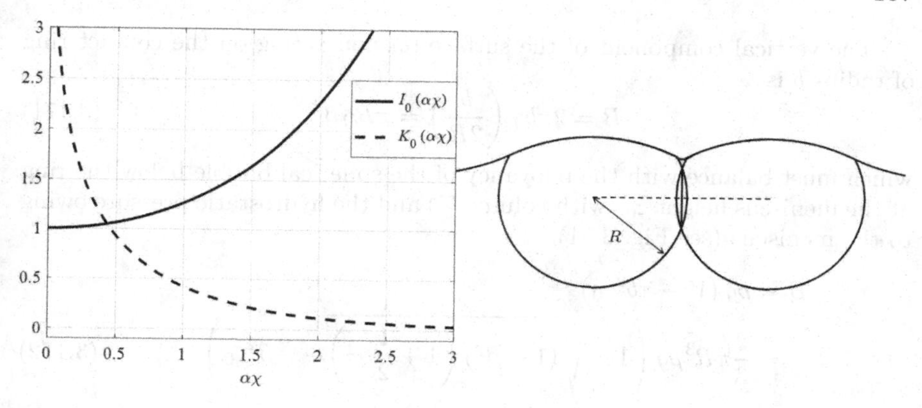

FIGURE 3.32
Bessel function I_0 and K_0 (left) and illustration of two floating bubbles in contact (right).

The solution of Eq. (3.164) takes the form

$$\zeta = C_1 I_0(\alpha\chi) + C_2 K_0(\alpha\chi) \tag{3.166}$$

where I_0 and K_0 are the modified Bessel function of the first and second kinds of order zero. Considering the boundary condition $\zeta \to 0$ at $\chi \to \infty$, the solution form of $I_0(\alpha\chi)$ is inappropriate as $(\alpha\chi) \to \infty$ at $\chi \to \infty$ (Fig. 3.32 left). To satisfy this far-field boundary condition, C_1 must be zero.

$$\zeta = C_2 K_0(\alpha\chi) \tag{3.167}$$

As another boundary condition, the surface gradient is imposed at the contact $r = b$ (see Fig. 3.31):

$$\frac{d\zeta}{d\chi} = \frac{\tan\psi - \tan\theta}{1 + \tan\psi\tan\theta} \qquad (\chi = \beta) \tag{3.168}$$

where $\beta = b/R$, θ is the contact angle, and ψ is the horizontal angle of the contact at the center of the bubble (see Fig. 3.31). Nicolson (1949) avoided the rigorous treatment of Eq. (3.168) and assumed that the contact point has the continuous surface slope of the bubble cap of radius $2R$:

$$\frac{d\zeta}{d\chi} = -\frac{\beta}{\sqrt{4-\beta^2}} \qquad (\chi = \beta) \tag{3.169}$$

Differentiating Eq. (3.167) to impose Eq. (3.169), the unknown constant C_2 can be specified. The surface form is finally given by

$$\zeta = \frac{\beta}{\alpha\sqrt{4-\beta^2}} \frac{K_0(\alpha\chi)}{K_1(\alpha\beta)} \tag{3.170}$$

in terms of β which is determined by the vertical mechanical balance.

The vertical component of the surface tension acting on the contact ring of radius b is

$$B = 2\pi b\gamma\left(\frac{b}{2R}\right) = \pi R\gamma\beta^2 \tag{3.171}$$

which must balance with the buoyancy of the spherical bubble below the ring at the meniscus height z_0 (with volume V) and the hydrostatic pressure owing to the meniscus (see Fig. 3.31).

$$\begin{aligned} B &= \rho g\left(V - \pi b^2 z_0\right) \\ &= \frac{2}{3}\pi R^3 \rho g\left(1 + \sqrt{(1-\beta^2)\left(1+\frac{1}{2}\beta^2\right)} - \frac{3}{2}\beta^2\zeta_0\right) \end{aligned} \tag{3.172}$$

where $\zeta_0 = z_0/R = \zeta\mid_{\chi=\beta}$ given by Eq. (3.170). Equating Eqs. (3.171) and (3.172), we find β satisfies the equation:

$$\beta^2 - \frac{2}{3}\alpha^2\left(1 + \sqrt{(1-\beta^2)\left(1+\frac{1}{2}\beta^2\right)} - \frac{3}{2}\frac{\beta^3}{\alpha\sqrt{4-\beta^2}}\frac{K_0(\alpha\beta)}{K_1(\alpha\beta)}\right) = 0 \tag{3.173}$$

This nonlinear equation for β can be numerically solved for given α. The equilibrium position of the floating bubble, ζ_0, and the surface form can be identified with the solution of β by Eq. (3.170).

The sloped surface, defined by Eq. (3.170), works as the apparent attraction of another bubble on the slope. When the same size of the bubble located on the meniscus surface, the horizontal component of the buoyancy is given using Eqs. (3.171) and (3.165) as

$$F_{atr} = B\frac{d\zeta}{d\chi} = -\pi R^3 \rho g \frac{\beta^3}{\alpha^2\sqrt{4-\beta^2}}\frac{K_1(\alpha\chi)}{K_1(\alpha\beta)} \tag{3.174}$$

Once the attracting two bubbles contact each other (when the relative distance $\chi = 2$), repulsive force, caused by the excess pressure, $2\gamma/R$, occurs on the area of contact with another one, $\pi\left(R^2 - (\chi R/2)^2\right)$, when $\chi \leq 2$ (see Fig. 3.32 right):

$$\begin{aligned} F_{rep} &= \pi R^2\left(1 - \frac{\chi^2}{4}\right)\frac{2\gamma}{R} \qquad &(3.175) \\ &\approx 2\pi R\gamma(2-\chi) = 2\pi R^3\rho g\frac{2-\chi}{\alpha^2} \qquad &(3.176) \end{aligned}$$

Accordingly, the horizontal force between two bubbles can be estimated by

$$F_h = \begin{cases} F_{atr} & (\chi > 2) \\ F_{atr} + F_{rep} & (\chi \leq 2) \end{cases} \tag{3.177}$$

In the surf zone, patches of bubbles remained on sea surfaces are often observed behind the breaking waves (Fig. 3.33 left). Also in open ocean, similar

FIGURE 3.33
Photos of a surf zone taken at Ishikari beach, Hokkaido (left), and open ocean taken at Shirahama oceanographic observation tower, Wakayama (right).

patches left behind wind wave breaking are observed to remain for long time (Fig. 3.33 right). The above microscopic mechanism to create bubble clusters (see Fig. 3.29) may provide an answer for a question how the patches are formed. On the one hand, entrapment of bubbles in vortices, disturbances by turbulence at surfaces (creating local surface slopes), in addition to surfactant effect, may affect the covering area and lifetimes of the patches.

3.6.4.3 Note on pressure in a bubble

Static pressure in a spherical bubble of radius R_1 located at the depth h of water pool is expressed by the sum of hydrostatic (ρgh), atmospheric (p_a), and capillary pressures, Eq. (3.29):

$$p_1 = p_a + \rho gh + \frac{2\gamma}{R_1} \tag{3.178}$$

When the bubble rises up and arrives a free surface, as it is released from the hydrostatic pressure, the pressure becomes:

$$p_2 = p_a + \frac{2\gamma}{R_2} \tag{3.179}$$

Assuming ideal gas inside the bubble, it satisfies the state equation:

$$\frac{p_1 V_1}{T_1} = \frac{p_2 V_2}{T_2} \tag{3.180}$$

The substitution of Eqs. (3.178) and (3.179) into Eq. (3.180) gives

$$\frac{1}{T_1}\left(p_a + \rho gh + \frac{2\gamma}{R_1}\right)\frac{4}{3}\pi R_1^3 = \frac{1}{T_2}\left(p_a + \frac{2\gamma}{R_2}\right)\frac{4}{3}\pi R_2^3$$

$$\therefore \left(\frac{R_2}{R_1}\right)^3 + \frac{2\gamma}{p_a R_1}\left(\left(\frac{R_2}{R_1}\right)^2 - \frac{T_2}{T_1}\right) = \frac{T_2}{T_1}\left(1 + \frac{\rho gh}{p_a}\right) \tag{3.181}$$

In general, the ratio of surface tension to atmospheric pressure $2\gamma/p_a R_1$ is negligibly small; e.g. for $R_1 = 1$ mm, $\gamma = 0.072$ Nm^{-1}, $p_a = 10^5$ Pa, $2\gamma/p_a R_1 = 1.4\times10^{-3} \ll 1$. Neglecting the second term of Eq. (3.181), the radius change can be estimated by

$$\frac{R_2}{R_1} = \left\{\frac{T_2}{T_1}\left(1 + \frac{\rho g h}{p_a}\right)\right\}^{1/3} \tag{3.182}$$

For instance, when the depth $h = 10$ m, water temperatures at the bottom and surface $T_1 = 17° = 290$ K and $T_2 = 27° = 300$ K, $R_2/R_1 \approx 1.27$ is estimated.

4

Linear Wave Theory

In this chapter, the Laplace equation for velocity potential is solved with specific boundary conditions describing water waves of interest. When the amplitude of oscillation is sufficiently small, nonlinear boundary conditions, considered in Section 3.2, can be linearized. This analytical approximation, linearization, provides the lowest order solutions of water waves and free-surface responses on various topographies, which describes major features of wave behaviors and has been widely applied to industries and engineering. In this chapter, the linearization procedure is first introduced, and analytical solutions for various linear water waves and surface oscillations are then considered in the following sections.

4.1 The Laplace equation

Fluid motions in water waves observed in any water environment are well-approximated as incompressible irrotational flows. On this assumption, the fluid velocity is defined in terms of velocity potential, $\boldsymbol{u} = \nabla\phi$, as introduced in Section 1.4. Substituting $\boldsymbol{u} = \boldsymbol{\nabla}\phi$ into the continuity equation $\boldsymbol{\nabla} \cdot \boldsymbol{u} = 0$, the Laplace equation for velocity potential is given by

$$\boldsymbol{\nabla}^2\phi = 0 \tag{4.1}$$

$$\frac{\partial^2\phi}{\partial x^2} + \frac{\partial^2\phi}{\partial y^2} + \frac{\partial^2\phi}{\partial z^2} = 0 \tag{4.2}$$

where $\boldsymbol{\nabla} = (\partial/\partial x, \partial/\partial y, \partial/\partial z)$ in the Cartesian coordinates (x, y, z).

In the cylindrical coordinate (r, θ, z), as $\boldsymbol{\nabla} = (\partial/\partial r, r^{-1}\partial/\partial\theta, \partial/\partial z)$, Eq. (1.6), the radial, azimuthal and axial velocities are given

$$\boldsymbol{u} = (u_r, u_\theta, u_z) = \boldsymbol{\nabla}\phi = \left(\frac{\partial\phi}{\partial r}, \frac{1}{r}\frac{\partial\phi}{\partial\theta}, \frac{\partial\phi}{\partial z}\right) \tag{4.3}$$

The substitution into the continuity equation $\boldsymbol{\nabla} \cdot \boldsymbol{u} = 0$, Eq. (1.70), gives the Laplace equation in the cylindrical system (see also Eq. (1.10)):

$$\frac{\partial^2\phi}{\partial r^2} + \frac{1}{r}\frac{\partial\phi}{\partial r} + \frac{1}{r^2}\frac{\partial^2\phi}{\partial\theta^2} + \frac{\partial^2\phi}{\partial z^2} = 0 \tag{4.4}$$

DOI: 10.1201/9781003140160-4

The Laplace equation is solved as a boundary value problem under suitable boundary conditions characterizing specific wave field to be solved.

4.2 Linear boundary conditions

The kinematic and dynamic boundary conditions, introduced in Section 3.2, are used for solving the Laplace equation. As these conditions are described as a form of a nonlinear differential equation, which is difficult to directly solve, they should be linearized with an appropriate assumption. In this chapter, a small amplitude of waves is assumed to reduce the nonlinear boundary conditions to linear ones on the basis of the perturbation method, which provides the lowest order (linear) solution of the Laplace equation. While the general perturbation method considering higher order quantities is introduced in Section 7.5 for deriving the nonlinear wave solution, in this section, we consider the boundary conditions for small values of surface elevation η and velocity potential ϕ. η and ϕ are replaced by the small variables η' and ϕ':

$$\eta = \epsilon\eta' \tag{4.5}$$

$$\phi = \epsilon\phi' \tag{4.6}$$

where dimensionless parameter $\epsilon \ll 1$ indicates the order of the small quantities. The substitution of Eq. (4.6) into the Laplace equation, Eq. (4.1), gives the Laplace equation at the order of ϵ:

$$\nabla^2\phi = \epsilon\nabla^2\phi' = 0 \tag{4.7}$$

Defining the coordinate with the origin at the still water level in the vertical axis (see the coordinate defined in Fig. 4.1 left), the kinematic boundary condition at the arbitrary form of bottom, Eq. (3.40), at this order is given:

$$\epsilon\frac{\partial\phi'}{\partial z} + \epsilon\frac{\partial h}{\partial x}\frac{\partial\phi'}{\partial x} + \epsilon\frac{\partial h}{\partial y}\frac{\partial\phi'}{\partial y} = 0 \quad (z = -h(x, y)) \tag{4.8}$$

In the special case of the flat bottom ($\partial h/\partial x = 0, \partial h/\partial y = 0$), Eq. (4.8) is reduced to

$$\epsilon\frac{\partial\phi'}{\partial z} = 0 \quad (z = -h) \tag{4.9}$$

Similarly the substitution of Eqs. (4.5) and (4.6) into the kinematic boundary condition at free surface ($z = \epsilon\eta'$), Eq. (3.41), gives

$$\epsilon\frac{\partial\phi'}{\partial z} = \epsilon\frac{\partial\eta'}{\partial t} + \epsilon^2\frac{\partial\phi'}{\partial x}\frac{\partial\eta'}{\partial x} + \epsilon^2\frac{\partial\phi'}{\partial y}\frac{\partial\eta'}{\partial y} \quad (z = \epsilon\eta') \tag{4.10}$$

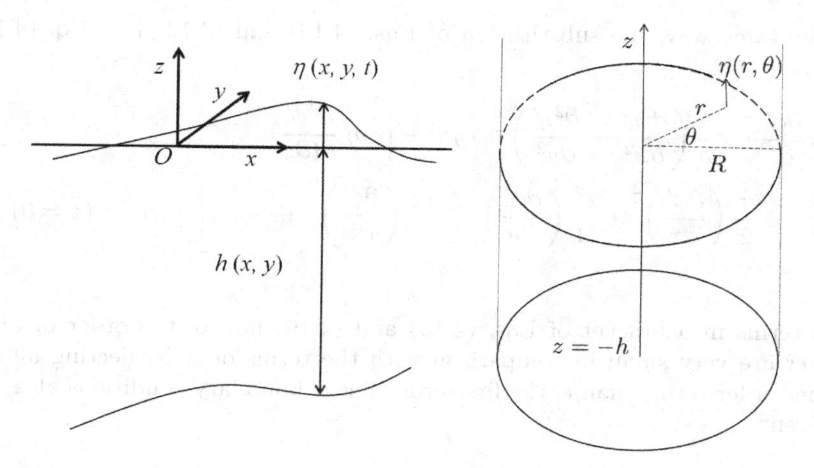

FIGURE 4.1
Free surfaces in Cartesian and cylindrical coordinates.

The dynamic condition at free surface, Eq. (3.56), is transformed in the same manner:

$$\epsilon\frac{\partial\phi'}{\partial t}-\epsilon\frac{\gamma}{\rho}\left(\frac{\partial^2\eta'}{\partial x^2}+\frac{\partial^2\eta'}{\partial y^2}\right)+gz$$
$$+\frac{\epsilon^2}{2}\left(\frac{\partial\phi'}{\partial x}\right)^2+\frac{\epsilon^2}{2}\left(\frac{\partial\phi'}{\partial y}\right)^2+\frac{\epsilon^2}{2}\left(\frac{\partial\phi'}{\partial z}\right)^2=\epsilon C(t)\quad(z=\epsilon\eta')\qquad(4.11)$$

The quantities at the surface location $z=\epsilon\eta'$ are expanded in the Taylor series about $z=0$ (still water level) with respect to the vertical z-axis:

$$\phi'\mid_{z=\epsilon\eta'}=\phi'\mid_{z=0}+\epsilon\eta'\left.\frac{\partial\phi'}{\partial z}\right|_{z=0}+O\left(\epsilon^2\right)\qquad(4.12)$$

$$gz\mid_{z=\epsilon\eta'}=gz\mid_{z=0}+\epsilon\eta' g=\epsilon\eta' g\qquad(4.13)$$

The substitution of Eq. (4.12) into Eq. (4.10) gives

$$\epsilon\frac{\partial}{\partial z}\left\{\phi'+\epsilon\eta'\frac{\partial\phi'}{\partial z}+O\left(\epsilon^2\right)\right\}=\epsilon\frac{\partial\eta'}{\partial t}+\epsilon^2\frac{\partial\eta'}{\partial x}\frac{\partial}{\partial x}\left\{\phi'+\epsilon\eta'\frac{\partial\phi'}{\partial z}+O\left(\epsilon^2\right)\right\}$$
$$+\epsilon^2\frac{\partial\eta'}{\partial y}\frac{\partial}{\partial y}\left\{\phi'+\epsilon\eta'\frac{\partial\phi'}{\partial z}+O\left(\epsilon^2\right)\right\}\quad(z=0)\qquad(4.14)$$

Rearranging the terms by the order,

$$\epsilon\frac{\partial\eta'}{\partial t}-\epsilon\frac{\partial\phi'}{\partial z}+\left[-\epsilon^2\frac{\partial}{\partial z}\left(\eta'\frac{\partial\phi'}{\partial z}\right)+\epsilon^2\frac{\partial\phi'}{\partial x}\frac{\partial\eta'}{\partial x}+\epsilon^2\frac{\partial\phi'}{\partial y}\frac{\partial\eta'}{\partial y}+O(\epsilon^3)\right]=0\quad(z=0)\qquad(4.15)$$

In the same way, the substitution of Eqs. (4.12) and (4.13) into Eq. (4.11) gives

$$\epsilon\frac{\partial \phi'}{\partial t} - \epsilon\frac{\gamma}{\rho}\left(\frac{\partial^2 \eta'}{\partial x^2} + \frac{\partial^2 \eta'}{\partial y^2}\right) + \epsilon g\eta' + \left[\epsilon^2\eta'\frac{\partial^2 \phi'}{\partial t \partial z}\right.$$
$$\left. + \frac{\epsilon^2}{2}\left(\frac{\partial \phi'}{\partial x}\right)^2 + \frac{\epsilon^2}{2}\left(\frac{\partial \phi'}{\partial y}\right)^2 + \frac{\epsilon^2}{2}\left(\frac{\partial \phi'}{\partial z}\right)^2 + O(\epsilon^3)\right] = \epsilon C' \quad (z=0) \tag{4.16}$$

The terms in a bracket of Eqs. (4.15) and (4.16) having the order of ϵ^2 or higher are very small in comparison with the terms of ϵ. Neglecting all the higher order terms than ϵ, the first-order linear boundary conditions at $z = 0$ is given

$$\epsilon\frac{\partial \eta'}{\partial t} - \epsilon\frac{\partial \phi'}{\partial z} = 0 \quad (z=0) \tag{4.17}$$

$$\epsilon\frac{\partial \phi'}{\partial t} - \epsilon\frac{\gamma}{\rho}\left(\frac{\partial^2 \eta'}{\partial x^2} + \frac{\partial^2 \eta'}{\partial y^2}\right) + \epsilon g\eta' = \epsilon C' \quad (z=0) \tag{4.18}$$

In summary, a system of the first-order equations is given by

$$\frac{\partial^2 \phi'}{\partial x^2} + \frac{\partial^2 \phi'}{\partial y^2} + \frac{\partial^2 \phi'}{\partial z^2} = 0 \quad (-h < z < 0) \tag{4.19}$$

$$\frac{\partial \eta'}{\partial t} - \frac{\partial \phi'}{\partial z} = 0 \quad (z=0) \tag{4.20}$$

$$\frac{\partial \phi'}{\partial t} - \frac{\gamma}{\rho}\left(\frac{\partial^2 \eta'}{\partial x^2} + \frac{\partial^2 \eta'}{\partial y^2}\right) + g\eta' = C' \quad (z=0) \tag{4.21}$$

$$\frac{\partial \phi'}{\partial z} + \frac{\partial h}{\partial x}\frac{\partial \phi'}{\partial x} + \frac{\partial h}{\partial y}\frac{\partial \phi'}{\partial y} = 0 \quad (z=-h(x,y)) \tag{4.22}$$

$$\text{or} \quad \frac{\partial \phi'}{\partial z} = 0 \quad (z=-h, \text{constant depth}) \tag{4.23}$$

The two linear boundary conditions at $z = 0$, Eqs. (4.20) and (4.21), may be combined; the differentiation of Eq. (4.21) with respect to t gives

$$\frac{\partial^2 \phi'}{\partial t^2} - \frac{\gamma}{\rho}\left(\frac{\partial^2}{\partial x^2} + \frac{\partial^2}{\partial y^2}\right)\frac{\partial \eta'}{\partial t} + g\frac{\partial \eta'}{\partial t} = 0 \tag{4.24}$$

The substitution of Eq. (4.20) into Eq. (4.24) gives the combined free surface condition:

$$\frac{\partial^2 \phi'}{\partial t^2} - \frac{\gamma}{\rho}\frac{\partial}{\partial z}\left(\frac{\partial^2 \phi'}{\partial x^2} + \frac{\partial^2 \phi}{\partial y^2}\right) + g\frac{\partial \phi'}{\partial z} = 0 \quad (z=0) \tag{4.25}$$

If the capillary effect is negligibly smaller than the gravity, Eq. (4.25) is reduced to

$$\frac{\partial^2 \phi'}{\partial t^2} + g\frac{\partial \phi'}{\partial z} = 0 \quad (z = 0) \tag{4.26}$$

In the case of fluid flows in a rectangular container, the wall boundary condition Eq. (3.39) (impermeable wall), defining no normal velocity to the wall, is also given in the same procedure. In the simplest case of a vertical wall (see the coordinate Fig. 3.6 left),

$$\frac{\partial \phi'}{\partial x} = 0 \quad \text{(at a vertical wall)} \tag{4.27}$$

For the flow in a circular container of radius R (see Fig. 4.1 right), the identical procedure provides the first-order equation system:

$$\frac{1}{r}\frac{\partial}{\partial r}\left(r\frac{\partial \phi'}{\partial r}\right) + \frac{1}{r^2}\frac{\partial^2 \phi'}{\partial \theta^2} + \frac{\partial \phi'}{\partial z^2} = 0 \quad (-h < z < 0, r < R) \tag{4.28}$$

$$\frac{\partial \eta'}{\partial t} - \frac{\partial \phi'}{\partial z} = 0 \quad (z = 0) \tag{4.29}$$

$$\frac{\partial \phi'}{\partial t} - \frac{\gamma}{\rho}\left(\frac{1}{r}\frac{\partial \eta'}{\partial r} + \frac{\partial^2 \eta'}{\partial r^2} + \frac{1}{r^2}\frac{\partial^2 \eta'}{\partial \theta^2}\right) + g\eta' = C' \quad (z = 0) \tag{4.30}$$

$$\frac{\partial \phi'}{\partial z} = 0 \quad (z = -h) \tag{4.31}$$

$$\frac{\partial \phi'}{\partial r} = 0 \quad (r = R) \tag{4.32}$$

The free-surface condition combined with Eqs. (4.29) and (4.30) is given by

$$\frac{\partial^2 \phi'}{\partial t^2} - \frac{\gamma}{\rho}\frac{\partial}{\partial z}\left(\frac{1}{r}\frac{\partial \phi'}{\partial r} + \frac{\partial^2 \phi'}{\partial r^2} + \frac{1}{r^2}\frac{\partial^2 \phi}{\partial \theta^2}\right) + g\frac{\partial \phi'}{\partial z} = 0 \tag{4.33}$$

4.3 Progressive waves

Consider one-dimensional horizontal propagation of regular periodic waves in constant depth h. We explore the solution of the Laplace equation for $\phi(x, z, t)$ for the waves:

$$\frac{\partial^2 \phi}{\partial x^2} + \frac{\partial^2 \phi}{\partial z^2} = 0 \tag{4.34}$$

A general method for solving this type of the differential equation is a separation of variable; ϕ is assumed to be expressed as $\phi = \xi(x)\zeta(z)\tau(t)$ where $\xi(x), \zeta(z), \tau(t)$ are functions depending only x, z, and t, respectively. As we look for the temporally oscillating solution with the angular frequency σ, we may assume $\tau = \mathrm{e}^{-i\sigma t}$; $\phi = \xi(x)\zeta(z)\mathrm{e}^{-i\sigma t}$. The substitution of ϕ into Eq. (4.34) gives

$$\frac{1}{\xi}\frac{\partial^2 \xi}{\partial x^2} + \frac{1}{\zeta}\frac{\partial^2 \zeta}{\partial z^2} = 0 \tag{4.35}$$

The first term depends only on x, which never affects the second one, and the second one depending on only z never affects the first one. Therefore, to satisfy Eq. (4.35), both terms are identically given by constant with opposite sign:

$$\frac{1}{\xi}\frac{d^2 \xi}{dx^2} = -\frac{1}{\zeta}\frac{d^2 \zeta}{dz^2} = \lambda \tag{4.36}$$

which yields the two equations:

$$\frac{d^2 \xi}{dx^2} - \lambda\xi = 0 \tag{4.37}$$

$$\frac{d^2 \zeta}{dz^2} + \lambda\zeta = 0 \tag{4.38}$$

If the eigenvalue λ is positive, the general solutions for Eqs. (4.37) and (4.38) take the forms

$$\xi = C_1\mathrm{e}^{\sqrt{\lambda}x} + C_2\mathrm{e}^{-\sqrt{\lambda}x} \tag{4.39}$$

and

$$\zeta = C_3\mathrm{e}^{i\sqrt{\lambda}z} + C_4\mathrm{e}^{-i\sqrt{\lambda}z} \tag{4.40}$$

This form of solution, describing exponential variation in horizontal x-axis and vertical oscillation in z, explains so-called evanescent waves, which is introduced in Section 4.8.

For the negative λ, Eqs. (4.37) and (4.38) may be written

$$\frac{d^2 \xi}{dx^2} + |\lambda|\xi = 0 \tag{4.41}$$

$$\frac{d^2 \zeta}{dz^2} - |\lambda|\zeta = 0 \tag{4.42}$$

The solutions for these equations take the forms describing horizontal oscillation

$$\xi = C_1\mathrm{e}^{i\sqrt{|\lambda|}x} + C_2\mathrm{e}^{-i\sqrt{|\lambda|}x} \tag{4.43}$$

and vertically exponential variation

$$\zeta = C_3\mathrm{e}^{\sqrt{|\lambda|}z} + C_4\mathrm{e}^{-\sqrt{|\lambda|}z} \tag{4.44}$$

As we look for the progressive wave solution with horizontal oscillation, Eqs. (4.43) and (4.44) are used as the general solutions. Introducing wave number defined by $k = \sqrt{|\lambda|}$, ϕ is thus expressed as

$$\begin{aligned}\phi = \xi\zeta\mathrm{e}^{-i\sigma t} &= \left(C_1\mathrm{e}^{ikx} + C_2\mathrm{e}^{-ikx}\right)\left(C_3\mathrm{e}^{kz} + C_4\mathrm{e}^{-kz}\right)\mathrm{e}^{-i\sigma t} \\ &= \left(C_3\mathrm{e}^{kz} + C_4\mathrm{e}^{-kz}\right)\left(C_1\mathrm{e}^{i(kx-\sigma t)} + C_2\mathrm{e}^{-i(kx+\sigma t)}\right)\end{aligned} \tag{4.45}$$

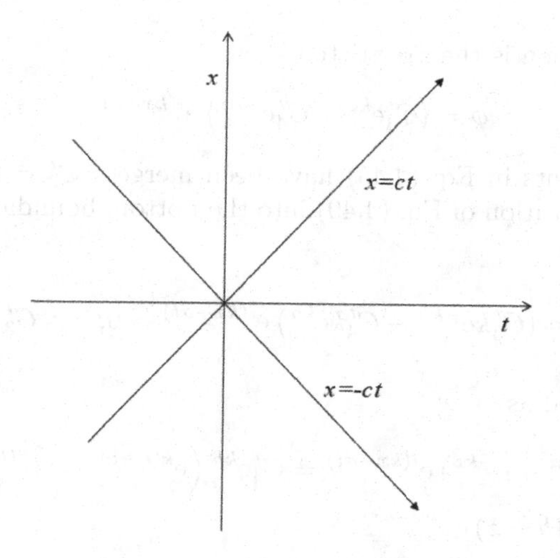

FIGURE 4.2
Characteristic lines.

The oscillation term, $e^{i(kx-\sigma t)}$ in Eq. (4.45), is often expressed in terms of phase θ as $e^{i\theta}$ where

$$\theta = kx - \sigma t \tag{4.46}$$

When the surface elevation of cosine waves is expressed by $\eta = Re[e^{i\theta}] = \cos\theta$, the wave crest achieving the maximum elevation is given at the phase $\theta = 0, 2\pi, \ldots$, and the trough appears at $\theta = \pi, 3\pi/2, \ldots$.

Eq. (4.46) may be rewritten as

$$\theta/k = x - ct \tag{4.47}$$

where $c = \sigma/k$ defines the wave speed. Eq. (4.47) indicates the specific phase moves in the x-direction with the speed c. For instance, considering the case of $\theta = 0$ (phase of the crest) and the constant wave speed, the position of the wave crest moves following the linear equation $x = ct$ (see Fig. 4.2). Accordingly, the term $e^{i(kx-\sigma t)}$ describes the wave propagation in positive the x-direction.

For another oscillation term $e^{-i(kx+\sigma t)}$ in Eq. (4.45), the waves with the phase $\theta = -(kx + \sigma t)$ travel following

$$-\theta/k = x + ct \tag{4.48}$$

Accordingly, the wave crest ($\theta = 0$) moves the opposite direction (negative x) along the characteristic line $x = -ct$ (see Fig. 4.2). While the both wave components are the solutions of the Laplace equation, as we look for the solution for the uni-directional progressive wave in positive the x-direction, the latter component of $e^{-i(kx+\sigma t)}$ is neglected, that is, $C_2 = 0$ in Eq. (4.45).

The general solution is then rewritten

$$\phi = \left(C_3'\mathrm{e}^{kz} + C_4'\mathrm{e}^{-kz}\right)\mathrm{e}^{i(kx-\sigma t)} \tag{4.49}$$

where the constants in Eq. (4.45) have been merged; $C_3' = C_3C_1$ and $C_4' = C_4C_1$. The substitution of Eq. (4.49) into the bottom boundary condition Eq. (4.23) gives

$$\left.\frac{\partial\phi}{\partial z}\right|_{z=-h} = \left(C_3'k\mathrm{e}^{-kh} - C_4'k\mathrm{e}^{kh}\right)\mathrm{e}^{i(kx-\sigma t)} = 0, \quad \therefore C_3' = C_4'\mathrm{e}^{2kh}$$

Thus ϕ is updated as

$$\begin{aligned}\phi &= C_4'\left(\mathrm{e}^{2kh}\mathrm{e}^{kz} + \mathrm{e}^{-kz}\right)\mathrm{e}^{i(kx-\sigma t)} = C_4'\mathrm{e}^{kh}\left(\mathrm{e}^{k(h+z)} + \mathrm{e}^{-k(h+z)}\right)\mathrm{e}^{i(kx-\sigma t)} \\ &= C_4''\cosh k\,(h+z)\,\mathrm{e}^{i(kx-\sigma t)}\end{aligned} \tag{4.50}$$

where the hyperbolic function $\cosh k\,(h+z) = \left(\mathrm{e}^{k(h+z)} + \mathrm{e}^{-k(h+z)}\right)/2$ (see Section 1.1.3). The substitution of Eq. (4.50) into the combined free-surface boundary condition, Eq. (4.25), gives the dispersion relation, which defines the relation between the angular frequency $\sigma(= 2\pi/T)$ and wave number $k(= 2\pi/L)$, where T is the wave period and L is the wavelength:

$$\sigma^2 = \left(gk + \frac{\gamma}{\rho}k^3\right)\tanh kh \tag{4.51}$$

The wave speed, defined by $c = \sigma/k = L/T$, is thus given as

$$c = \sqrt{\left(\frac{g}{k} + \frac{\gamma}{\rho}k\right)\tanh kh} \tag{4.52}$$

If we assume the progressive oscillation form of the surface elevation with amplitude A,

$$\eta = A\mathrm{e}^{i(kx-\sigma t)} \tag{4.53}$$

The substitution of Eqs. (4.53) and (4.50) into the kinematic boundary condition at a free-surface, Eq. (4.20), determines the unknown constant C_4'':

$$-iA\sigma = C_4''k\sinh kh, \quad \therefore C_4'' = -i\frac{A\sigma}{k\sinh kh}$$

Finally the velocity potential for the progressive small amplitude waves is obtained as

$$\phi = -iA\frac{\sigma}{k}\frac{\cosh k\,(h+z)}{\sinh kh}\mathrm{e}^{i(kx-\sigma t)} \tag{4.54}$$

Substituting Euler's formula (Section 1.1.3), $\mathrm{e}^{i(kx-\sigma t)} = \cos(kx-\sigma t) + i\sin(kx-\sigma t)$, into Eq. (4.54) and considering the real part, the trigonometric representation of ϕ is given

$$\phi = Re\left[-iA\frac{\sigma}{k}\frac{\cosh k\,(h+z)}{\sinh kh}\left(\cos\,(kx-\sigma t)+i\sin\,(kx-\sigma t)\right)\right]$$
$$= A\frac{\sigma}{k}\frac{\cosh k\,(h+z)}{\sinh kh}\sin\,(kx-\sigma t) \tag{4.55}$$

Following the definition of velocity potential, Eq. (1.132), horizontal and vertical velocities are then given by

$$u = \frac{\partial \phi}{\partial x} = A\sigma\frac{\cosh k(h+z)}{\sinh kh}\cos(kx-\sigma t) \tag{4.56}$$

$$w = \frac{\partial \phi}{\partial z} = A\sigma\frac{\sinh k(h+z)}{\sinh kh}\sin(kx-\sigma t) \tag{4.57}$$

Here the derivatives of the hyperbolic functions should be referred to Section A.2 in Appendix. The pressure under waves is derived from the linearized Bernoulli equation Eq. (1.163) as

$$p = -\rho g z + \rho A\frac{\sigma^2}{k}\frac{\cosh k(h+z)}{\sinh kh}\cos(kx-\sigma t) \tag{4.58}$$

The first term on right-hand side of Eq. (4.58) indicates static pressure, and the second one is termed dynamic pressure.

If the surface tension is negligibly small ($\gamma \to 0$), the dispersion relation Eq. (4.51) and wave velocity Eq. (4.52) are reduced to

$$\sigma^2 = gk\tanh kh \tag{4.59}$$

$$c = \sqrt{\frac{g}{k}\tanh kh} = \frac{g}{\sigma}\tanh kh \tag{4.60}$$

In gravity free field ($g \to 0$ in Eqs. (4.51) and (4.52)), the dispersion relation and wave speed for capillary waves are given by

$$\sigma^2 = \frac{\gamma}{\rho}k^3\tanh kh \tag{4.61}$$

$$c = \sqrt{\frac{\gamma}{\rho}k\tanh kh} \tag{4.62}$$

Fig. 4.3 shows the dispersion relations (left), Eqs. (4.51), (4.59), and (4.61), and the wave speeds (right), Eqs. (4.52), (4.60), and (4.62) in deep water ($\tanh kh \to 1$, see Section 1.1.3). Eq. (4.51) is well approximated by Eq. (4.59) when k is sufficiently small, while the deviation increases with k. The wave number at the intersection of Eqs. (4.59) and (4.61), $k_c = \sqrt{\rho g/\gamma} \sim 367.0$ m^{-1} (corresponding wavelength $L_c = 17.1$ mm) when $\rho = 1000.0$ kg/m^3, $\gamma =$ 0.07275 N/m. The wave velocity in the gravity-dominant regime ($k \ll k_c$), well approximated by Eq. (4.60), decreases with k, while the capillary wave velocity Eq. (4.62) in $k \gg k_c$ increases with k. Accordingly, water waves have minimum wave velocity $c_m = (4g\gamma/\rho)^{1/4} \sim 0.231$ m/s at $k = k_c$ in the deep water regime.

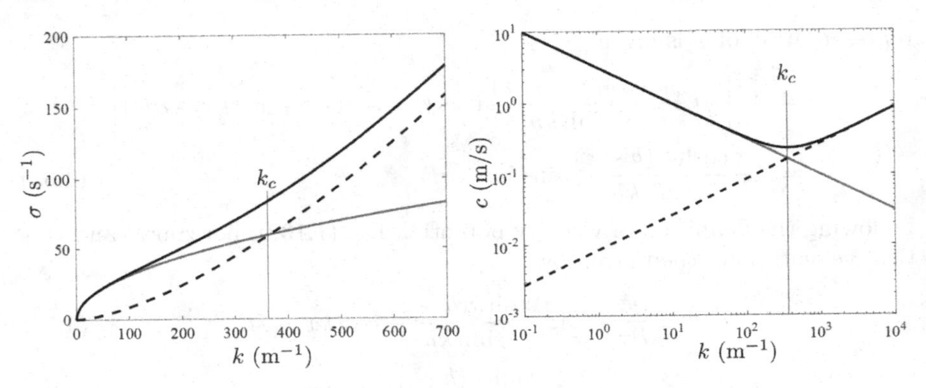

FIGURE 4.3
Dispersion relations (left, black line (capillary+gravity): Eq. (4.51), gray line (gravity only): Eq. (4.59), broken line (capillary only): Eq. (4.61)) and wave velocities (right, black line (capillary+gravity): Eq. (4.52), gray line (gravity only): Eq. (4.60), broken line (capillary only): Eq. (4.62)).

4.4 Waves on thin sheets of fluid

Formations of sheet or film of liquid are observed in various flows of the natural environment: cavity walls at crown splash [125], foams floating on surfaces [55], ascending jet produced at wave impacts on vertical wall [121], and overturning jets of breaking waves [1]. The local oscillation on such thin sheets is governed by surface tension rather than gravity, when a reference Eötvös number, Eq. (3.149), $Eo = gh^2\rho/\gamma \ll 1$ for very thin thickness h. In this section, we consider the capillary oscillation of a thin liquid sheet in two-dimensional space (x, y) in gravity-free field.

There are two oscillation modes: if the displacement of opposite surfaces of the sheet is in phase, the waves are called symmetrical waves, and if it is in opposite phase, antisymmetrical waves [96] (see Fig. 4.4). These waves with different modes have the different dispersion relation and thus propagate with different wave speeds on the sheets.

Consider the harmonic surface displacements of the symmetric waves horizontally propagating on a sheet of thickness $2h$ (see Fig. 4.4). The origin of the vertical z-axis is taken at the center of the sheet. The top surface is located at

$$z = \eta_{s1} = h - Aie^{i(kx-\sigma t)} \quad (\mathrm{Re}[\eta_{s1}] = h + A\sin(kx - \sigma t))$$

and the bottom one at

$$z = \eta_{s2} = -h + Aie^{i(kx-\sigma t)} \quad (\mathrm{Re}[\eta_{s2}] = -h - A\sin(kx - \sigma t))$$

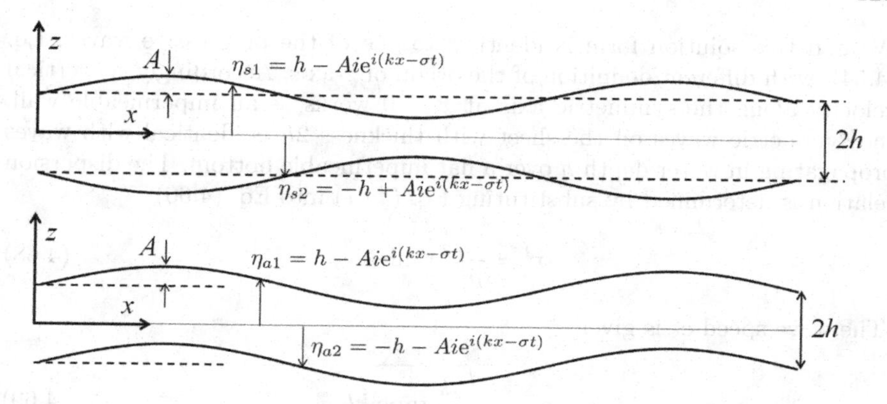

FIGURE 4.4
Symmetrical waves (top) and antisymmetrical waves (bottom) propagating on a thin sheet of fluid.

where A is the amplitude. The Laplace equation, kinematic boundary condition and combined surface condition in this case are written as

$$\frac{\partial^2 \phi}{\partial x^2} + \frac{\partial^2 \phi}{\partial z^2} = 0 \quad (-h < z < h) \tag{4.63}$$

$$\frac{\partial \eta_{s1}}{\partial t} - \frac{\partial \phi}{\partial z} = 0 \quad (z = h) \tag{4.64}$$

$$\frac{\partial \eta_{s2}}{\partial t} - \frac{\partial \phi}{\partial z} = 0 \quad (z = -h) \tag{4.65}$$

$$\frac{\partial^2 \phi}{\partial t^2} - \frac{\gamma}{\rho}\frac{\partial}{\partial z}\frac{\partial^2 \phi}{\partial x^2} = 0 \quad (z = \pm h) \tag{4.66}$$

As given in Section 4.3, the general solution of the Laplace equation Eq. (4.63), $\phi = \left(C_1 \mathrm{e}^{kz} + C_2 \mathrm{e}^{-kz}\right) \mathrm{e}^{i(kx-\sigma t)}$ describes the wave phase propagation in positive x direction. The constants C_1 and C_2 are determined by the linearized kinematic boundary conditions at both surfaces, Eqs. (4.64) and (4.65)

$$C_1 \mathrm{e}^{kh} - C_2 \mathrm{e}^{-kh} = -A\sigma/k, \quad C_1 \mathrm{e}^{-kh} - C_2 \mathrm{e}^{kh} = A\sigma/k$$

$$\therefore C_1 = -A\frac{\sigma}{k}\frac{\cosh(kh)}{\sinh(2kh)} = C_2$$

The velocity potential for the symmetrical waves is thus given as (see Section A.2 in Appendix for the properties of hyperbolic functions)

$$\begin{aligned}\phi &= -A\frac{\sigma}{k}\frac{\cosh kh}{\sinh 2kh}\left(\mathrm{e}^{kz} + \mathrm{e}^{-kz}\right)\mathrm{e}^{i(kx-\sigma t)} \\ &= -A\frac{\sigma}{k}\frac{\cosh kz}{\sinh kh}\mathrm{e}^{i(kx-\sigma t)}\end{aligned} \tag{4.67}$$

We find this solution form is identical to one of the progressive waves, Eq. (4.54), with different definition of the origin of z-axis. Accordingly, as vertical velocity along the symmetric axis at $z = 0$ works as an impermeable wall, the symmetric waves on the sheet with thickness $2h$ is identical with waves propagating in water depth h over a flat impermeable bottom. The dispersion relation is determined by substituting Eq. (4.67) into Eq. (4.66):

$$\sigma^2 = \frac{\gamma}{\rho} k^3 \tanh kh \tag{4.68}$$

The wave speed c_s is given

$$c_s = \sqrt{\frac{\gamma k}{\rho} \tanh kh} \tag{4.69}$$

When kh is small, using the shallow water approximation Eqs. (1.33), (4.68), and (4.69) are approximated as

$$\sigma^2 \approx \frac{\gamma}{\rho} k^4 h \tag{4.70}$$

and

$$c_s \approx k \sqrt{\frac{\gamma}{\rho} h} \tag{4.71}$$

The identical procedure is taken for the antisymmetrical waves having the top and bottom surfaces at

$$z = \eta_{a1} = h - Ai\mathrm{e}^{i(kx-\sigma t)} \quad \text{and} \quad z = \eta_{a2} = -h - Ai\mathrm{e}^{i(kx-\sigma t)}$$

The kinematic boundary conditions, Eqs. (4.64) and (4.65), are used to determine the solution of the Laplace equation Eq. (4.63), $\phi = \left(C_1 \mathrm{e}^{kz} + C_2 \mathrm{e}^{-kz}\right) \mathrm{e}^{i(kx-\sigma t)}$:

$$C_1 \mathrm{e}^{kh} - C_2 \mathrm{e}^{-kh} = -A\sigma/k, \quad C_1 \mathrm{e}^{-kh} - C_2 \mathrm{e}^{kh} = -A\sigma/k$$
$$\therefore C_1 = -A \frac{\sigma}{k} \frac{\sinh kh}{\sinh 2kh} = -C_2$$

Therefore the solution for the antisymmetric waves is given

$$\phi = -A \frac{\sigma}{k} \frac{\sinh kz}{\cosh kh} \mathrm{e}^{i(kx-\sigma t)} \tag{4.72}$$

where

$$\sigma^2 = \frac{\gamma k^3}{\rho} \coth kh \tag{4.73}$$

and the wave speed

$$c_a = \sqrt{\frac{\gamma k}{\rho} \coth kh} \tag{4.74}$$

The shallow water approximation reduces them as

$$\sigma^2 \approx \frac{\gamma k^2}{\rho h} \tag{4.75}$$

and

$$c_a \approx \sqrt{\frac{\gamma}{\rho h}} \tag{4.76}$$

The wave speed of symmetric waves c_s increases with thickness, while the one of antisymmetrical waves c_a contrarily decreases. We also find that c_a is independent of the wavelength and is analogous to waves propagating a string stretched with tension γ and mass per unit length ρh.

4.5 Standing waves

The general function form of velocity potential to satisfy the Laplace equation, Eq. (4.45), involves two waves advancing in the opposite directions. In this section, we consider superposition of these waves. The surface elevations (η^+ and η^-) and velocity potentials (ϕ^+ and ϕ^-) for the wave train propagating in the positive and negative directions are given from Eqs. (4.53) and (4.54) by

$$\eta^{\pm} = A^{\pm} e^{i(\pm kx - \sigma t)} \tag{4.77}$$

$$\phi^{\pm} = -iA^{\pm}\frac{\sigma}{k}\frac{\cosh k(h+z)}{\sinh kh} e^{i(\pm kx - \sigma t)} \tag{4.78}$$

where A^+ and A^- are the amplitudes of η^+ and η^-, respectively. The superposed η is thus

$$\begin{aligned} \eta = \eta^+ + \eta^- &= \left(A^+ e^{ikx} + A^- e^{-ikx}\right) e^{-i\sigma t} \\ &= \left(A^+ (\cos kx + i \sin kx) + A^- (\cos(-kx) + i \sin(-kx))\right) \\ &\quad \times (\cos(-\sigma t) + i \sin(-\sigma t)) \\ &= \left(A^+ + A^-\right) \cos kx \cos \sigma t + \left(A^+ - A^-\right) \sin kx \sin \sigma t \\ &\quad + i\left\{\left(A^+ - A^-\right) \sin kx \cos \sigma t - \left(A^+ + A^-\right) \cos kx \sin \sigma t\right\} \end{aligned} \tag{4.79}$$

The real part of η is thus

$$\eta = Re[\eta] = \left(A^+ + A^-\right) \cos kx \cos \sigma t + \left(A^+ - A^-\right) \sin kx \sin \sigma t \tag{4.80}$$

If $A^+ = A^- = A$,

$$\eta_{st} = Re[\eta] = 2A \cos kx \cos \sigma t \tag{4.81}$$

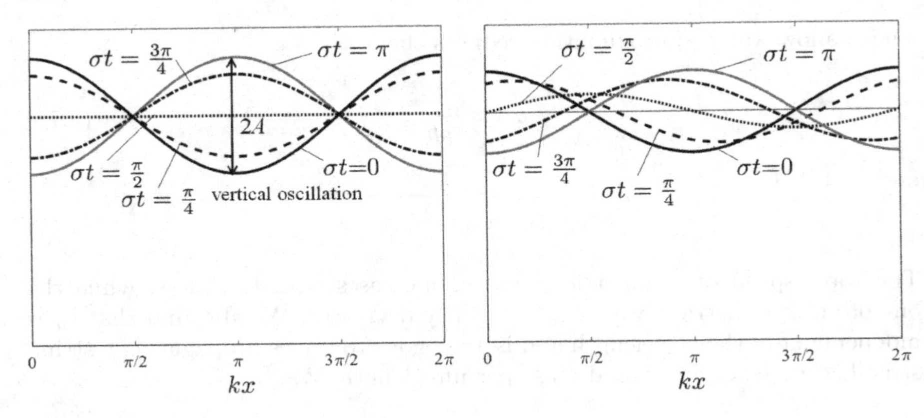

FIGURE 4.5
Standing wave oscillation, Eq. (4.81), (left) and partial standing wave, Eq. (4.80), (right).

In the same way, the superposed ϕ is given by

$$\phi_{st} = Re[\phi] = -2A\frac{\sigma}{k}\frac{\cosh k\,(h+z)}{\sinh kh}\cos kx \sin \sigma t \tag{4.82}$$

Here the dispersion relation is identical to the one of progressive waves Eq. (4.51).

When a train of waves travels in the normal direction to a vertical wall, the waves reflect on the wall and propagate in the opposite direction in phase. The in-phase wave oscillation created by superposition of trains of incident wave and reflected wave is called the standing wave (Fig. 4.5 left). Eq. (4.82) gives the velocities and pressure under the standing waves:

$$u_{st} = \frac{\partial \phi_s}{\partial x} = 2A\sigma\frac{\cosh k(h+z)}{\sinh kh}\sin kx \sin \sigma t \tag{4.83}$$

$$w_{st} = \frac{\partial \phi_s}{\partial z} = -2A\sigma\frac{\sinh k(h+z)}{\sinh kh}\cos kx \sin \sigma t \tag{4.84}$$

$$p_{st} = -\rho g z + 2\rho A\frac{\sigma^2}{k}\frac{\cosh k\,(h+z)}{\sinh kh}\cos kx\, \cos \sigma t \tag{4.85}$$

The wave force acting on the wall can be estimated by integrating p_{st} from the bottom to the free surface, which is a basis of design wave force for a vertical breakwater.

When the wave energy of incident waves is not perfectly reflected on a wall, $A^+ \neq A^-$, e.g. the case of rubble structure partially dissipates the wave energy. The waves described by Eq. (4.80) is called partial standing waves (see Fig. 4.5 right). The superposed surface form changes with the phase.

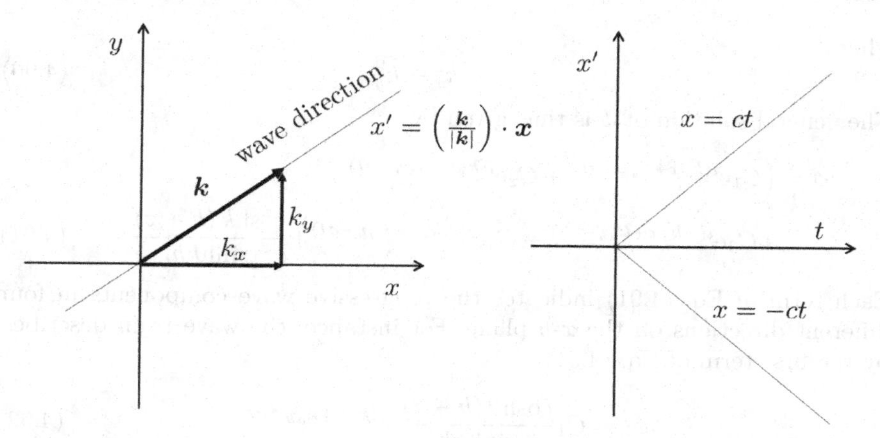

FIGURE 4.6
Wave direction defined by the wave number vector (left) and characteristic lines along the wave direction (right).

4.6 Planar wave propagation

Three-dimensional flows in planar wave propagation are considered in this section. We solve the Laplace equation in three-dimensional space, Eq. (4.19) by introducing the separation of variable

$$\phi(x, y, z, t) = \varphi(x, y)\zeta(z)\mathrm{e}^{-i\sigma t} \tag{4.86}$$

While waves travel in an arbitrary direction on a planar space (x, y), if the horizontal axis is appropriately transformed to be in the wave direction (x' in Fig. 4.6), the function form of $\zeta(z)$ along the wave direction must be identical to one of the one-dimensional progressive waves. We may, therefore, choose the identical function form of ζ to Eq. (4.54):

$$\phi(x, y, z, t) = \varphi(x, y)\frac{\cosh k\,(h+z)}{\sinh kh}\mathrm{e}^{-i\sigma t} \tag{4.87}$$

The substitution into the Laplace equation Eq. (4.19) yields

$$\frac{\partial^2 \varphi}{\partial x^2} + \frac{\partial^2 \varphi}{\partial y^2} + k^2\varphi = 0 \tag{4.88}$$

The general solution describing harmonic oscillations in x and y directions, satisfying Eq. (4.88), takes the form:

$$\varphi = C_1\mathrm{e}^{i(k_x x + k_y y)} + C_2\mathrm{e}^{i(k_x x - k_y y)} + C_3\mathrm{e}^{i(-k_x x + k_y y)} + C_4\mathrm{e}^{-i(k_x x + k_y y)} \tag{4.89}$$

where

$$k^2 = k_x^2 + k_y^2 \tag{4.90}$$

The general solution of ϕ is thus given by

$$\phi = \left(C_1 e^{i(k_x x + k_y y - \sigma t)} + C_2 e^{i(k_x x - k_y y - \sigma t)} + C_3 e^{i(-k_x x + k_y y - \sigma t)} + C_4 e^{i(-k_x x - k_y y - \sigma t)} \right) \frac{\cosh k(h+z)}{\sinh kh} \tag{4.91}$$

Each term of Eq. (4.91) indicates the progressive wave components in four different directions on the x-y plane. For instance, the wave train described by the first term of Eq. (4.91)

$$\phi_1 = C_1 \frac{\cosh k(h+z)}{\sinh kh} e^{i(k_x x + k_y y - \sigma t)} \tag{4.92}$$

has the wave phase expressed by

$$\theta = k_x x + k_y y - \sigma t = \boldsymbol{k} \cdot \boldsymbol{x} - \sigma t$$
$$\therefore \frac{\theta}{|\boldsymbol{k}|} = \left(\frac{\boldsymbol{k}}{|\boldsymbol{k}|} \right) \cdot \boldsymbol{x} - ct \tag{4.93}$$

where $\boldsymbol{k} = (k_x, k_y)$ is the wave number vector, $\boldsymbol{k}/|\boldsymbol{k}|$ is thus the unit wave number vector, the wave speed defined by $c = \sigma/|\boldsymbol{k}|$, and absolute value of the wave number vector $|\boldsymbol{k}| = k_x^2 + k_y^2 = k$ (see Fig. 4.6 left). Eq. (4.93) indicates the specific phase advances in the direction of $\boldsymbol{k}/|\boldsymbol{k}|$ with the speed c. Fig. 4.6 (right) illustrates the characteristic line of the wave crest ($\theta = 0$) advancing in the wave direction $x' = (\boldsymbol{k}/|\boldsymbol{k}|) \cdot \boldsymbol{x}$. We find the wave phase moves along the characteristic line of $x' = ct$ (see also Fig. 4.2). Accordingly, the solution of ϕ_1 along the wave direction is identical with Eq. (4.54) for the one-dimensional progressive wave:

$$\phi_1 = -iA \frac{\sigma}{k} \frac{\cosh k(h+z)}{\sinh kh} e^{i(k_x x + k_y y - \sigma t)} \tag{4.94}$$

The corresponding surface elevation is thus:

$$\eta_1 = A e^{i(k_x x + k_y y - \sigma t)} \tag{4.95}$$

The substitution of ϕ_1 into the surface boundary condition Eq. (4.25) gives the identical dispersion relation to Eq. (4.51):

$$\sigma^2 = \left(gk + \frac{\gamma}{\rho} k \left(k_x^2 + k_y^2 \right) \right) \tanh kh \tag{4.96}$$

with $k^2 = k_x^2 + k_y^2$.

It should be noted that all four wave components in Eq. (4.91) describe propagation in mutually perpendicular directions of (k_x, k_y), $(k_x, -k_y)$, $(-k_x, k_y)$, and $(-k_x, -k_y)$, which configures the planar standing wave system, which is introduced in the next section.

4.7 Free oscillation of water surfaces in containers

In a system performing small oscillations, there is a set of eigenfrequencies, called spectrum (see also Section 7.4.2). If we consider a system of N degrees of freedom, there are N normal modes with each eigenfrequency, which takes the form

$$\phi(x, y, z, t) = A_n \phi_n(x, y, z) \mathrm{e}^{i\sigma_n t} \; (n = 1, 2, \ldots, N) \tag{4.97}$$

The free motion is described by the sum of the normal modes:

$$\phi = \sum_{n=1}^{N} A_n \phi_n(x, y, z) \mathrm{e}^{i\sigma_n t} \tag{4.98}$$

The forced motion consists of free modes together with a periodic forced motion. When the frequency of forcing approaches one of the spectrum frequencies $\sigma_n/2\pi$, the amplitude of the mode becomes infinite owing to resonance. The resonant oscillation in a partially filled fluid tank is called liquid sloshing, which has caused accidents of seismically exited fuel tanks in past.

In this section, typical free oscillating motion in containers is explained before introducing the force motion in the later section.

4.7.1 Oscillation in a rectangle container

4.7.1.1 Two-dimensional tank

Consider the simplest case of the free oscillation of inviscid incompressible fluid between vertical walls with interval length a and water depth h in a two-dimensional (x, z) space. While waves reflect on tank walls, the reflected waves propagate in the opposite side and reflect again on another wall; that is, the standing wave system occurs in the tank. The solution requires the impermeable condition, Eq. (4.27), fulfilled at the walls:

$$u \,|_{x=0,\; a} = \left. \frac{\partial \phi}{\partial x} \right|_{x=0,\; a} = 0 \tag{4.99}$$

We readily find the horizontal velocity of standing waves, Eq. (4.83), satisfies Eq. (4.99) when the wave number $k_n = n\pi/a$ $(n = 1, 2, \ldots)$. In this case, the velocity potential and surface elevation in free oscillation are given as the sum of the normal mode (see Fig. 4.7):

$$\eta = \sum_{n=1} A_n \cos k_n x \cos \sigma_n t \tag{4.100}$$

$$\phi = -\sum_{n=1} A_n \frac{\sigma_n}{k_n} \frac{\cosh k_n (h + z)}{\sinh k_n h} \cos k_n x \sin \sigma_n t \tag{4.101}$$

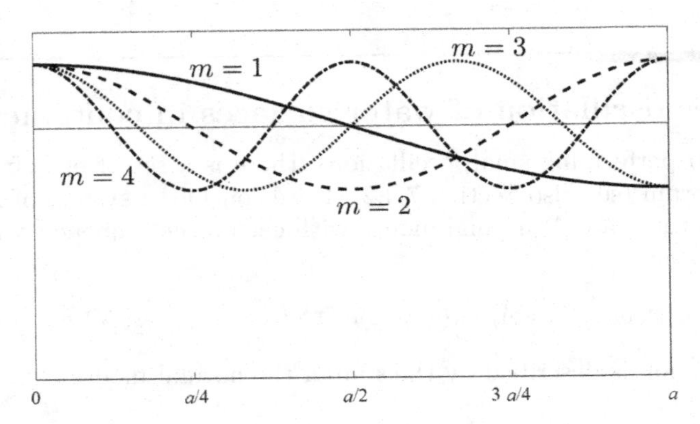

FIGURE 4.7
Oscillation modes in a two-dimensional tank.

where $k_n = n\pi/a$ $(n = 1, 2, \ldots)$, and the dispersion relation

$$\sigma_n^2 = \left(gk_n + \frac{\gamma}{\rho}k_n^3\right)\tanh k_n h \tag{4.102}$$

The period of natural oscillation is therefore given by

$$T_n = \frac{2\pi}{\sigma_n} = \frac{2\pi}{\sqrt{(gk_n + \gamma k_n^3/\rho)\tanh k_n h}} \tag{4.103}$$

4.7.1.2 Three-dimensional tank

Next we consider three-dimensional natural oscillations in a tank of finite length a, width b with constant water depth h. As noted in Section 4.5, a standing wave system is formed by the superposition of wave trains traveling to the mutually opposite directions. In the current planar wave case, all four wave components in Eq. (4.91), describing propagation in mutually perpendicular directions of (k_x, k_y), $(k_x, -k_y)$, $(-k_x, k_y)$, and $(-k_x, -k_y)$, are superposed to describe the planar standing wave system:

$$\begin{aligned}\phi &= -iA\frac{\sigma}{k}\frac{\cosh k(h+z)}{\sinh kh}\Big(e^{i(k_x x + k_y y - \sigma t)} + e^{i(k_x x - k_y y - \sigma t)} \\ &\quad + e^{i(-k_x x + k_y y - \sigma t)} + e^{i(-k_x x - k_y y - \sigma t)}\Big) \\ &= -iA\frac{\sigma}{k}\frac{\cosh k(h+z)}{\sinh kh}\left(e^{ik_x x} + e^{-ik_x x}\right)\left(e^{ik_y y} + e^{-ik_y y}\right)e^{-i\sigma t} \\ &= -4iA\frac{\sigma}{k}\frac{\cosh k(h+z)}{\sinh kh}\cos k_x x \cos k_y y\,(\cos\sigma t - i\sin\sigma t)\end{aligned}$$

where all wave components have the same amplitude A. The trigonometric representation of the superposed velocity potential is given by

$$\phi = Re[\phi] = -4A\frac{\sigma}{k}\frac{\cosh k\,(h+z)}{\sinh kh}\cos k_x x \cos k_y y \sin \sigma t \tag{4.104}$$

Similarly, the surface elevation of the planar standing waves is given

$$\eta = Re[\eta] = 4A\cos k_x x \cos k_y y \cos \sigma t \tag{4.105}$$

To achieve free oscillation in the rectangular tank, the standing wave solution Eq. (4.104) should satisfy the impermeable condition at the wall locations ($x = 0, a$ and $y = 0, b$):

$$u\,|_{x=0,\ a} = \left.\frac{\partial \phi}{\partial x}\right|_{x=0,\ a} = 0 \tag{4.106}$$

$$v\,|_{y=0,\ b} = \left.\frac{\partial \phi}{\partial y}\right|_{y=0,\ b} = 0 \tag{4.107}$$

They are attained when $k_{xn} = n\pi/a$ ($n = 1, 2, \ldots$) and $k_{ym} = m\pi/b$ ($m = 1, 2, \ldots$). The surface elevation and velocity potential are expressed by the sum of (n, m) mode oscillations:

$$\eta = \sum_m \sum_n A_{nm} \cos\left(\frac{n\pi}{a}x\right)\cos\left(\frac{m\pi}{b}y\right)\cos \sigma_{nm} t \tag{4.108}$$

$$\phi = -\sum_m \sum_n A_{nm}\frac{\sigma_{nm}}{k_{nm}}\frac{\cosh k_{nm}\,(h+z)}{\sinh k_{nm} h}\cos\left(\frac{n\pi}{a}x\right)\cos\left(\frac{m\pi}{b}y\right)\sin \sigma_{nm} t \tag{4.109}$$

where $k_{nm} = \sqrt{k_{xn}^2 + k_{ym}^2} = \sqrt{(n\pi/a)^2 + (m\pi/b)^2}$ and the dispersion relation

$$\sigma_{nm}^2 = \left(g k_{nm} + \frac{\gamma}{\rho} k_{nm}^3\right)\tanh k_{nm} h \tag{4.110}$$

The natural period of oscillation is identified as

$$T_{nm} = \frac{2\pi}{\sigma_{nm}} = \frac{2\pi}{\sqrt{(g k_{nm} + \gamma k_{nm}^3/\rho)\tanh\,(k_{nm} h)}} \tag{4.111}$$

For two-dimensional mode ($m = 0$), as $k_{nm} \to k_n$, Eq. (4.111) agrees with Eq. (4.103).

Fig. 4.8 shows the surface displacement for $(m, n) = (1, 1)$ and (3, 2) mode oscillations at $t = 0$ in the rectangular tank. When the forced oscillation with the frequency σ_f approaches the spectrum frequencies of Eq. (4.110), the amplitude of the free motion approaches infinity owing to a resonance.

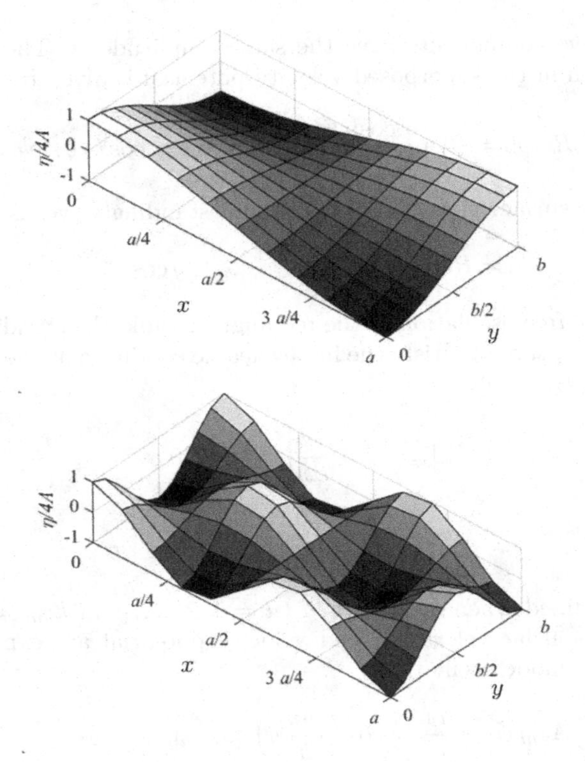

FIGURE 4.8
Surface elevations of $(m, n) = (1, 1)$ mode (top) and $(3, 2)$ mode (bottom) of free oscillation in the rectangular tank.

4.7.2 Oscillation in a cylindrical container

In this section, we consider free oscillation in a cylindrical container of radius R and water depth h (see Fig. 4.1 right). The equation system Eqs. (4.28)–(4.33) in the cylindrical coordinates (r, θ, z) is used to solve the problem. Assuming the harmonic water surface oscillation with angular frequency σ, the variable separation for the velocity potential may be given by the form:

$$\phi\left(r, \theta, z, t\right) = \varphi(r)\zeta(z)\cos n\theta e^{i\sigma t} \tag{4.112}$$

where φ and ζ are the functions of only r and z, respectively. The substitution of Eq. (4.112) into Eq. (4.28) gives

$$\frac{1}{\varphi}\left(\frac{\partial^2\varphi}{\partial r^2} + \frac{1}{r}\frac{\partial\varphi}{\partial r} - \frac{n^2}{r^2}\varphi\right) = -\frac{1}{\zeta}\frac{\partial^2\zeta}{\partial z^2} \tag{4.113}$$

As the left-hand side depends only on r and the right-hand side only on z, Eq. (4.113) can be decomposed into two equation with the constant value λ:

$$-\frac{1}{\varphi}\left(\frac{d^2\varphi}{dr^2}+\frac{1}{r}\frac{d\varphi}{dr}-\frac{n^2}{r^2}\varphi\right)=\lambda \tag{4.114}$$

$$\frac{1}{\zeta}\frac{d^2\zeta}{dz^2}=\lambda \tag{4.115}$$

As noted in Section 4.3, Eq. (4.115) has three solution forms for $\lambda > 0$, $\lambda = 0$, and $\lambda < 0$. In the case of negative eigenvalue $\lambda < 0$, $\zeta = C_1 e^{i\sqrt{|\lambda|}z} + C_2 e^{-i\sqrt{|\lambda|}z}$, indicating the vertical oscillation in the evanescent wave mode, which is introduced in the later section. For $\lambda = 0$, $\zeta = C_1 z + C_2$ and the solution for Eq. (4.114) $\varphi = C_1/r + C_2 r$, which is not an oscillation form of interest. The positive eigenvalue $\lambda > 0$ describes the free oscillation with the radial wave number $k = \sqrt{\lambda}$. The general solution of Eq. (4.115) is given

$$\zeta = C_1 e^{kz} + C_2 e^{-kz} \tag{4.116}$$

From the bottom boundary condition Eq. (4.31), $\partial\zeta/\partial z\,|_{z=-h}= 0$:

$$\left.\frac{\partial\zeta}{\partial z}\right|_{z=-h} = C_1 k e^{-kh} + C_2 k e^{kh} = 0, \rightarrow C_1 = C_2 e^{2kh} \tag{4.117}$$

$$\therefore \zeta = C_2 e^{kh}\left(e^{k(h+z)} + e^{-k(h+z)}\right) = C_3 \cosh k\,(h+z) \tag{4.118}$$

where C_3 is the constant.

Substituting $\lambda = k^2$ into Eq. (4.114), we find Eq. (4.114) takes the form of well-known Bessel's differential equation:

$$r^2\frac{d^2\varphi}{dr^2} + r\frac{d\varphi}{dr} + \left(k^2r^2 - n^2\right)\varphi = 0 \tag{4.119}$$

The solution of Eq. (4.119) is given by the Bessel functions of the first kind, J_n and the second kind, Y_n (see Section A.8.1 in Appendix):

$$\varphi = C_4 J_n\,(kr) + C_5 Y_1\,(kr) \tag{4.120}$$

While $J_n(kr)$ has finite values at $kr = 0$ and oscillatory behaviors in kr (see Fig. 4.9), $Y_n(kr)$ is singular at $kr = 0$. Since ϕ must be finite on the whole domain, Y_n is inappropriate and thus C_5 must be zero. Eq. (4.112) is then given with Eq. (4.118) by

$$\phi = C_0 J_n\,(kr)\cosh k\,(h+z)\cos n\theta e^{i\sigma t} \tag{4.121}$$

where C_0 is the constant to be determined later. The substitution into the wall boundary condition Eq. (4.32) gives

$$\left.\frac{d\phi}{dr}\right|_{r=R} = C_0 \left.\frac{dJ_n(kr)}{dr}\right|_{r=R} \cosh k(h+z)\cos n\theta\, e^{i\sigma t} = 0 \tag{4.122}$$

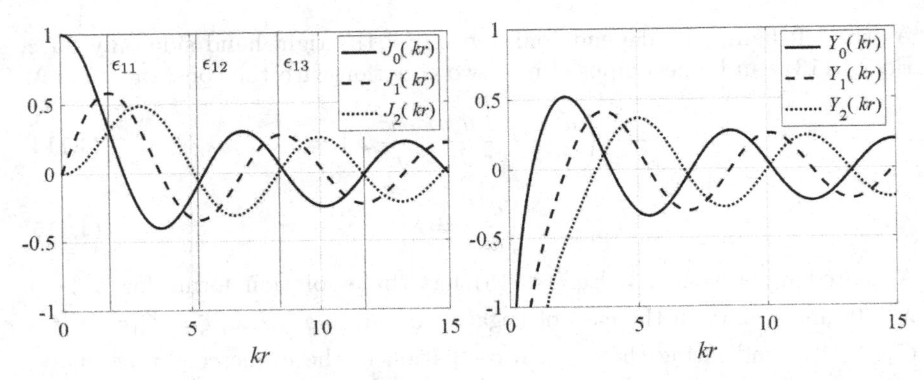

FIGURE 4.9
Bessel functions of the first kind J_n (left) and second kind Y_n (right); the first (ϵ_{11}), second (ϵ_{12}) and third (ϵ_{13}) root of $dJ_1(kr)/dr = 0$ are shown in the left panel.

TABLE 4.1
Roots ϵ_{nm} of $dJ_n(r)/dr = 0$

	$m = 1$	$m = 2$	$m = 3$	$m = 4$	$m = 5$
$n = 0$	3.8317	7.0155	10.1734	13.3236	16.4706
$n = 1$	1.8411	5.3314	8.5363	11.7060	14.8635
$n = 2$	3.0542	6.7061	9.9694	13.1703	16.3475
$n = 3$	4.2011	8.0152	11.3459	14.5858	17.7887
$n = 4$	5.3175	9.2823	12.6819	15.9641	19.1960

This condition is satisfied for arbitrary z, θ and t when $dJ_n(kr)/dr|_{r=R} = 0$. Given ϵ_{nm} as the mth root of $dJ_n(kr)/dr$ (see Fig. 4.9 left and Table 4.1), since $dJ_n(\epsilon_{nm})/dr = 0$, Eq. (4.122) is achieved when $kR = \epsilon_{nm}$. Eq. (4.121) is rewritten as

$$\phi_{nm} = C_0 J_n(k_{nm}r) \cosh k_{nm}(h+z) \cos n\theta e^{i\sigma t} \tag{4.123}$$

where $k_{nm} = \epsilon_{nm}/R$. The substitution of Eq. (4.123) into the free surface boundary condition Eq. (4.33),

$$\begin{aligned} &-\sigma_{nm}^2 J_n(k_{nm}r)\cosh k_{nm}h + gk_{nm}J_n(k_{nm}r)\sinh k_{nm}h \\ &-\frac{\gamma}{\rho}k_{nm}\sinh k_{nm}h\left(\frac{1}{r}\frac{dJ_n(k_{nm}r)}{dr} + \frac{d^2J_n(k_{nm}r)}{dr^2} - \frac{n^2}{r^2}J_n(k_{nm}r)\right) = 0 \end{aligned}$$

Since $J_n(k_{nm}r)$ satisfies Eq. (4.119), the terms in the bracket are reduced to $-k_{nm}^2 J_n(k_{nm}r)$. The dispersion relation for the free oscillation in a cylindrical

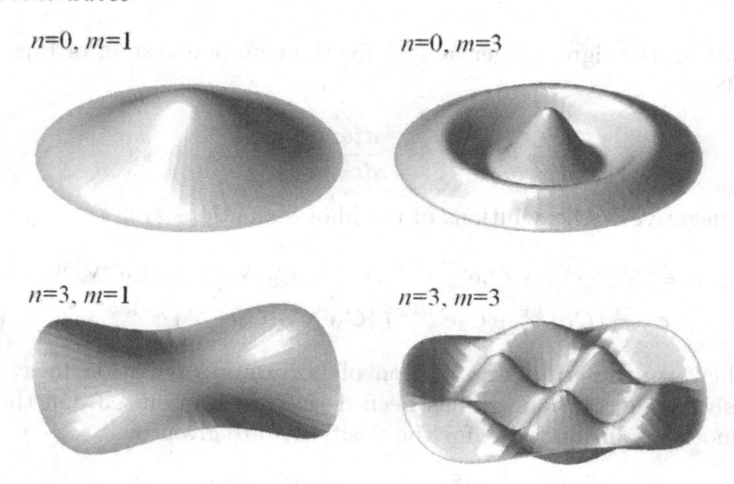

FIGURE 4.10
Free oscillations of the free surface at the different modes $(n,\ m)$.

tank is then given as

$$\sigma_{nm}^2 = \left(gk_{nm} + \frac{\gamma}{\rho} k_{nm}^3 \right) \tanh k_{nm} h \tag{4.124}$$

When the surface is given in terms of the amplitude of (n, m) mode A_{nm}:

$$\eta = \sum_{n=0} \sum_{m=1} A_{nm} J_n \left(k_{nm} r \right) \cos n\theta \mathrm{e}^{i\sigma t} \tag{4.125}$$

the kinematic boundary condition at the surface, Eq. (4.29), determines the coefficient in Eq. (4.123); $C_0 = iA_{nm}\sigma_{nm}/k_{nm} \sinh k_{nm}h$. The solution is thus given by

$$\phi = i \sum_{n=0} \sum_{m=1} A_{nm} \frac{\sigma_{nm}}{k_{nm}} J_n \left(k_{nm} r \right) \frac{\cosh k_{nm} \left(h + z \right)}{\sinh k_{nm} h} \cos n\theta \mathrm{e}^{i\sigma t} \tag{4.126}$$

Typical surface forms depending on the oscillation mode are shown in Fig. 4.10.

4.8 Evanescent waves

When we solve the Laplace equation by the variable separation assuming $\phi(x, y, t) = \xi(x)\zeta(z)\mathrm{e}^{-i\sigma t}$ (see Eqs. (4.35) and (4.36)), the solution form ξ and

ζ depends on the sign of eigenvalue λ for the equation system of Eqs. (4.37) and (4.38):

$$\frac{d^2\xi}{dx^2} - \lambda\xi = 0, \quad \frac{d^2\zeta}{dz^2} + \lambda\zeta = 0$$

For the negative λ, the solutions of the above equations are:

$$\xi = C_1 e^{i\sqrt{|\lambda|}x} + C_2 e^{-i\sqrt{|\lambda|}x}, \quad \zeta = C_3 e^{\sqrt{|\lambda|}z} + C_4 e^{-\sqrt{|\lambda|}z}$$
$$\therefore \phi_p = \left(C_1 e^{ikx} + C_2 e^{-ikx}\right)\left(C_3 e^{kz} + C_4 e^{-kz}\right) e^{-i\sigma t} \tag{4.127}$$

Since this form describes the motion of horizontal oscillation in x, as the progressive wave solution, ϕ_p has been derived in Section 4.3. On the other hand, another solution forms for the positive λ are given by

$$\xi = C_1 e^{\sqrt{\lambda}x} + C_2 e^{-\sqrt{\lambda}x}, \quad \zeta = C_3 e^{i\sqrt{\lambda}z} + C_4 e^{-i\sqrt{\lambda}z}$$
$$\therefore \phi_e = \left(C_1 e^{kx} + C_2 e^{-kx}\right)\left(C_3 e^{ikz} + C_4 e^{-ikz}\right) e^{-i\sigma t} \tag{4.128}$$

This type of the solution, featuring horizontal exponential variation and vertical oscillation, describes evanescent waves. In this section, we look for the wave solution of the evanescent mode.

First we consider a far-field boundary condition. Since we expect a finite wave solution, the exponential form e^{kx} in ϕ_e is inappropriate as $e^{kx} \to \infty$ at $x \to \infty$, and thus C_1 must be zero. Eq. (4.128) is rewritten

$$\phi_e = \left(C_3' e^{ikz} + C_4' e^{-ikz}\right) e^{-kx} e^{-i\sigma t} \tag{4.129}$$

Assuming the constant depth h, Eq. (4.129) is substituted into the bottom boundary condition Eq. (4.23) to have the relation of the coefficients C_3' and C_4'; $C_3' = C_4' e^{2ikh}$, Eq. (4.129) is updated as

$$\begin{aligned}\phi_e &= C_4' e^{ikh}\left(e^{ik(h+z)} + e^{-ik(h+z)}\right) e^{-kx} e^{-i\sigma t} \\ &= C_0 \cos k\,(h+z)\, e^{-kx} e^{-i\sigma t}\end{aligned} \tag{4.130}$$

where C_0 is the constant to be determined later. Assuming negligibly small surface tension, the substitution of Eq. (4.130) into the free-surface boundary condition Eq. (4.26) provides the dispersion relation for the evanescent waves:

$$\sigma^2 = -gk \tan kh \tag{4.131}$$

Assuming the surface elevation takes the consistent oscillation form with Eq. (4.130),

$$\eta = A e^{-kx} e^{-i\sigma t} \tag{4.132}$$

where A is the amplitude at $x = 0$. The kinematic boundary condition at the surface Eq. (4.20) determines the constant $C_0 = iA\sigma/k \sin kh$, and thus the

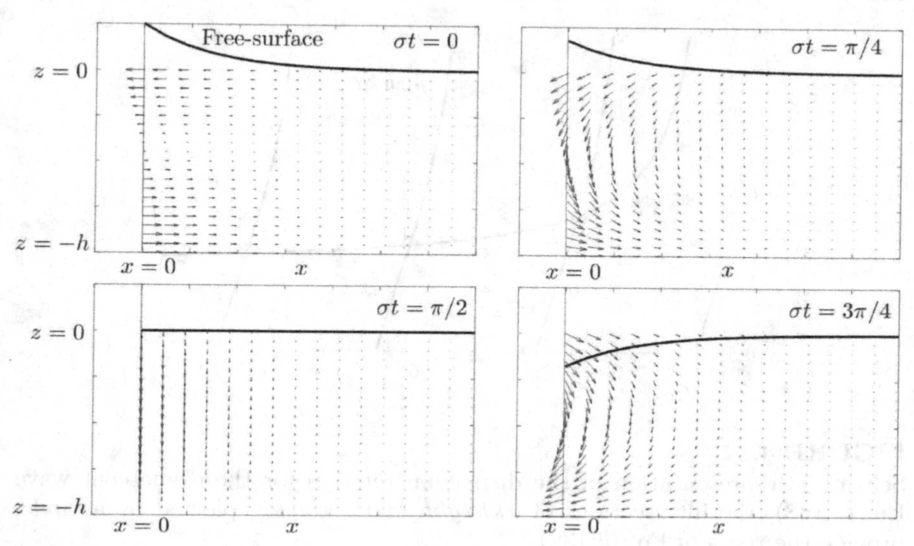

FIGURE 4.11
Free surface and fluid velocity of evanescent waves, estimated by Eqs. (4.132) and (4.133), at $\sigma t = 0,\ \pi/4,\ \pi/2,\ 3\pi/4$.

solution is derived as

$$\phi_e = iA\frac{\sigma}{k}\frac{\cos k\,(h+z)}{\sin kh}\mathrm{e}^{-kx}\mathrm{e}^{-i\sigma t} \tag{4.133}$$

The trigonometric forms of Eqs. (4.132) and (4.133) are given

$$\eta = Re\left[A\mathrm{e}^{-kx}\left(\cos\left(-\sigma t\right) + i\sin\left(-\sigma t\right)\right)\right] = A\mathrm{e}^{-kx}\cos\sigma t \tag{4.134}$$

$$\begin{aligned}\phi_e &= Re\left[iA\frac{\sigma}{k}\frac{\cos k\,(h+z)}{\sin kh}\mathrm{e}^{-kx}\left(\cos\left(-\sigma t\right) + i\sin\left(-\sigma t\right)\right)\right]\\ &= A\frac{\sigma}{k}\frac{\cos k\,(h+z)}{\sin kh}\mathrm{e}^{-kx}\sin\sigma t\end{aligned} \tag{4.135}$$

We find the horizontal velocity of this wave has temporally oscillating feature at $x = 0$

$$\left.\frac{\partial\phi_e}{\partial x}\right|_{x=0} = -A\sigma\frac{\cos k\,(h+z)}{\sin kh}\sin\sigma t = U(z)\sin\sigma t \tag{4.136}$$

where $U(z)$ is the amplitude of velocity at $x = 0$. Accordingly, the evanescent waves are induced by the forced oscillation with the frequency σ at $x = 0$, and exponentially decays in the x-direction (see Fig. 4.11), which has been referred in a wavemaker theory.

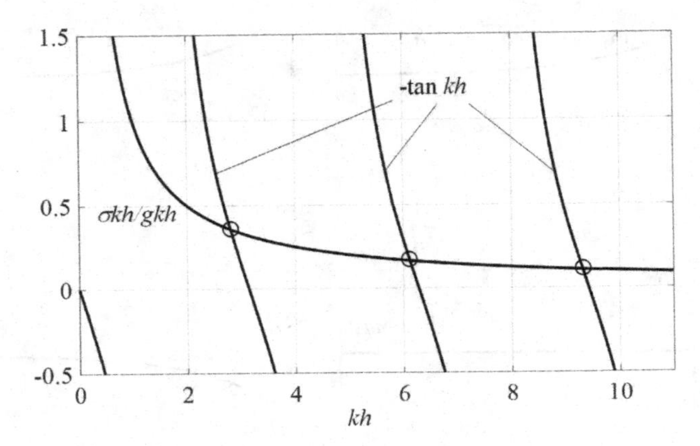

FIGURE 4.12
Schematic representation of the dispersion relation for the evanescent wave Eq. (4.138); the intersections of $\sigma kh/gkh$ and $-\tan kh$, plotted by a circle, provide the roots of Eq. (4.138).

Eq. (4.131) has infinite numbers of solutions for the single σ (see Fig. 4.12). Eq. (4.135) is thus generalized for the n-th mode of wave component:

$$\phi_{en} = A_n \frac{\sigma}{k_n} \frac{\cos k_n (h+z)}{\sin k_n h} \mathrm{e}^{-k_n x} \sin \sigma t \tag{4.137}$$

with the dispersion relation

$$\sigma^2 = -gk_n \tan k_n h \tag{4.138}$$

where k_n is the n th root of Eq. (4.138) $(n = 1, \ldots, \infty)$.

It should be noted that, since both ϕ_p, Eq. (4.127), and ϕ_e, Eq. (4.128), are the solutions of the Laplace equation, the flows defined by both ϕ_p and ϕ_e are simultaneously present. Therefore, the flow can be generally described by the superposition of the sum of all evanescent wave component and the progressive wave component, Eq. (4.55):

$$\phi = A_0 \frac{\sigma}{k_0} \frac{\cosh k_0 (h+z)}{\sinh k_0 h} \cos (k_0 x - \sigma t) + \sum_{n=1}^{\infty} A_n \frac{\sigma}{k_n} \frac{\cos k_n (h+z)}{\sin k_n h} \mathrm{e}^{-k_n x} \sin \sigma t \tag{4.139}$$

where the subscript 0 indicates the progressive wave component, following the dispersion relation $\sigma^2 = gk_0 \tanh k_0 h$. The amplitude of horizontal velocity at $x = 0$, Eq. (4.136), is now given by

$$U(z) = A_0 \sigma \frac{\cosh k_0 (h+z)}{\sinh k_0 h} - \sum_{n=1}^{\infty} A_n \sigma \frac{\cos k_n (h+z)}{\sin k_n h} \tag{4.140}$$

As the set $\{\cosh k_0(h+z), \cos k_n(h+z)\}$ is a complete orthogonal set[17],

$$\int_{-h}^{0} \cosh k_0(h+z)\cos k_n(h+z)dz = 0, \quad \int_{-h}^{0} \cos k_m(h+z)\cos k_n(h+z)dz = 0$$

If both sides of Eq. (4.140) are multiplied by $\cosh k_0\,(h+z)$ and integrated over the depth, as no contribution comes from the series of the evanescent components, the amplitude of progressive waves can be determined by

$$A_0 = \frac{4k_0 \sinh k_0 h \int_{-h}^{0} U(z)\cosh k_0\,(h+z)\,dz}{\sigma\,(\sinh 2k_0 h + 2k_0 h)} \tag{4.141}$$

Similarly, multiplying $\cos k_n\,(h+z)$ and integrating, the amplitudes of the evanescent modes are given

$$A_n = -\frac{4k_n \sin k_n h \int_{-h}^{0} U(z)\cos k_n\,(h+z)\,dz}{\sigma\,(\sin 2k_n h + 2k_n h)} \tag{4.142}$$

The wave field described by Eq. (4.139) is typically observed in a wave experiment using a wavemaker. The oscillating wave paddle creates both of the progressive and evanescent waves. For a piston type of wave paddle having a vertical-wall shape, as U in Eq. (4.141) is not a function of z, Eq. (4.141) gives

$$A_0 = \frac{2U\,(\cosh 2k_0 h - 1)}{\sigma\,(\sinh 2k_0 h + 2k_0 h)} \tag{4.143}$$

Using the definition of wave height, $H = 2A_0$, and stroke $S = 2U/\sigma$ of the wave paddle, the above relation is rewritten

$$\frac{H}{S} = \frac{2\,(\cosh 2k_0 h - 1)}{(\sinh 2k_0 h + 2k_0 h)} \tag{4.144}$$

which is called the transfer function that estimates a far-field wave height of generated progressive waves as a function of the stroke of a wavemaker. As found from Fig. 4.11, the evanescent waves affect the flow near the moving boundary, while this effect exponentially decays to be negligibly small far from the boundary. In laboratory experiments of progressive waves, the wave measurements should be performed far enough away from the wave paddle where is unaffected by the evanescent waves.

4.9 Edge waves

In the solution procedure of the Laplace equation, a general solution describing horizontal wave propagation, $\exp[i(kx - \sigma t)]$, is chosen to attain the solution for progressive waves in Section 4.3, and in Section 4.8, another solution form

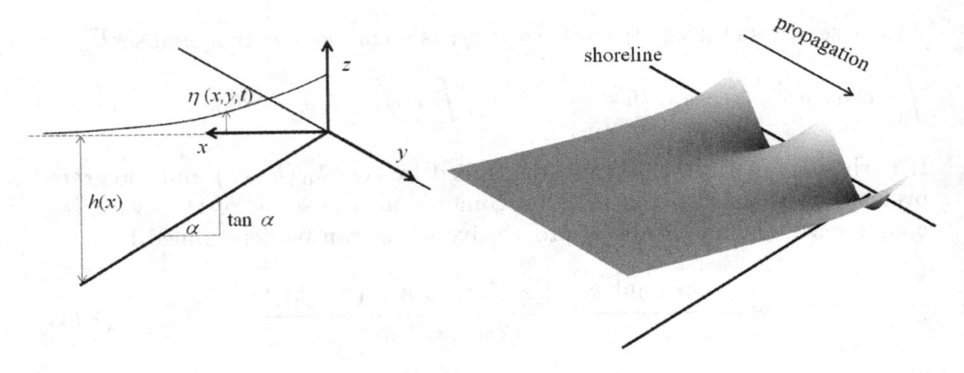

FIGURE 4.13
Coordinates (left) and surface elevation (right) of edge waves.

describing vertical propagation, $\exp[i(kz - \sigma t)]$, is considered for evanescent waves. In this section, we look for the other solution describing lateral propagation in the y-direction perpendicular to x and z axes; that is, the solution form of $\exp[i(ky - \sigma t)]$. The waves propagate along a shore of beach, called edge waves, was firstly derived by Stokes (1846)[93].

Following Stokes[93], we consider waves traveling in the long shore direction (y-axis) on the uniform sloping beach with the bottom angle α with respect to horizontal, offshore ward x-axis (Fig. 4.13). The origin of x-axis (cross-shore direction) is taken at the shore location. The water depth $h(x) = x \tan \alpha$, assuming no bottom slope in the y-direction. The vertical z-axis has the origin at a still water level.

As noted, a general solution describing the lateral propagation in y-axis is chosen in this case:

$$\phi = \varphi(x, z)\mathrm{e}^{i(ky - \sigma t)} \tag{4.145}$$

The substitution into the Laplace equation gives

$$\frac{\partial^2 \varphi}{\partial x^2} + \frac{\partial^2 \varphi}{\partial z^2} - k^2 \varphi = 0 \tag{4.146}$$

Considering a finite value of the solution at $x \to \infty$, we readily find φ takes the form:

$$\varphi = C\mathrm{e}^{-k_x x + k_z z} \tag{4.147}$$

where the wave number k_x in the x-direction , k_z in z direction, and

$$k^2 = k_x^2 + k_z^2 \tag{4.148}$$

Eq. (4.145) is thus rewritten

$$\phi = C\mathrm{e}^{-k_x x + k_z z}\mathrm{e}^{i(ky - \sigma t)} \tag{4.149}$$

The substitution into the sloped bottom boundary condition, Eq. (4.22), gives

$$k_z = \frac{\partial h}{\partial x} k_x = \tan \alpha k_x \tag{4.150}$$

From Eqs. (4.148) and (4.150), k_z and k_x are expressed in terms of k:

$$k_z^2 = \frac{k^2 \tan^2 \alpha}{1 + \tan^2 \alpha}$$

$$\therefore k_z = \frac{k \tan \alpha}{\sqrt{1 + \tan \alpha}} = k \sin \alpha, \quad k_x = \sqrt{k^2 - k_z^2} = k \cos \alpha \tag{4.151}$$

Eq. (4.149) is thus written in terms of k:

$$\phi = C\mathrm{e}^{-k \cos \alpha x + k \sin \alpha z} \mathrm{e}^{i(ky - \sigma t)} \tag{4.152}$$

Assuming negligibly small surface tension in this case, the combined free-surface condition, Eq. (4.26), provides the dispersion relation for the edge waves:

$$\sigma^2 = gk \sin \alpha \tag{4.153}$$

Assuming the surface form with the amplitude A given by

$$\eta = iA\mathrm{e}^{-k \cos \alpha x} \mathrm{e}^{i(ky - \sigma t)} \tag{4.154}$$

the kinematic boundary condition at free surface, Eq. (4.20), determines $C = A\sigma/k \sin \alpha$. The solution for the Stokes' edge wave is finally given by

$$\phi = A \frac{\sigma}{k \sin \alpha} \mathrm{e}^{-k \cos \alpha x + k \sin \alpha z} \mathrm{e}^{i(ky - \sigma t)} \tag{4.155}$$

Ursell (1952) [103)] found the additional, series of discrete modes of the edge waves. The proposed solution is:

$$\phi = \mathrm{e}^{i(ky - \sigma t)} \left[A_s \mathrm{e}^{-k(\cos \alpha x - k \sin \alpha z)} + \sum_{m=1}^{n} A_{mn} \left\{ \mathrm{e}^{-k(x \cos(2m-1)\alpha - zk \sin(2m-1)\alpha)} \right.\right.$$
$$\left.\left. + \mathrm{e}^{-k(x \cos(2m+1)\alpha + zk \sin(2m+1)\alpha)} \right\} \right] \tag{4.156}$$

where $A_s = A\sigma/k \sin \alpha$ is the amplitude of the Stokes edge wave. The above solution satisfies the bottom boundary condition, Eq. (4.22), and the free-surface boundary condition, Eq. (4.26), when A_{mn} is given by

$$A_{mn} = (-1)^m \prod_{r=1}^{m} \frac{\tan(n - r + 1)\alpha}{\tan(n + r)\alpha} \tag{4.157}$$

and the dispersion relation

$$\sigma^2 = gk \sin(2n + 1)\alpha \tag{4.158}$$

FIGURE 4.14
Regular pattern of wave run-up on a beach, induced by edge waves.

The Ursell's edge wave Eq. (4.156) with $n = 0$ agrees with the Stokes one Eq. (4.155).

The edge waves may affect coastal process. The surface oscillation along a shore modifies the wave height of run-up waves on the shore and local run-up distance, resulting in the regular pattern of wave run-up on the beach (Fig. 4.14). This effect may influence local beach erosion and transformation of shores, resulting in a formation of beach cusp. The edge waves also affect evolution of tsunami, which is introduced in Section 5.4.2.

5

Shallow Water Equation

The shallow water equation describes fluid flows in a thin layer bounded by the bottom and free surface. The equation is derived by integrating the Navier-Stokes equation over the layer under a hydrostatic assumption, so that three-dimensional system of the Navier-Stokes equation can be reduced to the two-dimensional planar system, which is advantageous to understand major features of the layer flows in the simple system. The equation has covered various scales of the flows dealt with over industrial, engineering and scientific fields; film flows in coating process, breakup of liquid sheet and jet in inkjet technology, open channel flows, tides and tsunami in ocean.

In this chapter, starting from derivation of the shallow water equations, important features of shallow water waves are addressed through some applications.

5.1 Derivation of shallow water equation

We consider the flow with depth $h(x, y)$ and surface elevation $\eta(x, y, t)$ in the Cartesian coordinate shown in Fig. 5.1 (left). Note that the origin of vertical axis z is the still water level. In the framework of shallow water equation, any quantities are integrated over a fluid layer. Assuming the fluid velocity $\boldsymbol{u} =$

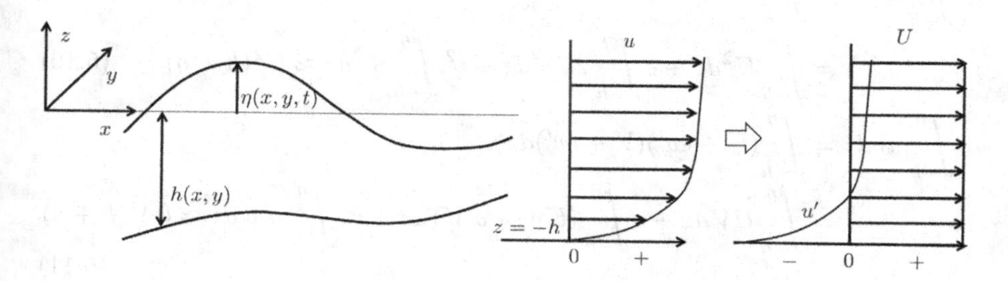

FIGURE 5.1
Coordinate system (left) and velocity profile decomposed into the depth mean U and the deviation u' (right).

DOI: 10.1201/9781003140160-5

(u, v, w) is decomposed into the depth uniform (mean) velocity $\boldsymbol{U} = (U, V, W)$ and the minor deviation $\boldsymbol{u}' = (u', v', w')$ in the layer (see Fig. 5.1 right).

$$u(x, y, z, t) = U(x, y, t) + \epsilon u'(x, y, z, t) \tag{5.1}$$
$$v(x, y, z, t) = V(x, y, t) + \epsilon v'(x, y, z, t) \tag{5.2}$$
$$w(x, y, z, t) = \epsilon w'(x, y, z, t) \tag{5.3}$$

where dimensionless parameter $\epsilon \ll 1$ indicates a small quantity. The above decomposition assumes the horizontal mean velocity (U, V) governs the flow and no mean vertical velocity, $W = 0$, within a thin layer. The integration of the above velocities over the depth gives

$$\int_{-h}^{\eta} u dz = U(h + \eta) + \epsilon \int_{-h}^{\eta} u' dz \tag{5.4}$$
$$\int_{-h}^{\eta} v dz = V(h + \eta) + \epsilon \int_{-h}^{\eta} v' dz \tag{5.5}$$
$$\int_{-h}^{\eta} w dz = \epsilon \int_{-h}^{\eta} w' dz \tag{5.6}$$

If the small quantities with ϵ is ignored, the approximation of velocity at the zero-th order, $O(\epsilon^0)$, is given

$$\int_{-h}^{\eta} u dz = U(h + \eta) \tag{5.7}$$

$$\int_{-h}^{\eta} v dz = V(h + \eta) \tag{5.8}$$

$$\int_{-h}^{\eta} w dz = 0 \tag{5.9}$$

Similarly, the integrated nonlinear terms are approximated as

$$\begin{aligned} \int_{-h}^{\eta} u^2 dz &= \int_{-h}^{\eta} (U + \epsilon u')^2 dz \\ &= \int_{-h}^{\eta} U^2 dz + \epsilon \int_{-h}^{\eta} 2Uu' dz + \epsilon^2 \int_{-h}^{\eta} u'^2 dz \approx U^2(h + \eta) \end{aligned} \tag{5.10}$$

$$\begin{aligned} \int_{-h}^{\eta} uv dz &= \int_{-h}^{\eta} (U + \epsilon u')(V + \epsilon v') dz \\ &= \int_{-h}^{\eta} UV dz + \epsilon \int_{-h}^{\eta} (Uv' + u'V)\, dz + \epsilon^2 \int_{-h}^{\eta} u'v' dz \approx UV(h + \eta) \end{aligned} \tag{5.11}$$

$$\int_{-h}^{\eta} uw dz = \int_{-h}^{\eta} (U + \epsilon u')(\epsilon w') dz = \epsilon \int_{-h}^{\eta} Uw' dz + \epsilon^2 \int_{-h}^{\eta} u'w' dz \approx 0 \tag{5.12}$$
$$\int_{-h}^{\eta} vw dz = \int_{-h}^{\eta} (V + \epsilon v')(\epsilon w') dz = \epsilon \int_{-h}^{\eta} Vw' dz + \epsilon^2 \int_{-h}^{\eta} v'w' dz \approx 0 \tag{5.13}$$

These approximations are considered in the derivation procedure of the shallow water equation.

5.1.1 Continuity equation

The depth-averaged continuity equation of incompressible fluid, ensuring mass conservation, is considered. The integration of the continuity equation, Eq. (1.68), over the depth gives

$$\int_{-h}^{\eta}\left(\frac{\partial u}{\partial x}+\frac{\partial v}{\partial y}+\frac{\partial w}{\partial z}\right)dz=\int_{-h}^{\eta}\frac{\partial u}{\partial x}dz+\int_{-h}^{\eta}\frac{\partial v}{\partial y}dz+w_{\eta}-w_{h}=0 \tag{5.14}$$

where the subscripts η and h indicate quantities at $z=\eta$ (free-surface) and $z=-h$ (bottom), respectively. Using the Leibniz rule, Eq. (A.45) in Appendix, Eq. (5.14) is rewritten

$$\frac{\partial}{\partial x}\int_{-h}^{\eta}udz-u_{\eta}\frac{\partial \eta}{\partial x}-u_{h}\frac{\partial h}{\partial x}+\frac{\partial}{\partial y}\int_{-h}^{\eta}vdz-v_{\eta}\frac{\partial \eta}{\partial y}-v_{h}\frac{\partial h}{\partial y}+w_{\eta}-w_{h}=0 \tag{5.15}$$

The substitution of the integrated velocity Eqs. (5.7) and (5.8) gives

$$\begin{aligned}&\frac{\partial}{\partial x}U\left(h+\eta\right)+\frac{\partial}{\partial y}V\left(h+\eta\right)\\&\quad+\left[w_{\eta}-u_{\eta}\frac{\partial \eta}{\partial x}-v_{\eta}\frac{\partial \eta}{\partial y}\right]_{\eta}-\left[w_{h}+u_{h}\frac{\partial h}{\partial x}+v_{h}\frac{\partial h}{\partial y}\right]_{h}=0\end{aligned} \tag{5.16}$$

The kinematic boundary condition Eq. (3.41); $\partial\eta/\partial t+u\partial\eta/\partial x+v\partial\eta/\partial y=w$ at $z=\eta$, reduces the terms of the square bracket $[\,]_{\eta}$ to $\partial\eta/\partial t$, and the terms of the bracket $[\,]_{h}$ are eliminated by the bottom boundary condition Eq. (3.40); $u\partial h/\partial x+v\partial h/\partial y+w=0$ at $z=-h$. Eq. (5.16) is rewritten to give the depth-averaged continuity equation:

$$\frac{\partial \eta}{\partial t}+\frac{\partial}{\partial x}U\left(h+\eta\right)+\frac{\partial}{\partial y}V\left(h+\eta\right)=0 \tag{5.17}$$

In the vector form,

$$\frac{\partial \eta}{\partial t}+\boldsymbol{\nabla}\cdot\left(\boldsymbol{U}\left(h+\eta\right)\right)=0, \tag{5.18}$$

where $\boldsymbol{U}=(U,V)$ and the gradient operator $\boldsymbol{\nabla}$ is redefined in this chapter as $\boldsymbol{\nabla}=(\partial/\partial x,\partial/\partial y)$.

5.1.2 Momentum equation

We next derive the depth-averaged momentum equation from the conservative form of Navier-Stokes equations, Eqs. (1.98)–(1.100):

$$\frac{\partial u}{\partial t}+\frac{\partial u^2}{\partial x}+\frac{\partial uv}{\partial y}+\frac{\partial uw}{\partial z}=-\frac{1}{\rho}\frac{\partial p}{\partial x}+\frac{1}{\rho}\left(\frac{\partial \tau_{xx}}{\partial x}+\frac{\partial \tau_{xy}}{\partial y}+\frac{\partial \tau_{xz}}{\partial z}\right) \tag{5.19}$$

$$\frac{\partial v}{\partial t} + \frac{\partial uv}{\partial x} + \frac{\partial v^2}{\partial y} + \frac{\partial vw}{\partial z} = -\frac{1}{\rho}\frac{\partial p}{\partial y} + \frac{1}{\rho}\left(\frac{\partial \tau_{yx}}{\partial x} + \frac{\partial \tau_{yy}}{\partial y} + \frac{\partial \tau_{yz}}{\partial z}\right) \tag{5.20}$$

$$\frac{\partial w}{\partial t} + \frac{\partial uw}{\partial x} + \frac{\partial vw}{\partial y} + \frac{\partial w^2}{\partial z} = -\frac{1}{\rho}\frac{\partial p}{\partial z} + \frac{1}{\rho}\left(\frac{\partial \tau_{zx}}{\partial x} + \frac{\partial \tau_{zy}}{\partial y} + \frac{\partial \tau_{zz}}{\partial z}\right) - g \tag{5.21}$$

where $\tau_{xx} \ldots \tau_{zz}$, in the second terms on the right-hand sides of these equations, are the viscous stress components defined in Eq. (1.52). The integration of Eq. (5.21) from arbitrary level z to η gives

$$\begin{aligned}\int_z^\eta \frac{\partial w}{\partial t}dz + \int_z^\eta \frac{\partial uw}{\partial x}dz + \int_z^\eta \frac{\partial vw}{\partial y}dz + w_\eta^2 - w_z^2 = -\frac{1}{\rho}\left(p_\eta - p_z\right) \\ + \frac{1}{\rho}\left(\int_z^\eta \frac{\partial \tau_{zx}}{\partial x}dz + \int_z^\eta \frac{\partial \tau_{zy}}{\partial y}dz + \tau_{zz}\mid_\eta - \tau_{zz}\mid_z\right) - (\eta - z)\, g\end{aligned} \tag{5.22}$$

Adopting the Leibniz rule, Eq. (A.45) in Appendix, Eq. (5.22) is transformed to

$$\begin{aligned}\frac{\partial}{\partial t}\int_z^\eta w dz + \frac{\partial}{\partial x}\int_z^\eta uw dz + \frac{\partial}{\partial y}\int_z^\eta vw dz - w_z^2 \\ - w_\eta\left[\frac{\partial \eta}{\partial t} + u_\eta \frac{\partial \eta}{\partial x} + v_\eta \frac{\partial \eta}{\partial y} - w_\eta\right]_\eta = -\frac{1}{\rho}\left(p_\eta - p_z\right) \\ + \frac{1}{\rho}\left(\int_z^\eta \frac{\partial \tau_{zx}}{\partial x}dz + \int_z^\eta \frac{\partial \tau_{zy}}{\partial y}dz + \tau_{zz}\mid_\eta - \tau_{zz}\mid_z\right) - (\eta - z)\, g\end{aligned} \tag{5.23}$$

Here the terms in $[\ ]_\eta$ are eliminated by the free-surface kinematic condition, Eq. (3.41). The approximations of Eqs. (5.9), (5.12), and (5.13) remove the first to third terms of Eq. (5.23). If Eqs. (5.1)–(5.3) are introduced, the fourth term, $w_z^2 = \epsilon^2 w_z'^2$, is the order of ϵ^2 and can be neglected. Since the viscous stress (the second term on the right-hand side), $\tau_{zx} = \mu(\epsilon\partial w'/\partial x + \cancel{\partial U/\partial z} + \epsilon\partial u'/\partial z)$ (U is not a function of z, see Eq. (5.1)), $\tau_{zy} = \mu(\epsilon\partial w'/\partial y + \cancel{\partial V/\partial z} + \epsilon\partial v'/\partial z)$ and $\tau_{zz} = 2\epsilon\mu\partial w'/\partial z$ (see definition of the stress in Eq. (1.52)), are the order of ϵ, all the terms can be ignored. Eq. (5.23) is thus reduced to

$$\frac{p_z}{\rho} \approx (\eta - z)\, g + p_\eta, \tag{5.24}$$

which indicates hydrostatic pressure. The gradient of Eq. (5.24) gives

$$\frac{1}{\rho}\frac{\partial p_z}{\partial x} = g\frac{\partial \eta}{\partial x} + \frac{1}{\rho}\frac{\partial p_\eta}{\partial x} \tag{5.25}$$

Eq. (5.25) is substituted into Eq. (5.19) and integrated over the depth:

$$\begin{aligned}\int_{-h}^\eta \frac{\partial u}{\partial t}dz + \int_{-h}^\eta \frac{\partial u^2}{\partial x}dz + \int_{-h}^\eta \frac{\partial uv}{\partial y}dz + u_\eta w_\eta - u_h w_h = -(h + \eta)\, g\frac{\partial \eta}{\partial x} \\ - \frac{h + \eta}{\rho}\frac{\partial p_\eta}{\partial x} + \frac{1}{\rho}\left(\int_{-h}^\eta \frac{\partial \tau_{xx}}{\partial x}dz + \int_{-h}^\eta \frac{\partial \tau_{xy}}{\partial y}dz + \tau_{xz}\mid_\eta - \tau_{xz}\mid_h\right)\end{aligned} \tag{5.26}$$

This equation is modified by using the Leibniz rule again and the kinematic boundary conditions at free surface and bottom:

$$\frac{\partial}{\partial t}\int_{-h}^{\eta} udz+\frac{\partial}{\partial x}\int_{-h}^{\eta} u^2dz+\frac{\partial}{\partial y}\int_{-h}^{\eta} uvdz$$
$$=-\left(h+\eta\right)g\frac{\partial \eta}{\partial x}-\frac{h+\eta}{\rho}\frac{\partial p_\eta}{\partial x}+\frac{1}{\rho}\left(\frac{\partial}{\partial x}\int_{-h}^{\eta}\tau_{xx}dz+\frac{\partial}{\partial y}\int_{-h}^{\eta}\tau_{xy}dz\right)$$
$$-\frac{1}{\rho}\left(\tau_{xx}\mid_\eta\frac{\partial \eta}{\partial x}+\tau_{xy}\mid_\eta\frac{\partial \eta}{\partial y}+\tau_{xx}\mid_h\frac{\partial h}{\partial x}+\tau_{xy}\mid_h\frac{\partial h}{\partial y}\right)+\frac{1}{\rho}\left(\tau_{xz}\mid_\eta-\tau_{xz}\mid_h\right) \tag{5.27}$$

The approximations of Eqs. (5.7), (5.10), and (5.11) provides the equation form in terms of the depth mean velocity:

$$\frac{\partial}{\partial t}U\left(h+\eta\right)+\frac{\partial}{\partial x}U^2\left(h+\eta\right)+\frac{\partial}{\partial y}UV\left(h+\eta\right)$$
$$=-\left(h+\eta\right)g\frac{\partial \eta}{\partial x}-\frac{h+\eta}{\rho}\frac{\partial p_\eta}{\partial x}+\frac{h+\eta}{\rho}\left(\frac{\partial \tau_{xx}}{\partial x}+\frac{\partial \tau_{xy}}{\partial y}\right)+\frac{\tau_{xz}\mid_\eta-\tau_{xz}\mid_h}{\rho} \tag{5.28}$$

It should be noted that τ_{xx} and τ_{xy} are approximated to be uniform over the depth as $\tau_{xx}=2\mu\partial/\partial x(U+\epsilon u')\approx 2\mu\partial U/\partial x$, $\tau_{xy}=\mu(\partial/\partial y(U+\epsilon u')+\partial/\partial x(V+\epsilon v'))\approx(\partial U/\partial y+\partial V/\partial x)$.

The same operation for Eq. (5.20) gives

$$\frac{\partial}{\partial t}V\left(h+\eta\right)+\frac{\partial}{\partial x}UV\left(h+\eta\right)+\frac{\partial}{\partial y}V^2\left(h+\eta\right)$$
$$=-\left(h+\eta\right)g\frac{\partial \eta}{\partial y}-\frac{h+\eta}{\rho}\frac{\partial p_\eta}{\partial y}+\frac{h+\eta}{\rho}\left(\frac{\partial \tau_{yx}}{\partial x}+\frac{\partial \tau_{yy}}{\partial y}\right)+\frac{\tau_{yz}\mid_\eta-\tau_{yz}\mid_h}{\rho} \tag{5.29}$$

Using the chain rule and the mass conservation given in Eq. (5.17) into Eqs. (5.28) and (5.29), the depth-averaged momentum equation is given

$$\frac{\partial U}{\partial t}+U\frac{\partial U}{\partial x}+V\frac{\partial U}{\partial y}=-g\frac{\partial \eta}{\partial x}-\frac{1}{\rho}\frac{\partial p_\eta}{\partial x}+\frac{1}{\rho}\left(\frac{\partial \tau_{xx}}{\partial x}+\frac{\partial \tau_{xy}}{\partial y}\right)+\frac{\tau_{xz}\mid_\eta-\tau_{xz}\mid_h}{\rho\left(h+\eta\right)} \tag{5.30}$$

$$\frac{\partial V}{\partial t}+U\frac{\partial V}{\partial x}+V\frac{\partial V}{\partial y}=-g\frac{\partial \eta}{\partial y}-\frac{1}{\rho}\frac{\partial p_\eta}{\partial y}+\frac{1}{\rho}\left(\frac{\partial \tau_{yx}}{\partial x}+\frac{\partial \tau_{yy}}{\partial y}\right)+\frac{\tau_{yz}\mid_\eta-\tau_{yz}\mid_h}{\rho\left(h+\eta\right)} \tag{5.31}$$

The vector form of the momentum equation is given

$$\frac{\partial \boldsymbol{U}}{\partial t}+\left(\boldsymbol{U}\cdot\boldsymbol{\nabla}\right)\boldsymbol{U}=-g\boldsymbol{\nabla}\eta-\frac{1}{\rho}\boldsymbol{\nabla}p_\eta+\frac{1}{\rho}\boldsymbol{\nabla}\cdot\boldsymbol{\tau}+\frac{\tau_b}{\rho\left(h+\eta\right)} \tag{5.32}$$

where $\boldsymbol{\tau}_b$ is the shear stress at the boundaries. The pressure at the free surface

p_η is given from Eq. (3.29) by

$$p_\eta = \gamma\kappa + p_a \approx -\gamma\nabla^2\eta + p_a \tag{5.33}$$

$$= -\gamma\left(\frac{\partial^2\eta}{\partial x^2} + \frac{\partial^2\eta}{\partial y^2}\right) + p_a \tag{5.34}$$

where p_a is the atmospheric pressure. Assuming the constant p_a in space, Eq. (5.32) is then rewritten

$$\frac{\partial \boldsymbol{U}}{\partial t} + (\boldsymbol{U}\cdot\nabla)\boldsymbol{U} = -g\nabla\eta + \frac{\gamma}{\rho}\nabla\left(\nabla^2\eta\right) + \frac{1}{\rho}\nabla\cdot\tau + \frac{\tau_b}{\rho(h+\eta)} \tag{5.35}$$

For inviscid fluid, Eq. (5.35) is reduced

$$\frac{\partial \boldsymbol{U}}{\partial t} + (\boldsymbol{U}\cdot\nabla)\boldsymbol{U} = -g\nabla\eta + \frac{\gamma}{\rho}\nabla\left(\nabla^2\eta\right) \tag{5.36}$$

If both capillary and viscous effects are negligibly smaller than the gravity, Eq. (5.35) is simplified

$$\frac{\partial \boldsymbol{U}}{\partial t} + (\boldsymbol{U}\cdot\nabla)\boldsymbol{U} = -g\nabla\eta \tag{5.37}$$

5.2 Linear shallow water waves

In order to understand fundamental properties of the shallow water equation, we consider the simplest case of one-dimensional inviscid flow, assuming small velocity and surface elevation, without capillary and viscous effects. In this case, Eqs. (5.17) and (5.37) are linearized:

$$\frac{\partial\eta}{\partial t} + \frac{\partial}{\partial x}Uh = 0 \tag{5.38}$$

$$\frac{\partial U}{\partial t} + g\frac{\partial\eta}{\partial x} = 0 \tag{5.39}$$

The differentiation of Eq. (5.38) with respect to t gives

$$\frac{\partial^2\eta}{\partial t^2} + h\frac{\partial}{\partial x}\frac{\partial U}{\partial t} + \frac{\partial U}{\partial t}\frac{\partial h}{\partial x} = 0 \tag{5.40}$$

Eq. (5.39) is substituted into Eq. (5.40) to give

$$\frac{\partial^2\eta}{\partial t^2} - gh\frac{\partial^2\eta}{\partial x^2} - g\frac{\partial\eta}{\partial x}\frac{\partial h}{\partial x} = 0 \tag{5.41}$$

In the case of constant depth, $\partial h/\partial x = 0$, Eq. (5.41) is reduced to

$$\frac{\partial^2\eta}{\partial t^2} - gh\frac{\partial^2\eta}{\partial x^2} = 0 \tag{5.42}$$

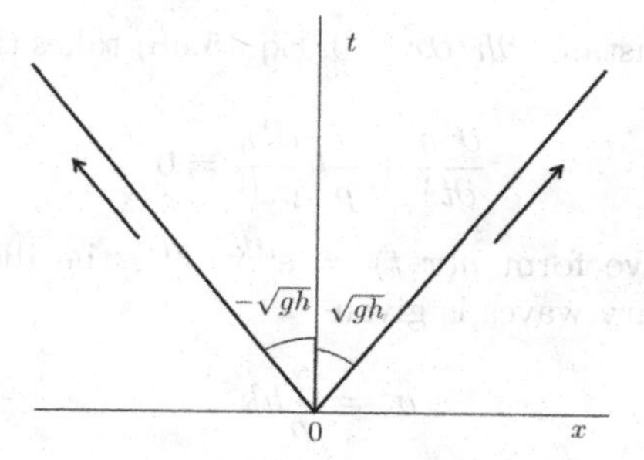

FIGURE 5.2
Characteristic lines of Eq. (5.42)D

Eq. (5.42) may be rewritten as

$$\left(\frac{\partial}{\partial t}-\sqrt{gh}\frac{\partial}{\partial x}\right)\left(\frac{\partial}{\partial t}+\sqrt{gh}\frac{\partial}{\partial x}\right)\eta=0 \tag{5.43}$$

Eq. (5.43) is satisfied when either bracket is zero:

$$\frac{d\eta}{dt}=\frac{\partial\eta}{\partial t}+\frac{dx}{dt}\frac{\partial\eta}{\partial x}=0 \tag{5.44}$$

where

$$\frac{dx}{dt}=\begin{cases}\sqrt{gh}\\-\sqrt{gh}\end{cases} \tag{5.45}$$

This equation indicates the surface forms traveling with the speed $\sqrt{gh}$ and $-\sqrt{gh}$ to the mutually opposite directions (Fig. 5.2). $\sqrt{gh}$ is known as the wave speed of long wave. Assuming a progressive wave form of the surface $\eta(x,t)=e^{i(kx-\sigma t)}$, the substitution into Eq. (5.42) gives the dispersion relation:

$$\sigma^2=ghk^2 \tag{5.46}$$

which is identical to the shallow water approximation of the dispersion relation for small amplitude wave, Eq. (7.29). Eq. (5.46) also describes the wave speed $c=\sigma/k=\pm\sqrt{gh}$.

We next consider another limit $g\to 0$ for Eq. (5.36), that is, capillary wave propagation on a thin liquid sheet of thickness h. The linearized Eq. (5.36) in this case is:

$$\frac{\partial \mathrm{U}}{\partial \mathrm{t}}=\frac{\gamma}{\rho}\mathrm{h}\frac{\partial^3\eta}{\partial \mathrm{x}^3} \tag{5.47}$$

The substitution into Eq. (5.40) gives

$$\frac{\partial^2\eta}{\partial t^2}+\frac{\gamma}{\rho}h\frac{\partial^4\eta}{\partial x^4}+\frac{\gamma}{\rho}\frac{\partial^3\eta}{\partial x^3}\frac{\partial h}{\partial x}=0 \tag{5.48}$$

If the thickness is constant, $\partial h/\partial x = 0$, Eq. (5.48) takes the form of a dynamic beam equation:

$$\frac{\partial^2 \eta}{\partial t^2} + \frac{\gamma}{\rho} h \frac{\partial^4 \eta}{\partial x^4} = 0 \tag{5.49}$$

Substituting the wave form $\eta(x,t) = \mathrm{e}^{i(kx-\sigma t)}$, the dispersion relation for shallow-water capillary waves is given

$$\sigma^2 = \frac{\gamma}{\rho} h k^4 \tag{5.50}$$

which defines the wave speed on the sheet:

$$c = \frac{\sigma}{k} = \pm k \sqrt{\frac{\gamma}{\rho} h} \tag{5.51}$$

Eqs. (5.50) and (5.51) coincide with Eqs. (4.70) and (4.71), which describes the shallow water properties of the potential waves propagating on a liquid sheet.

5.3 Method of characteristics

The method of characteristics is introduced for solving nonlinear shallow water equation in this section. We use a system of one-dimensional shallow water equation for inviscid fluid without surface tension, Eq. (5.37):

$$\frac{\partial \eta}{\partial t} + \frac{\partial}{\partial x} U (h + \eta) = 0 \tag{5.52}$$

$$\frac{\partial U}{\partial t} + U \frac{\partial U}{\partial x} + g \frac{\partial \eta}{\partial x} = 0 \tag{5.53}$$

Eq. (5.52) is transformed by a chain rule as

$$\frac{\partial}{\partial t} (\eta + h) + U \frac{\partial}{\partial x} (\eta + h) + (\eta + h) \frac{\partial U}{\partial x} = 0 \tag{5.54}$$

If we define $c^2 = g(\eta + h)$, Eq. (5.54) takes the form

$$\frac{\partial c^2}{\partial t} + U \frac{\partial c^2}{\partial x} + c^2 \frac{\partial U}{\partial x} = 0 \tag{5.55}$$

$$\therefore 2 \frac{\partial c}{\partial t} + 2U \frac{\partial c}{\partial x} + c \frac{\partial U}{\partial x} = 0 \tag{5.56}$$

The momentum equation, Eq. (5.53), may be expressed by

$$\frac{\partial U}{\partial t} + U \frac{\partial U}{\partial x} + g \frac{\partial}{\partial x} (\eta + h) = g \frac{\partial h}{\partial x} \tag{5.57}$$

The identical transformation by c gives

$$\frac{\partial U}{\partial t} + U\frac{\partial U}{\partial x} + 2gc\frac{\partial c}{\partial x} = g\frac{\partial h}{\partial x} \tag{5.58}$$

The sum of Eqs. (5.56) and (5.58) gives

$$\frac{\partial}{\partial t}(U + 2c) + (U + c)\frac{\partial}{\partial x}(U + 2c) = g\frac{\partial h}{\partial x} \tag{5.59}$$

and the difference gives

$$\frac{\partial}{\partial t}(U - 2c) + (U - c)\frac{\partial}{\partial x}(U - 2c) = g\frac{\partial h}{\partial x} \tag{5.60}$$

Assuming the constant depth ($\partial h/\partial x = 0$), and introducing the characteristic valuables:

$$\alpha = U + 2c \tag{5.61}$$

$$\beta = U - 2c \tag{5.62}$$

Eqs. (5.59) and (5.60) can be transformed to the ordinary differential equations along the characteristics C_+ and C_- given by $dx/dt\,|_+ = U + c$ and $dx/dt\,|_- = U - c$:

$$\left(\frac{\partial}{\partial t} + \left.\frac{dx}{dt}\right|_+ \frac{\partial}{\partial x}\right)\alpha = \frac{d\alpha}{dt} = 0 \tag{5.63}$$

$$\left(\frac{\partial}{\partial t} + \left.\frac{dx}{dt}\right|_- \frac{\partial}{\partial x}\right)\beta = \frac{d\beta}{dt} = 0 \tag{5.64}$$

α and β are constant along the characteristic curves C_+ and C_-, which is known as Riemann invariants. In the case of uniform sloping bottom described by $h = ax + b$ with constant a and b, as $\partial h/\partial x = a$, integrating right-hand side of Eqs. (5.59) and (5.60), the characteristic variables $\alpha = U + 2c - gat$ and $\beta = U - 2c - gat$ are chosen for Eqs. (5.63) and (5.64).

Fig. 5.3 (left) shows typical characteristic lines of a simple wave train advancing with the constant depth h, the constant fluid velocity U, and the constant wave speed c in one-dimensional space. In this case, the characteristics are described by straight lines. The initial wave propagates into still water (the region of quiet) with $c_0 = \sqrt{gh}$, which is described by the straight line $x = c_0 t$ of the initial characteristic C_1^0. The set of straight characteristics above C_1^0 (family of characteristics C_1^0) is determined by the prescribed conditions at $x = 0$ along t axis.

We then consider a straight characteristic extending from $t = \tau$ at $x = 0$:

$$\frac{dx}{dt} = U(\tau) + c(\tau) \tag{5.65}$$

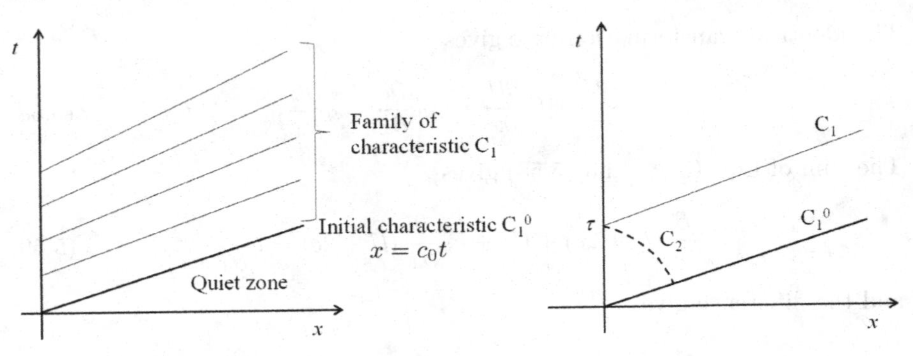

FIGURE 5.3
Characteristic lines of a simple wave train (left) and relation of characteristic family to the initial characteristic (right).

Supposing a characteristic C_2 extending from $t = \tau$ to the initial characteristic C_1^0 (see Fig. 5.3 right), Eq. (5.62) is constant along C_2:

$$U(\tau) - 2c(\tau) = U_0 - 2c_0 \tag{5.66}$$

Substituting into Eq. (5.65), the C_1 characteristics are provided as

$$\frac{dx}{dt} = \frac{1}{2}(3U(\tau) - U_0) + c(\tau) \tag{5.67}$$

$$= 3c(\tau) - 2c_0 + U_0 \tag{5.68}$$

We find, if either $U(\tau)$ or $c(\tau)$ is given, the characteristics C_1 can be determined.

As an example of the solution procedure, we consider a dam-break flow of a semi-infinite water column with constant depth h, initially located in $x > 0$ (see Fig. 5.4 bottom). While the initial depression wave propagate along the C_1^0characteristic straight line $x = c_0 t$ in positive the x-direction, the wave front (characteristic C_2^0) extends with velocity W in negative direction (see Fig. 5.4 top). Because of the initially still water state of the column (in the quiet zone), $U_0 = 0$ and thus Eqs. (5.66) and (5.68) for C_1 characteristic give

$$c = \frac{1}{2}U + c_0 \tag{5.69}$$

$$\frac{dx}{dt} = \frac{3}{2}U + c_0 \tag{5.70}$$

As the characteristic is described by the straight line (see Fig. 5.4 top), $dx/dt = x/t$,

$$\frac{x}{t} = \frac{3}{2}U + c_0 \tag{5.71}$$

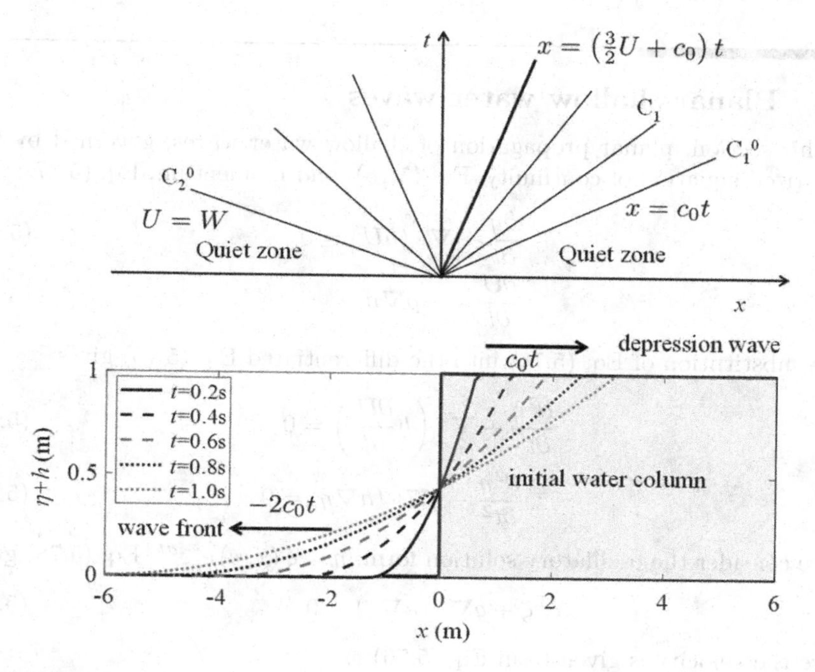

FIGURE 5.4
Characteristic lines for the dam-break flow (top) and temporal variations of the surface forms (bottom).

The substitution of Eqs. (5.71) into (5.69) gives

$$U = \frac{2}{3}\left(\frac{x}{t} - c_0\right) \tag{5.72}$$

$$c = \frac{1}{3}\left(\frac{x}{t} + 2c_0\right) \tag{5.73}$$

Since Eq. (5.73) states that $x/t \geq -2c_0$ as $c \geq 0$ (see Fig. 5.4 top), the minimum velocity is achieved at the front of wave propagating in negative direction; $W = -2c_0$. Accordingly, the back ward terminal characteristic C_2^0 is given by $x = Wt = -2c_0 t$. Substituting $U = W = -2c_0$ into Eq. (5.69), we find $c = 0$ on C_2^0. Since $c = \sqrt{g(h+\eta)} = 0$, we find the water surface is located at the bottom surface $\eta = -h$, which describes the front of wave advancing with the velocity $-2c_0$. The surface elevation is given from $c = \sqrt{g(h+\eta)}$ with Eq. (5.73)

$$\eta = \frac{1}{9g}\left(\frac{x}{t} + 2c_0\right)^2 - h \tag{5.74}$$

Fig. 5.4 (bottom) shows the temporal variations of the surface form of bad-break flow. Many other flows solved by the method of characteristics are introduced in Stoker[92].

5.4 Planar shallow water waves

In this section, planar propagation of shallow water waves, governed by the linearized equation of continuity, Eq. (5.18), and momentum, Eq. (5.37):

$$\frac{\partial \eta}{\partial t} + \boldsymbol{\nabla} \cdot (h\boldsymbol{U}) = 0 \tag{5.75}$$

$$\frac{\partial \boldsymbol{U}}{\partial t} = -g\boldsymbol{\nabla}\eta \tag{5.76}$$

The substitution of Eq. (5.76) into the differentiated Eq. (5.75) gives

$$\frac{\partial^2 \eta}{\partial t^2} + \boldsymbol{\nabla} \cdot \left(h\frac{\partial \boldsymbol{U}}{\partial t} \right) = 0 \tag{5.77}$$

$$\therefore \frac{\partial^2 \eta}{\partial t^2} - g\boldsymbol{\nabla} \cdot (h\boldsymbol{\nabla}\eta) = 0 \tag{5.78}$$

If we consider the oscillatory solution form, $\eta = \zeta\,(x, y)\,\mathrm{e}^{-i\sigma t}$, Eq. (5.78) gives

$$\sigma^2\zeta + g\boldsymbol{\nabla} \cdot (h\boldsymbol{\nabla}\zeta) = 0 \tag{5.79}$$

Here the velocity is given from Eq. (5.76) as

$$\boldsymbol{U} = -\frac{ig}{\sigma}\boldsymbol{\nabla}\zeta\mathrm{e}^{-i\sigma t} \tag{5.80}$$

For the simplest case of the constant water depth ($\boldsymbol{\nabla}h = 0$), Eq. (5.79) is reduced to

$$\sigma^2\zeta + gh\boldsymbol{\nabla}^2\zeta = 0 \tag{5.81}$$

specifically described by

$$\sigma^2\zeta + gh\left(\frac{\partial^2 \zeta}{\partial x^2} + \frac{\partial^2 \zeta}{\partial y^2} \right) = 0 \tag{5.82}$$

Assuming the solution takes the form of horizontal oscillation (see Section 4.6):

$$\zeta = A\mathrm{e}^{i\boldsymbol{k}\cdot\boldsymbol{x}} = A\mathrm{e}^{i(k_x x + k_y y)} \tag{5.83}$$

the dispersion relation is determined by substituting into Eq. (5.82):

$$\sigma^2 = gh\left(k_x^2 + k_y^2\right) \tag{5.84}$$

The wave speed is thus defined by

$$C = \frac{\sigma}{k} = \pm\sqrt{gh} \tag{5.85}$$

where $k = \sqrt{k_x^2 + k_y^2}$.

5.4.1 Seiching

Consider the shallow water surface oscillation in a rectangle tank with length a and width b in (x, y) space, which is identical setup to Section 4.7.1.2. The impermeable boundary condition is imposed at the tank walls:

$$\left.\frac{\partial \zeta}{\partial x}\right|_{x=0,\ a} = 0, \quad \left.\frac{\partial \zeta}{\partial y}\right|_{y=0,\ b} = 0 \tag{5.86}$$

which satisfies $U\mid_{x=0,\ a}= 0$ and $V\mid_{y=0,\ b}$ (see Eq. (5.80)). The solution of Eq. (5.82) to satisfy this condition is given by

$$\zeta = A_{nm} \cos(k_{xn} x) \cos(k_{ym} y) \tag{5.87}$$

where $k_{xn} = n\pi/a$ $(n = 1, 2, \ldots)$ and $k_{ym} = m\pi/b$ $(m = 1, 2, \ldots)$, and A_{nm} is the amplitude of (n, m) modes. The dispersion relation is given by

$$\sigma_{nm} = \sqrt{ghk_{nm}} \tag{5.88}$$

where $k_{nm} = \sqrt{k_{xn}^2 + k_{ym}^2} = \sqrt{(n\pi/a)^2 + (m\pi/b)^2}$.

We find the general form of the dispersion relation of free oscillation, derived in Eq. (4.110), $\sigma_{nm}^2 = gk_{nm} \tanh k_{nm}h$, approximated in shallow water regime ($\tanh k_{nm}h \to k_{nm}h$, see Section 1.1.3) agrees with Eq. (5.88).

5.4.2 Edge waves

The edge waves traveling along the straight beach of the constant slope is introduced in Section 4.9. A shallow water representation of the edge waves is considered here. The identical coordinate, introduced in Section 4.9, is considered (see Fig. 4.13). Eq. (5.78) in this case (uniform slope in the x-direction and uniform depth in the y-direction) is given

$$\frac{\partial^2 \eta}{\partial t^2} - g\left\{\frac{\partial h}{\partial x}\frac{\partial \eta}{\partial x} + h\frac{\partial^2 \eta}{\partial x^2} + h\frac{\partial^2 \eta}{\partial y^2}\right\} = 0 \tag{5.89}$$

Assuming the solution form describing the wave propagation in the y-direction,

$$\eta = \zeta(x) e^{i(ky - \sigma t)} \tag{5.90}$$

Eq. (5.89) gives

$$\begin{aligned} &\sigma^2 \zeta + g\left\{\tan\alpha \frac{\partial \zeta}{\partial x} + x\tan\alpha \frac{\partial^2 \zeta}{\partial x^2} - k^2 x \tan\alpha \zeta\right\} = 0 \\ &\therefore x\frac{\partial^2 \zeta}{\partial x^2} + \frac{\partial \zeta}{\partial x} + \left(\frac{\sigma^2}{g\tan\alpha} - k^2 x\right)\zeta = 0 \end{aligned} \tag{5.91}$$

where the water depth $h = x \tan\alpha$ and α is the slope angle. Using the transformations of $\xi = 2kx$ and $\zeta(x) = \mathrm{e}^{-kx}\chi(\xi)$, Eq. (5.91) is expressed by

$$\xi \frac{\partial^2 \chi}{\partial \xi^2} + (1 - \xi) \frac{\partial \chi}{\partial x} + \beta \chi = 0 \tag{5.92}$$

where

$$\beta = \frac{1}{2} \left(\frac{\sigma^2}{gk \tan\alpha} - 1 \right) \tag{5.93}$$

Eq. (5.92) with non-negative integer of β, i.e. $\beta = n$ $(n = 0, 1, 2, \ldots)$, is called the Laguerre's differential equation that has the solution of the Laguerre polynomials,

$$L_n(\xi) = \sum_{m=0}^{n} \frac{(-1)^m}{m!} \frac{(n!)^2}{m!\,(n-m)!} \xi^n \tag{5.94}$$

The first few Laguerre polynomials are $L_0(\xi) = 1$, $L_1(\xi) = -\xi + 1$, $L_2(\xi) = (\xi^2 - 4\xi + 2)/2$. The surface form is given

$$\zeta_n = A_n L_n(2kx) \mathrm{e}^{-kx} \tag{5.95}$$

where A_n is the amplitude of the n mode. The dispersion relation is given from Eq. (5.93) by

$$\sigma^2 = (1 + 2n) gk \tan\alpha \tag{5.96}$$

Eq. (5.96) has an analogy with the dispersion relation of Ursell's edge waves, Eq. (4.158), which are in agreement when $\alpha \to 0$.

While the presence of small-scale edge waves can be confirmed by the regular pattern of wave run-up on a beach (Fig. 4.14), large-scale edge waves have been also observed when tsunami arrives coast[109)]. In the 2011 Tohoku tsunami, the sea level oscillation with wave period of 30–90 min, depending on the local slope of beaches, has been recorded at tidal stations along costs of Japan after the first arrival of tsunami.

Fig. 5.5 shows the typical surface forms of the edge waves with the period of 42 min. We find that the wavelength increases with the mode (43 km for the 0th mode and 128 km for the first one) and that large area of sea surface is deformed. During the propagation of the edge waves along the shore, the successive wave crests arrive many times at the fixed location on coast, which may extend tsunami flooding on land area as the highest surface elevation is attained at the shore location (see the surface level at $x = 0$ in Fig. 5.5).

5.5 Flows in a rotating system

The effects of rotation of the Earth to flows are described by an apparent force known as the Coriolis force that deflects ocean currents, tides and infra-gravity

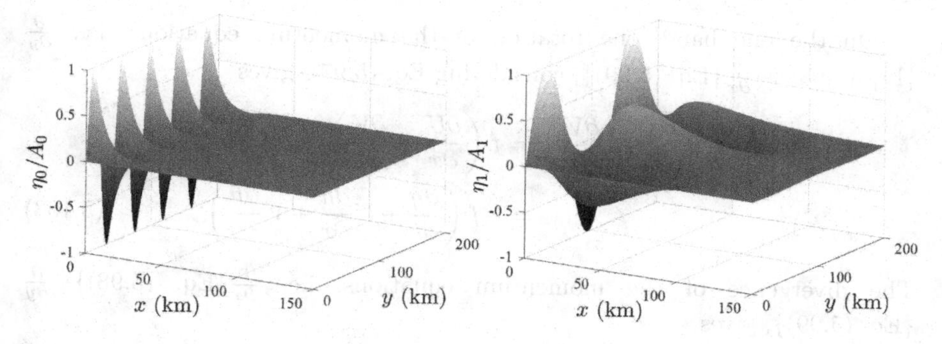

FIGURE 5.5
Surface elevations of edge waves ($\tan\alpha = 4.4 \times 10^{-3}$, $\sigma = 2.5 \times 10^{-3}$ s^{-1}); mode 0 (left) and mode 1 (right).

waves advancing over latitudes. The Coriolis force is described in terms of the Coriolis coefficient $f = 2\Omega \sin\theta$ where Ω is the angular rotation rate of the Earth and θ is the latitude (see Fig. 5.8). The waves in the rotating system are introduced in this section.

The linearized shallow water equation system including the Coriolis term is given by

$$\frac{\partial \eta}{\partial t} + h\left(\frac{\partial U}{\partial x} + \frac{\partial V}{\partial y}\right) + U\frac{\partial h}{\partial x} + V\frac{\partial h}{\partial y} = 0 \tag{5.97}$$

$$\frac{\partial U}{\partial t} - fV = -g\frac{\partial \eta}{\partial x} \tag{5.98}$$

$$\frac{\partial V}{\partial t} + fU = -g\frac{\partial \eta}{\partial y} \tag{5.99}$$

The differentiation of Eq. (5.97) with respect to t twice gives

$$\frac{\partial^3 \eta}{\partial t^3} + h\frac{\partial^2}{\partial t^2}\left(\frac{\partial U}{\partial x} + \frac{\partial V}{\partial y}\right) + \frac{\partial^2 U}{\partial t^2}\frac{\partial h}{\partial x} + \frac{\partial^2 V}{\partial t^2}\frac{\partial h}{\partial y} = 0 \tag{5.100}$$

Differentiating the momentum equations, Eqs. (5.98) and (5.99), with respect to t and substituting each other,

$$\begin{aligned}\frac{\partial^2 U}{\partial t^2} &= f\frac{\partial V}{\partial t} - g\frac{\partial^2 \eta}{\partial t \partial x} \\ &= -f^2 U - gf\frac{\partial \eta}{\partial y} - g\frac{\partial^2 \eta}{\partial t \partial x}\end{aligned} \tag{5.101}$$

$$\begin{aligned}\frac{\partial^2 V}{\partial t^2} &= f\frac{\partial U}{\partial t} - g\frac{\partial^2 \eta}{\partial t \partial y} \\ &= -f^2 V + gf\frac{\partial \eta}{\partial x} - g\frac{\partial^2 \eta}{\partial t \partial y}\end{aligned} \tag{5.102}$$

On the one hand, the rotation of the momentum equations, i.e. $\frac{\partial}{\partial y}$ {Eq. (5.98)}$-\frac{\partial}{\partial x}${Eq. (5.99)}, considering Eq. (5.97), gives

$$\begin{aligned}\frac{\partial}{\partial t}\left(\frac{\partial U}{\partial y}-\frac{\partial V}{\partial x}\right)&=f\left(\frac{\partial U}{\partial x}+\frac{\partial V}{\partial y}\right)\\&=-f\left(\frac{\partial \eta}{\partial t}+U\frac{\partial h}{\partial x}+V\frac{\partial h}{\partial y}\right)\end{aligned}\tag{5.103}$$

The divergence of the momentum equations, i.e. $\frac{\partial}{\partial x}${Eq. (5.98)}$+\frac{\partial}{\partial y}$ {Eq. (5.99)}, gives

$$\frac{\partial}{\partial t}\left(\frac{\partial U}{\partial x}+\frac{\partial V}{\partial y}\right)=-f\left(\frac{\partial U}{\partial y}-\frac{\partial V}{\partial x}\right)-g\left(\frac{\partial^2 \eta}{\partial x^2}+\frac{\partial^2 \eta}{\partial y^2}\right)\tag{5.104}$$

The substitution of Eq. (5.104) into Eq. (5.103) differentiated with respect to t gives

$$\frac{\partial^2}{\partial t^2}\left(\frac{\partial U}{\partial x}+\frac{\partial V}{\partial y}\right)=f^2\left(\frac{\partial \eta}{\partial t}+U\frac{\partial h}{\partial x}+V\frac{\partial h}{\partial y}\right)-g\frac{\partial}{\partial t}\left(\frac{\partial^2 \eta}{\partial x^2}+\frac{\partial^2 \eta}{\partial y^2}\right)\tag{5.105}$$

Substituting Eqs. (5.105), (5.101), and (5.102) into Eq. (5.100), the combined shallow water equation in the rotating system is derived as

$$\frac{\partial^3 \eta}{\partial t^3}-g\left(\frac{\partial}{\partial x}h\frac{\partial^2 \eta}{\partial t\partial x}+\frac{\partial}{\partial y}h\frac{\partial^2 \eta}{\partial t\partial y}\right)+f^2\frac{\partial \eta}{\partial t}+gf\left(\frac{\partial h}{\partial y}\frac{\partial \eta}{\partial x}-\frac{\partial h}{\partial x}\frac{\partial \eta}{\partial y}\right)=0\tag{5.106}$$

5.5.1 Poincaré waves

Considering the oscillatory function form of the surface elevation $\eta(x,y,t)=\zeta(x,y)\mathrm{e}^{-i\sigma t}$, Eq. (5.106) gives

$$\begin{aligned}&\left(\sigma^2-f^2\right)\zeta+g\left\{\left(\frac{\partial h}{\partial x}-i\frac{f}{\sigma}\frac{\partial h}{\partial y}\right)\frac{\partial \zeta}{\partial x}+\left(\frac{\partial h}{\partial y}+i\frac{f}{\sigma}\frac{\partial h}{\partial x}\right)\frac{\partial \zeta}{\partial y}\right\}\\&\quad+gh\left(\frac{\partial^2 \zeta}{\partial x^2}+\frac{\partial^2 \zeta}{\partial y^2}\right)=0\end{aligned}\tag{5.107}$$

When the constant depth is assumed, the equation is reduced to

$$\left(\sigma^2-f^2\right)\zeta+gh\left(\frac{\partial^2 \zeta}{\partial x^2}+\frac{\partial^2 \zeta}{\partial y^2}\right)=0\tag{5.108}$$

Considering the surface form described by $\zeta=A\mathrm{e}^{i(k_x x+k_y y)}$, Eq. (5.108) gives the dispersion relation:

$$\sigma^2=f^2+gh\left(k_x^2+k_y^2\right)\tag{5.109}$$

$$=f^2+ghk^2\tag{5.110}$$

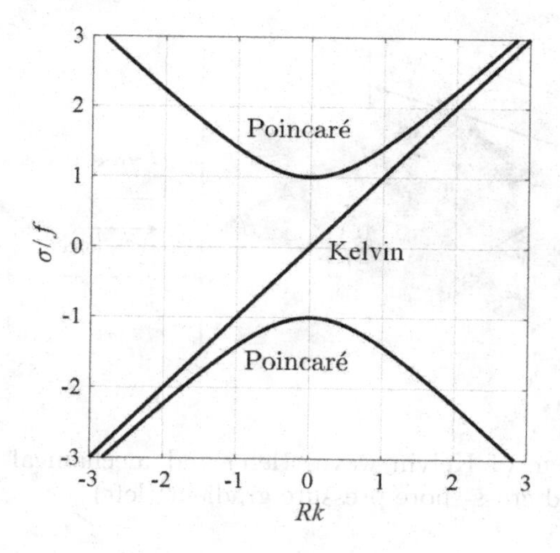

FIGURE 5.6
Dispersion relations of Poincaré and Kelvin waves.

where $k^2 = k_x^2 + k_y^2$ (see Fig. 5.6). This wave oscillation is known as a Poincaré wave.

Assuming the velocity having the consistent exponential form to η:

$$\boldsymbol{U} = \hat{\boldsymbol{U}} e^{i(k_x x + k_y y - \sigma t)} \tag{5.111}$$

The substitution into Eqs. (5.101) and (5.102) gives the amplitudes of the velocities:

$$\hat{U} = Ag\frac{ifk_y + \sigma k_x}{\sigma^2 - f^2} \tag{5.112}$$

$$\hat{V} = Ag\frac{-ifk_x + \sigma k_y}{\sigma^2 - f^2} \tag{5.113}$$

Introducing the Rossby radius of deformation defined by $R_o = \sqrt{gh}/f$, the dispersion relation, Eq. (5.110), is transformed as

$$\sigma^2 = gh\left(R_o^{-2} + k^2\right) \tag{5.114}$$

If $k \gg R_o^{-1}$ (short wave), the Coriolis effect is much smaller than the gravity, and thus dispersion relation is identical with Eq. (5.84), $\sigma \approx \sqrt{gh}k$. If $k \ll R_o^{-1}$, $\sigma \approx f$. At this limit, as $k \to 0$ (very long wave), the surface elevation $\eta = Ae^{i(k_x x + k_y y - ft)} \to Ae^{-ift}$, that is, η has no spatial gradient, and thus Eqs. (5.98) and (5.99) are reduced to

$$\frac{\partial U}{\partial t} - fV = 0 \tag{5.115}$$

$$\frac{\partial V}{\partial t} + fU = 0 \tag{5.116}$$

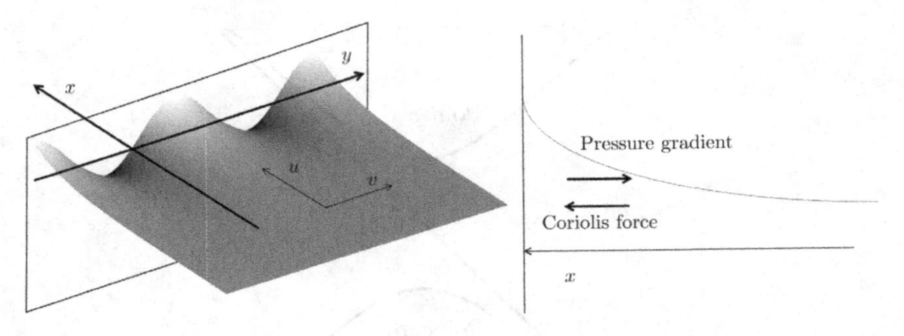

FIGURE 5.7
Coordinate system of Kelvin waves (left) and mechanical balance between Coriolis force and cross-shore pressure gradient (left).

The solution may be given as

$$U = Ce^{-ift} \tag{5.117}$$

$$V = -Cie^{-ift} \tag{5.118}$$

where C is constant. Accordingly, the flow is driven on a circle by the Coriolis force, known as inertial oscillation, which may be regarded as the natural oscillation of ocean.

5.5.2 Kelvin waves

Consider the linear shallow water flow where the Coriolis force balances with the pressure gradient perpendicular to the coast of constant depth h (see Fig. 5.7). Introducing the coordinate x in the cross-shore direction and y in the longshore direction (Fig. 5.7 left). In this equilibrium state, the cross-shore velocity U is zero anywhere. The equation system, Eqs. (5.97), (5.98), and (5.99), is reduced in this case:

$$\frac{\partial \eta}{\partial t} + h\frac{\partial V}{\partial y} = 0 \tag{5.119}$$

$$fV = g\frac{\partial \eta}{\partial x} \tag{5.120}$$

$$\frac{\partial V}{\partial t} = -g\frac{\partial \eta}{\partial y} \tag{5.121}$$

While Eq. (5.120) states that the cross-shore mechanical balance, Eqs. (5.119) and (5.121) give

$$\frac{\partial^2 \eta}{\partial t^2} - gh\frac{\partial^2 \eta}{\partial y^2} = 0 \tag{5.122}$$

indicating the wave propagation in the y-direction with the wave speed $\pm\sqrt{gh}$ (see Section 5.2). The surface elevation takes the form $\eta = \zeta(x)\,e^{i(ky-\sigma t)}$ with dispersion relation $\sigma = \sqrt{gh}k$ (see Fig. 5.6). The substitution η and $V = \hat{V}e^{i(ky-\sigma t)}$ into Eqs. (5.120) and (5.121) gives

$$\frac{f}{\sigma}k\zeta - \frac{\partial\zeta}{\partial x} = 0 \tag{5.123}$$

The solution, known as a coastal Kelvin wave, is given

$$\begin{aligned}\zeta &= Ae^{-fkx/\sigma} = Ae^{-fx/\sqrt{gh}} \\ &= Ae^{-x/R_o}\end{aligned} \tag{5.124}$$

where the Rossby radius of deformation $R_o = \sqrt{gh}/f$. The coastal Kelvin wave has similar features to the edge waves; propagation along a coast and exponential sea-level decay away from the coast, while properties of the wave are governed by the Coriolis forces.

5.6 Tsunami

Tsunami is well approximated as a shallow water wave because of the very long wavelength. Therefore fundamental features of the shallow water flow, described in the previous sections, are observed during evolution of tsunami. Since the equations are difficult to analytically solve for complex bathymetry, shore forms and the initial sea-level change, the prediction of tsunami on coasts has been generally made by computations of the shallow water equation system on the real bathymetry. In this section, the computational procedure of tsunami is first briefly introduced, and features of the 2011 Tohoku tsunami observed in the computed results are then introduced.

5.6.1 Shallow water equation in a spherical coordinate

Since tsunami may spread a large domain over ocean, the spherical coordinate system (r, θ, φ) to express location on the Earth is generally introduced in a computation of tsunami. Since the location on the Earth is conventionally described by latitude and longitude, the corresponding coordinates, latitude θ and longitude φ have been used in geophysics, instead of the polar angle defined in the normal spherical coordinate (see Section 1.1.1.3). In the spherical Earth coordinate, $\sin\theta$ in differential operators provided in Section 1.1.1.3 is replaced by $\cos\theta$. The longitudinal (U_φ) and latitudinal (U_θ) components of the depth mean velocity are redefined as

$$U_\varphi = \frac{1}{h+\eta}\int_{R-h}^{R+\eta} u_\varphi dr, \quad U_\theta = \frac{1}{h+\eta}\int_{R-h}^{R+\eta} u_\theta dr \tag{5.125}$$

where R is the radius of the Earth (Fig. 5.8).

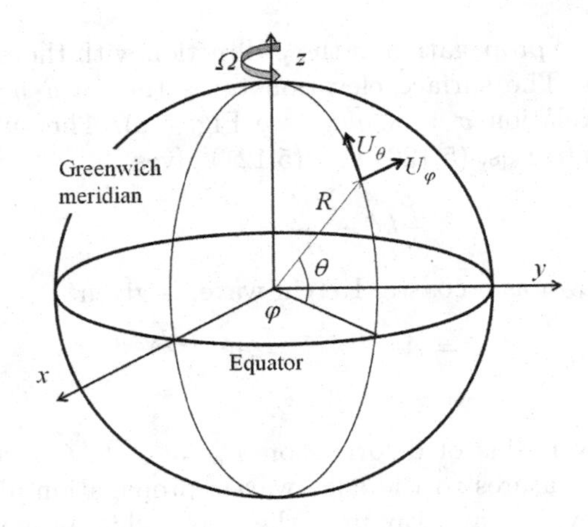

FIGURE 5.8
Spherical earth coordinate system.

The mass conservation, Eq. (5.18), and the momentum equations, Eq. (5.35), without the capillary terms, including the Coriolis force, are transformed as

$$\frac{\partial \eta}{\partial t}+\frac{1}{R\cos\theta}\frac{\partial}{\partial \varphi}U_\varphi\left(h+\eta\right)+\frac{1}{R\cos\theta}\frac{\partial}{\partial \theta}U_\theta\cos\theta\left(h+\eta\right)=0 \tag{5.126}$$

$$\begin{aligned}\frac{\partial U_\varphi}{\partial t}&+\frac{U_\varphi}{R\cos\theta}\frac{\partial U_\varphi}{\partial \varphi}+\frac{U_\theta}{R}\frac{\partial U_\varphi}{\partial \theta}=-\frac{1}{\rho R\cos\theta}\frac{\partial p_a}{\partial \varphi}-\frac{g}{R\cos\theta}\frac{\partial \eta}{\partial \varphi}+\frac{\tau_\varphi\mid_\eta-\tau_\varphi\mid_h}{\rho\left(h+\eta\right)}\\&+\nu\left(\frac{1}{R^2\cos^2\theta}\frac{\partial^2 U_\varphi}{\partial \varphi^2}+\frac{1}{R^2\cos\theta}\frac{\partial}{\partial \theta}\left(\cos\theta\frac{\partial U_\varphi}{\partial \theta}\right)\right)\\&+\left(2\Omega+\frac{U_\varphi}{R\cos\theta}\right)U_\theta\sin\theta\end{aligned} \tag{5.127}$$

$$\begin{aligned}\frac{\partial U_\theta}{\partial t}&+\frac{U_\varphi}{R\cos\theta}\frac{\partial U_\theta}{\partial \varphi}+\frac{U_\theta}{R}\frac{\partial U_\theta}{\partial \theta}=-\frac{1}{\rho R\cos\theta}\frac{\partial p_a}{\partial \theta}-\frac{g}{R\cos\theta}\frac{\partial \eta}{\partial \theta}+\frac{\tau_\theta\mid_\eta-\tau_\theta\mid_h}{\rho\left(h+\eta\right)}\\&+\nu\left(\frac{1}{R^2\cos^2\theta}\frac{\partial^2 U_\theta}{\partial \varphi^2}+\frac{1}{R^2\cos\theta}\frac{\partial}{\partial \theta}\left(\cos\theta\frac{\partial U_\theta}{\partial \theta}\right)\right)\\&+\left(2\Omega+\frac{U_\varphi}{R\cos\theta}\right)U_\theta\sin\theta\end{aligned} \tag{5.128}$$

where Ω is the earth's rate of rotation, p_a atmospheric pressure, η surface elevation. In Eqs. (5.127) and (5.128), the left-hand side expresses the advection of flow. The first and second terms on the right-hand side indicate contributions of pressure gradients of air and water, respectively. The third term of the

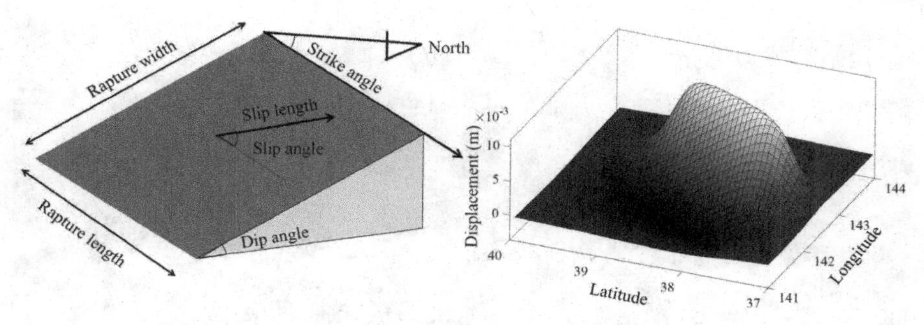

FIGURE 5.9
Fault parameters (left), and the vertical displacement estimated by the Okada's fault model[79] for the 2003 Tokachi-off tsunami.

right-hand side indicates the friction at the surface and bottom. The fourth and last terms indicate viscous and Coriolis effects, respectively.

5.6.2 Generation of tsunami

When stress occurring at the boundary of moving tectonic plates overcomes the friction between them, the plate is ruptured and slipped to deform the sea floor. The upward displacement of the floor lifts water above results in sea level rise and the downward movement depresses the sea surface. The local displacement of the floor thus determines the initial surface form of tsunami.

An elastic fault model has been commonly used to estimate the displacement, which describes geometrical properties of the rapture plane by the fault parameters; rupture length, rupture width, slip length, rake angle, strike angle, dip angle, and fault depth, shown in Fig. 5.9. Okada (1985)[79] has analytically derived three-dimensional elastic displacements at arbitrary location above rectangle deformation sources. Fig. 5.9 (right) illustrates the vertical displacement estimated by the Okada's fault model[79] for the 2003 Tokachi-off tsunami. Tanioka and Satake (1995)[95] proposed the tsunami generation model to use as an initial sea level (η_0), based on bulk water displacement due to horizontal sea floor displacement (u_x, u_y) and direct upward water displacement by vertical sea floor displacement (u_z):

$$\eta_0 = u_z + u_x \frac{\partial h}{\partial x} + u_y \frac{\partial h}{\partial y} \tag{5.129}$$

5.6.3 The 2011 Tohoku tsunami

The 9.1 magnitude plate boundary earthquake occurred off Tohoku, Japan on March 11, 2011 brought great tsunami disaster to Japan. The evolution of tsunami can be evaluated by computing the nonlinear shallow water equation

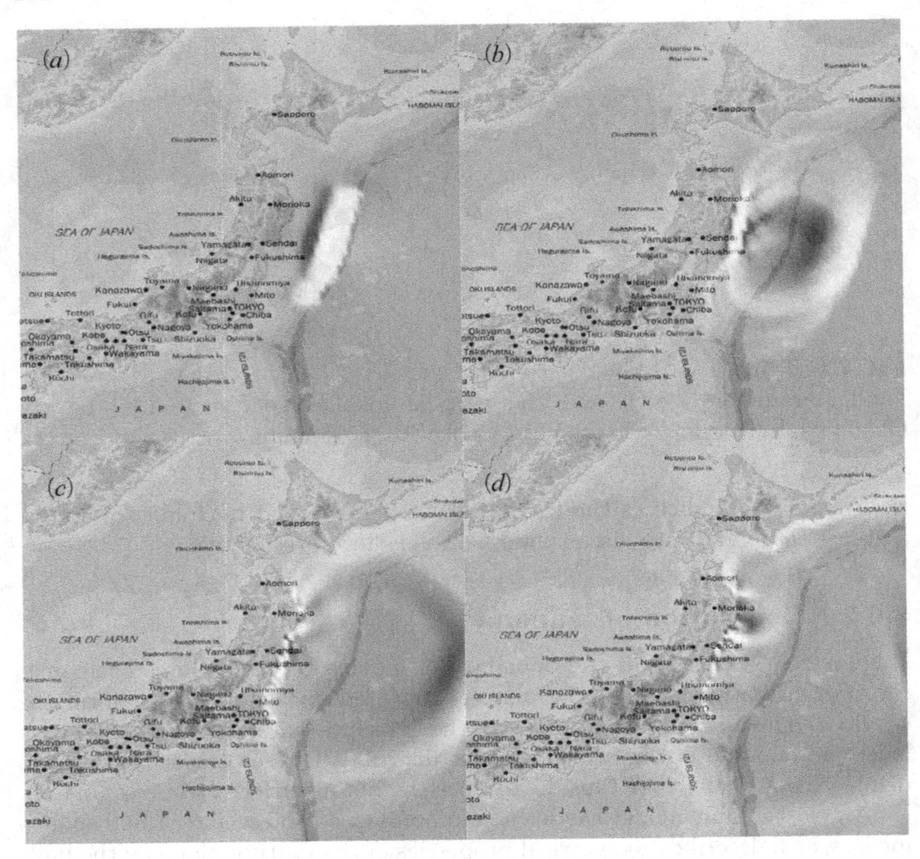

FIGURE 5.10
Evolution of the 2011 Tohoku tsunami; (*a*) initial surface elevation ($t = 0$), estimated by Tanioka and Satake[95] and Okada (1985)[79], (*b*) $t = 1000$ s, (*c*) $t = 2000$ s, (*c*) $t = 3000$ s.

system with the models of the fault rupture and initial wave source introduced in Section 5.6.1 and Section 5.6.2[74]. The specific computational methods should be referred to a specialized book such as Gotoh et al. (2013)[29].

Fig. 5.10 shows the computed surface elevations of the 2011 Tohoku tsunami. The initial sea level displacement extending along the Japan trench (*a*) radically spreads on the Pacific ocean (*b*). The tsunami grows near the coast during shoaling process, introduced in Chapter 7 (see *c*). We observe the positive and negative surface elevations alternately align to the coast line of Tohoku and propagate along the shore; that is, the edge waves, noted in Section 5.4.2, are generated after the tsunami arrives the coast. As Tohoku has ria shaped coast, seiching in bays (Section 5.4.1) has been also observed. Accordingly, multiple eigenfrequencies of free oscillations were present in bay-shaped coast of Tohoku, as also noted by Ursell (1952) [103], which may create extraordinary high tide up to 40 m [71] by the superposition.

6

Stability of Flows

A stable flow retrieves the former state even if it is disturbed. In unstable flows, the disturbance is amplified to change the flow state. The stability analysis examines the growth of disturbance and explains the mechanisms of the state change. In this chapter, instabilities of surface flows and waves owing to shear, acceleration and surface tension are studied through a linear stability analysis. Mechanisms of wave generation and breakup of free surface triggered by the instabilities are also introduced.

6.1 Gas-liquid two-layer flow

Stability of a two-layer flow with different densities and different flow velocities across an interface is mathematically studied, which refers to mechanical conditions of wind-wave generation and propagation. The well-known Kelvin-Helmholtz and Rayleigh-Taylor instabilities are introduced in this section.

6.1.1 Solution of the flow

Consider gas and liquid flows between two horizontal walls. The horizontal x-axis and the vertical z-axis of the origin at the still level of the interface are defined in Fig. 6.1. The layer thicknesses (depths) of the liquid, of density ρ_l, and the gas, of ρ_g, are h_l and h_g, respectively. We consider the displacement of the interface $\eta(x,t)$ in the uniform flows of velocity U_l in the liquid layer and U_g in the gas layer. Assuming inviscid irrotational flow, the velocity potential is given as the superposition of the uniform flow (Section 1.6.1) and the perturbations described by $\phi_l(x,z,t)$ for liquid phase and $\phi_g(x,z,t)$ for gas phase:

$$\phi(x,z,t) = \begin{cases} U_l x + \epsilon\phi_l & (-h_l < z < 0) \\ U_g x + \epsilon\phi_g & (0 < z < h_g) \end{cases} \tag{6.1}$$

where the dimensionless small parameter $\epsilon \ll 1$ indicates the order of perturbation. The substitution of Eq. (6.1) into the Laplace equation gives

$$\epsilon\nabla^2\phi_l = 0 \quad (-h_l < z < 0) \tag{6.2}$$

$$\epsilon\nabla^2\phi_g = 0 \quad (0 < z < h_g) \tag{6.3}$$

DOI: 10.1201/9781003140160-6

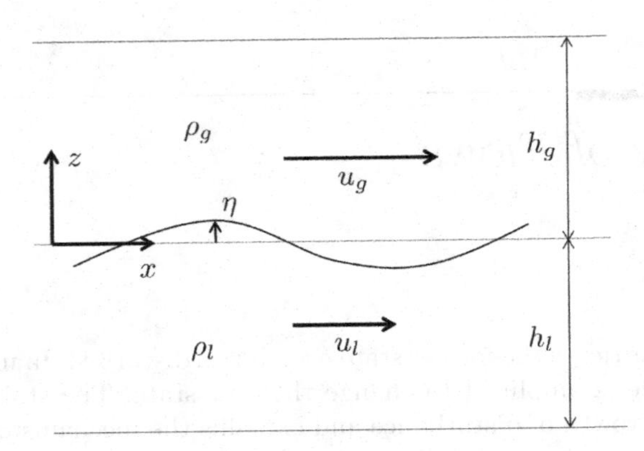

FIGURE 6.1
Illustration of gas-liquid two layer flow.

The kinematic boundary conditions at the flat impermeable bottom and top walls, Eq. (3.38), are given by

$$\epsilon \frac{\partial \phi_l}{\partial z} = 0 \quad (z = -h_l) \tag{6.4}$$

$$\epsilon \frac{\partial \phi_g}{\partial z} = 0 \quad (z = h_g) \tag{6.5}$$

Assuming the perturbed surface elevation $\eta = \epsilon\eta$, the kinematic boundary conditions at the liquid and gas phase sides of the interfaces, Eq. (3.41), are given

$$\frac{\partial \phi}{\partial z} - \frac{\partial \eta}{\partial t} - \frac{\partial \phi}{\partial x}\frac{\partial \eta}{\partial x} = \begin{cases} \epsilon \dfrac{\partial \phi_l}{\partial z} - \epsilon \dfrac{\partial \eta}{\partial t} - \left(U_l + \epsilon \dfrac{\partial \phi_l}{\partial x}\right) \epsilon \dfrac{\partial \eta}{\partial x} = 0 & (z = \epsilon\eta) \\ \epsilon \dfrac{\partial \phi_g}{\partial z} - \epsilon \dfrac{\partial \eta}{\partial t} - \left(U_g + \epsilon \dfrac{\partial \phi_g}{\partial x}\right) \epsilon \dfrac{\partial \eta}{\partial x} = 0 & (z = \epsilon\eta) \end{cases} \tag{6.6}$$

These conditions are linearized in the identical procedure to Section 4.2. The Taylor approximation for ϕ_l about $z = 0$, $\phi_l \mid_{z=\epsilon\eta} = \phi_l \mid_{z=0} + \epsilon\eta(\partial\phi/\partial z) \mid_{z=0} + O(\epsilon^2)$ (see Eq. (4.12)) is substituted into Eq. (6.6):

$$\frac{\partial}{\partial z}\left(\epsilon\phi_l + \epsilon^2 \eta \frac{\partial \phi}{\partial z} + O(\epsilon^3)\right) - \epsilon \frac{\partial \eta}{\partial t}$$
$$- \left(U_l + \epsilon\phi_l + \epsilon^2 \eta \frac{\partial \phi}{\partial z} + O(\epsilon^3)\right) \epsilon \frac{\partial \eta}{\partial x} = 0 \quad (z = 0)$$

As we are interested in the solution for the first-order Laplace equation Eq. (6.2), the terms of the order of ϵ are used as the linearized boundary condition:

$$\frac{\partial \phi_l}{\partial z} - \frac{\partial \eta}{\partial t} - U_l \frac{\partial \eta}{\partial x} = 0 \quad (z = 0) \tag{6.7}$$

The same operation for ϕ_g gives

$$\frac{\partial \phi_g}{\partial z} - \frac{\partial \eta}{\partial t} - U_g \frac{\partial \eta}{\partial x} = 0 \quad (z = 0) \tag{6.8}$$

The dynamic boundary condition, given from the Bernoulli equation, Eq. (1.163), at the interface ($z = \eta$) is also linearized in the same manner (see also Section 4.2). The first-order dynamic boundary conditions at $z = 0$ are given

$$\frac{\partial \phi_l}{\partial t} + U_l \frac{\partial \phi_l}{\partial x} + \frac{p_l}{\rho_l} + g\eta = 0 \quad (z = 0) \tag{6.9}$$

$$\frac{\partial \phi_g}{\partial t} + U_g \frac{\partial \phi_g}{\partial x} + \frac{p_g}{\rho_g} + g\eta = 0 \quad (z = 0) \tag{6.10}$$

where p_l and p_g are the pressure at the liquid and gas sides on the interface. The jump condition of normal stress of inviscid fluid is given from Eq. (3.49); $p_l = p_g + \gamma\kappa$ (see Section 3.2.2), which connects Eqs. (6.9) and (6.10) at $z = 0$:

$$\rho_l \left(\frac{\partial \phi_l}{\partial t} + U_l \frac{\partial \phi_l}{\partial x} + g\eta \right) = \rho_g \left(\frac{\partial \phi_g}{\partial t} + U_g \frac{\partial \phi_g}{\partial x} + g\eta \right) - \gamma\kappa \tag{6.11}$$

where γ is the surface tension. From Eq. (3.19), the curvature κ is approximated as

$$\kappa \approx -\frac{\partial^2 \eta}{\partial x^2} \tag{6.12}$$

In summary, a system of the first-order equations is given by

$$\frac{\partial^2 \phi_l}{\partial x^2} + \frac{\partial^2 \phi_l}{\partial z^2} = 0 \quad (-h_l < z < 0), \qquad \frac{\partial^2 \phi_g}{\partial x^2} + \frac{\partial^2 \phi_g}{\partial z^2} = 0 \quad (0 < z < h_g) \tag{6.13}$$

$$\frac{\partial \phi_l}{\partial z} = 0 \quad (z = -h_l), \qquad \frac{\partial \phi_g}{\partial z} = 0 \quad (z = h_g) \tag{6.14}$$

$$\frac{\partial \phi_l}{\partial z} - \frac{\partial \eta}{\partial t} - U_l \frac{\partial \eta}{\partial x} = 0, \qquad \frac{\partial \phi_g}{\partial z} - \frac{\partial \eta}{\partial t} - U_g \frac{\partial \eta}{\partial x} = 0 \quad (z = 0) \tag{6.15}$$

$$\rho_l \left(\frac{\partial \phi_l}{\partial t} + U_l \frac{\partial \phi_l}{\partial x} + g\eta \right) = \rho_g \left(\frac{\partial \phi_g}{\partial t} + U_g \frac{\partial \phi_g}{\partial x} + g\eta \right) + \gamma \frac{\partial^2 \eta}{\partial x^2} \quad (z = 0) \tag{6.16}$$

When the progressive wave form of the solution (see Section 4.3),

$$(\eta(x,t), \phi_l(x,z,t), \phi_g(x,z,t)) = (\eta', \phi_l'(z), \phi_g'(z))\, e^{i(kx - \sigma t)} \tag{6.17}$$

is substituted into Eq. (6.13), the amplitudes of velocity potential, ϕ_l' and ϕ_g', are determined by Eqs. (6.14) and (6.15) in the same procedure with Section 4.3:

$$\phi_l' = -i\eta' \frac{\sigma - U_l k}{k} \frac{\cosh k(h_l + z)}{\sinh kh_l} \tag{6.18}$$

$$\phi'_g = i\eta' \frac{\sigma - U_g k}{k} \frac{\cosh k (h_g - z)}{\sinh k h_g} \tag{6.19}$$

Accordingly, the flow solutions are given by the form

$$\phi_l = -iA \frac{\sigma - U_l k}{k} \frac{\cosh k (h_l + z)}{\sinh k h_l} e^{i(kx - \sigma t)} \tag{6.20}$$

$$\phi_g = iA \frac{\sigma - U_g k}{k} \frac{\cosh k (h_g - z)}{\sinh k h_g} e^{i(kx - \sigma t)} \tag{6.21}$$

$$\eta = A e^{i(kx - \sigma t)} \tag{6.22}$$

where A is the amplitude of surface elevation.

The eigenvalue equation of this system (dispersion relation) is derived by substituting these solutions into Eq. (6.16):

$$\rho_l (\sigma - U_l k)^2 \coth k h_l + \rho_g (\sigma - U_g k)^2 \coth k h_g = gk (\rho_l - \rho_g) + \gamma k^3 \tag{6.23}$$

For a free surface flow, i.e. no effect of gas phase ($\rho_g = 0$ and $U_g = 0$), the dispersion relation in wave-current coexisting field, as introduced in Section 7.3, can be given as

$$(\sigma - U_l k)^2 = \left(gk + \frac{\gamma}{\rho_l} k^3 \right) \tanh k h_l \tag{6.24}$$

If no current, $U_l = 0$, Eq. (6.24) expresses the dispersion of progressive waves, Eq. (4.51), where the frequency is given by

$$\sigma = \sqrt{\left(gk + \frac{\gamma}{\rho_l} k^3 \right) \tanh k h_l} \tag{6.25}$$

The general form of the frequency of the perturbed waves in the system of uniform flows is given from Eq. (6.23) as

$$\sigma = k \, \frac{\rho_l U_l \tanh k h_g + \rho U_g \tanh k h_l}{\rho_l \tanh k h_g + \rho_g \tanh k h_l} \pm \left\{ \tanh k h_l \tanh k h_g \left(\frac{(\rho_l - \rho_g) gk + \gamma k^3}{\rho_l \tanh k h_g + \rho_g \tanh k h_l} - \frac{\rho_l \rho_g (U_l - U_g)^2 k^2}{(\rho_l \tanh k h_g + \rho_g \tanh k h_l)^2} \right) \right\}^{1/2} \tag{6.26}$$

In the case of no top wall ($h_g \to \infty$, and thus $\tanh k h_g \approx 1$, see Fig. 1.2) and finite liquid depth (like as atmosphere above coastal region), Eq. (6.26) is reduced to

$$\sigma = k \frac{\rho_l U_l + \rho U_g \tanh k h_l}{\rho_l + \rho_g \tanh k h_l} \pm \left(\tanh k h_l \left(\frac{(\rho_l - \rho_g) gk + \gamma k^3}{\rho_l + \rho_g \tanh k h_l} - \frac{\rho_l \rho_g (U_l - U_g)^2 k^2}{(\rho_l + \rho_g \tanh k h_l)^2} \right) \right)^{1/2} \tag{6.27}$$

The deep water approximation for the both phases ($\tanh kh_g \approx 1$ and $\tanh kh_l \approx 1$) provides the dispersion in deep water:

$$\sigma = k\frac{\rho_l U_l + \rho U_g}{\rho_l + \rho_g} \pm \left(\frac{(\rho_l - \rho_g) gk + \gamma k^3}{\rho_l + \rho_g} - \frac{\rho_l \rho_g}{(\rho_l + \rho_g)^2} (U_l - U_g)^2 k^2 \right)^{1/2} \tag{6.28}$$

6.1.2 Stability

When we consider simple progressive waves, Eq. (6.25) gives a real value of the frequency σ for any positive k, and thus Eqs. (6.20)–(6.22) provide finite wave solutions with exponential oscillation $e^{-i\sigma t}$. However, if σ has imaginary value, the solutions Eqs. (6.20)–(6.22) exponentially increase in time and provide infinite values. For instance, if $\sigma = i\omega$ is given (when ω is a positive real value), $e^{-i\sigma t} = e^{\omega t}$, that is, the exponential oscillation in Eqs. (6.20)–(6.22) turns into the exponential growth. In this case ω is termed a growth rate (of instability). A flow described by a solution diverging to infinity is unstable. For Eq. (6.28), the system is unstable if the sum of the terms inside the bracket (square root) are negative:

$$\alpha = \frac{(\rho_1 - \rho_2) gk + \gamma k^3}{\rho_1 + \rho_2} - \frac{\rho_1 \rho_2}{(\rho_1 + \rho_2)^2} (U_1 - U_2)^2 k^2 < 0 \tag{6.29}$$

where the subscripts 1 and 2 are used here to indicate the quantities of the lower and upper layers, instead of l and g, for introducing arbitrary arrangement of fluids. In the simplest case of $\gamma = 0$ and $U_1 = U_2 = 0$:

$$\alpha = \frac{(\rho_1 - \rho_2) gk}{\rho_1 + \rho_2} \tag{6.30}$$

If the fluid of the upper layer is heavier than the lower one, $\rho_2 > \rho_1$ (like as liquid lied on gas layer), α must be negative, $\alpha < 0$, for any positive k, that is, the system is unstable. This is well known as Rayleigh-Taylor instability, which often triggers an overturning replacement of the upper and lower fluids. The upper, heavier fluid comes down and the under lighter one moves up with the characteristic wavelength $\lambda = 2\pi/k$. The characteristic time of the instability evolution for free surface (i.e. $\rho_1 = 0$) estimated by the inverse of a growth rate $\omega = \sqrt{-\alpha}$:

$$\tau = \frac{1}{\omega} = \sqrt{\frac{\lambda}{2\pi g}} \tag{6.31}$$

When the perturbed wavelength $\lambda = 1$ cm, Eq. (6.31) estimates $\tau = 0.013$s, indicating that the instability rapidly develops within very short time.

We find from the first term of Eq. (6.29) that the surface tension suppresses the Rayleigh-Taylor instability. The free surface ($\rho_1 = 0$) is stable in

FIGURE 6.2
Electrolytically generated microbubbles; microbubbles are vertically ejected at regular spacing on a cylindrical electrode[110]. Rayleigh-Taylor instability creates the regular array of typical mushroom-like bubble plumes.

$k_{rt} > \sqrt{\rho_2 g/\gamma}$ or $\lambda_{rt} < 2\pi\sqrt{\gamma/\rho_2 g} \approx 17.1$ mm (when $\rho_2 = 1000.0$ kg/m^3, $\gamma = 0.07275$ N/m for water), which is coincident with the wavelength (L_c) of progressive waves with minimum wave speed noted in Section 4.3. In natural environment, fluid replacements owing to the Rayleigh-Taylor instability are observed when temperature in a lower layer of atmosphere is higher than the upper one, ocean surface is cooled down, sediments transported from river to the ocean surface, an air tube is entrapped by breaking waves and so on. Watanabe et al. (2021)[110] found from a laboratory experiment of electrolytically generated micro-bubbles that density difference between the liquid and bubble layers on electrode induces a regular pattern in bubble ejections with spacing of λ through the Rayleigh-Taylor instability (see Fig. 6.2 and also Section 3.6.3).

Next we consider another extreme case of gravity free ($g = 0$) and surface tension free ($\gamma = 0$) in Eq. (6.29):

$$\alpha = -\frac{\rho_1 \rho_2}{(\rho_1 + \rho_2)^2}(U_1 - U_2)^2 k^2 \tag{6.32}$$

We find, if $U_1 \neq U_2$, α is always negative for any $k > 0$ regardless of the densities. This is well known as Kelvin-Helmholtz instability, caused by discontinuous velocity of the flows across an interface. In general case including gravity and capillary effects in stable densities of the fluids ($\rho_1 > \rho_2$), Eq. (6.29) indicates that gravity and surface tension stabilize the system against Kelvin-Helmholtz instability. At the neutral state, $\alpha = 0$, Eq. (6.29) is reduced

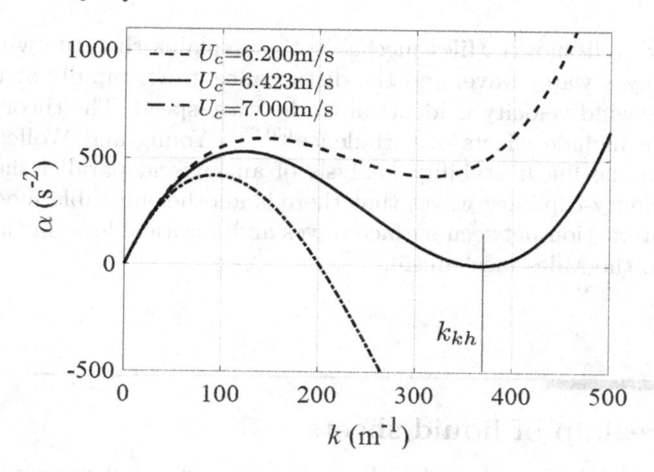

FIGURE 6.3
Growth of the Kelvin-Helmholtz instability. Negative α indicates instability growth. The system is stable when $U_c < 6.423$ m/s, while the range of unstable wave number extends in $U_c > 6.423$ m/s.

to a quadratic equation for k

$$\left(\rho_1^2 - \rho_2^2\right) g + \left(\rho_1 + \rho_2\right) \gamma k^2 - \rho_1 \rho_2 U_c^2 k = 0 \tag{6.33}$$

The critical velocity $U_c = |U_1 - U_2|$ can be derived from discriminant of the above equation; that is, the system is stable for all k if

$$U_c < 2^{1/2} \gamma^{1/4} g^{1/4} \left(\rho_1 + \rho_2\right)^{1/2} \left(\rho_1 - \rho_2\right)^{1/4} \rho_1^{-1/2} \rho_2^{-1/2} \tag{6.34}$$

For air and water two-layer flow ($\rho_1 = 1000.0$ kg/m^3, $\gamma = 0.07275$ N/m for water and $\rho_2 = 1.293$ kg/m^3 for air), $U_c \approx 6.423$ m/s. Fig. 6.3 shows the variation of α as a function of the wave number near the critical wind velocity U_c. For higher wind velocity than U_c, the waves in the higher wave numbers where $\alpha < 0$ are amplified. The critical wave number $k_{kh} = \sqrt{\left(\rho_1 - \rho_2\right) g / \gamma}$ and wavelength $\lambda_{kh} = 2\pi \sqrt{\gamma / \left(\rho_1 - \rho_2\right) g} \approx 17.1$ mm, which is again coincident with L_c and λ_{rt}.

Generation of ocean waves by wind has attracted great interest from ocean researchers from past to present. While the Kelvin-Helmholtz instability is the important mechanism of wave generation, waves are observed at wind velocity sufficiently lower than U_c in real ocean. While the theory introduced above assumes the inviscid and irrotational flow, viscous stress and turbulence in a boundary layer formed on the interface affects response of the interface and stability of the flow. According to Philips (1957) [81], pressure fluctuations of wind turbulence resonate with free surface when the wave frequency matches with the temporal fluctuation of wind $\sigma = \boldsymbol{U_g} \cdot \boldsymbol{k}$, which can initiate small waves. The well-known Miles mechanism[64] explains that the wind shear is disturbed by a water wave, and the disturbance grows rapidly at the critical level where wind velocity is identical to the wave speed. The theory has been modified to include effects of turbulence[65][66]. Young and Wolfe (2014)[127] found from the linear stability analysis of an inviscid parallel shear flow of air over gravity-capillary waves that there is another unstable mode resulted from an interaction between surface waves and a critical level in the water, in addition to the Miles mechanism.

6.2 Breakup of liquid sheets

When a liquid sheet is ejected with a certain speed, the flow may be unstable to amplify the waves on the sheet, resulting in breakup into ligaments and droplets. This breakup behavior may be observed when wind tears a crest of ocean waves to produce spume droplets.

Two types of capillary waves propagating on sheets, symmetric and antisymmetric waves have been introduced in Section 4.4. In this section, growths of disturbances on the both waves on moving sheets are studied.

6.2.1 Antisymmetric waves

Squire (1953)[91] first provided the instability analysis of the antisymmetric waves on a moving sheet of inviscid liquid. The theory was confirmed to satisfactory describe features observed in experimental waves. Following Squire (1953), we consider antisymmetric waves, having the top surface at $z = \eta_{a+} = h - Aie^{i(kx-\sigma t)}$ and the bottom one at $z = \eta_{a-} = -h - Aie^{i(kx-\sigma t)}$, propagate on horizontally moving sheet with thickness $2h$ at the velocity U (see Fig. 6.4 bottom). The solution procedure is similar to that of Section 6.1, but the indication of small parameter ϵ is omitted for simplicity in this section. The velocity potential takes the form

$$\phi(x,z,t) = \begin{cases} Ux + \phi_l(x,z,t) & (-h < z < h) \\ \phi_g(x,z,t) & (z < -h, z > h) \end{cases} \tag{6.35}$$

where the velocity potential ϕ_l for liquid in the sheet and ϕ_g for the ambient gas must satisfy the Laplace equation

$$\frac{\partial^2 \phi_l}{\partial x^2} + \frac{\partial^2 \phi_l}{\partial z^2} = 0 \quad (-h < z < h) \tag{6.36}$$

$$\frac{\partial^2 \phi_g}{\partial x^2} + \frac{\partial^2 \phi_g}{\partial z^2} = 0 \quad (z < -h, z > h) \tag{6.37}$$

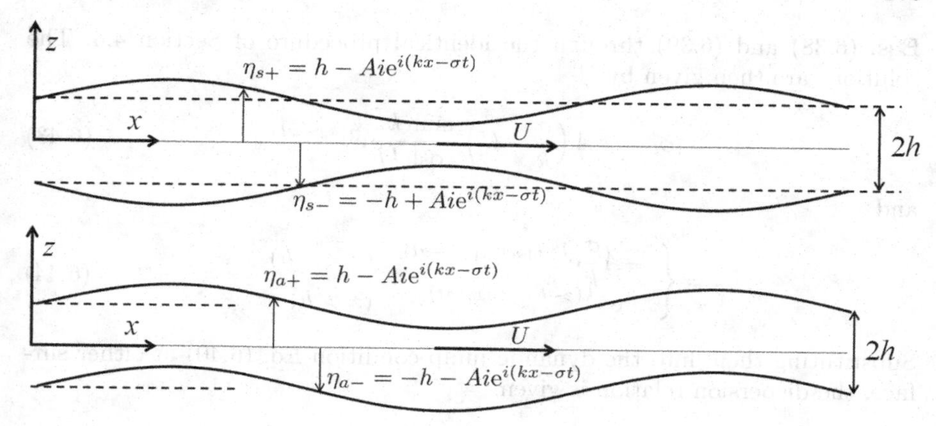

FIGURE 6.4
Symmetric waves (top) and antisymmetric waves (bottom) of liquid sheets moving with speed U.

The kinematic boundary conditions at the first order, $O(\epsilon)$, corresponding to Eqs. (6.7) and (6.8), are given by

$$\frac{\partial \phi_l}{\partial z} - \frac{\partial \eta_{a\pm}}{\partial t} - U\frac{\partial \eta_{a\pm}}{\partial x} = 0 \quad (z = \pm h) \tag{6.38}$$

$$\frac{\partial \phi_g}{\partial z} - \frac{\partial \eta_{a\pm}}{\partial t} = 0 \quad (z = \pm h) \tag{6.39}$$

The jump condition of the normal stress, corresponding to Eq. (6.11), at the boundaries $z = \pm h$ is given

$$\rho_l \left(\frac{\partial \phi_l}{\partial t} + U\frac{\partial \phi_l}{\partial x} \right) = \rho_g \frac{\partial \phi_g}{\partial t} + \gamma \frac{\partial^2 \eta_{a\pm}}{\partial x^2} \quad (z = \pm h) \tag{6.40}$$

The progressive wave form of the solution is given by

$$\phi_l = \left(C_1 e^{kz} + C_2 e^{-kz} \right) e^{i(kx-\sigma t)} \tag{6.41}$$

Since air is at rest far from the sheet, the far-field boundary condition, $\nabla \phi_g \to 0$ at $z \to \pm\infty$, requires the solution form of ϕ_g to be finite at $z = \pm\infty$, which gives

$$\phi_g = \begin{cases} C_3 e^{kz} e^{i(kx-\sigma t)} & (z < -h) \\ C_4 e^{-kz} e^{i(kx-\sigma t)} & (z > h) \end{cases} \tag{6.42}$$

The coefficients are determined by substituting Eqs. (6.41) and (6.42) into

Eqs. (6.38) and (6.39) through the identical procedure of Section 4.4. The solutions are then given by

$$\phi_l = -A\left(\frac{\sigma}{k} - U\right)\frac{\sinh kz}{\cosh kh}\mathrm{e}^{i(kx-\sigma t)} \tag{6.43}$$

and

$$\phi_g = \begin{cases} -A\frac{\sigma}{k}\mathrm{e}^{k(z+h)}\mathrm{e}^{i(kx-\sigma t)} & (z < -h) \\ A\mathrm{e}^{-k(z-h)}\mathrm{e}^{i(kx-\sigma t)} & (z > h) \end{cases} \tag{6.44}$$

Substituting them into the dynamic jump condition Eq. (6.40) at either surface, the dispersion relation is given

$$\sigma = \frac{1}{\rho_l \tanh kh + \rho_g}\Big\{ Uk\rho_l \tanh kh \pm\sqrt{U^2k^2\rho_l^2\tanh^2 kh - (\rho_l \tanh kh + \rho_g)(\rho_l \tanh khU^2k^2 - \gamma k^3)}\Big\} \tag{6.45}$$

If $\rho_g = 0$ and $U = 0$, Eq. (6.45) agrees with the dispersion relation for the antisymmetric free surface flow, Eq. (4.73). If the sum of the terms in the square root is negative, the system is unstable. Assuming that instability is present and σ is complex, $\sigma = \sigma_r + i\sigma_i$, σ_r defines the wave velocity and σ_i indicates instability growth:

$$\frac{\sigma_r}{k} = \frac{U}{1 + \rho'\mathrm{coth}kh} \tag{6.46}$$

$$\frac{h\sigma_i}{U} = \frac{kh\sqrt{\mathrm{coth}kh}}{1 + \rho'\mathrm{coth}kh}\left(\rho' - We\,(kh + kh\rho'\mathrm{coth}kh)\right)^{1/2} \tag{6.47}$$

where $\rho' = \rho_g/\rho_l$ and the Weber number $We = \gamma/\rho_l U^2 h$.

A shallow water approximation, Eq. (1.33), can be applied for the waves on a thin sheet. In this case, Eq. (6.47) may be reduced to

$$\frac{h\sigma_i}{U} \approx \frac{(kh)^{3/2}}{kh + \rho'}\left(\rho'(1 - We) - khWe\right)^{1/2} \tag{6.48}$$

When $\rho' \ll 1$ and small We, we find the maximum $\sigma_i \approx \rho'/2\sqrt{We}$ is achieved at $kh \approx \rho'/2We$. Fig. 6.5 illustrates the instability growth as a function of kh for $\rho' = 10^{-3}$. According to Squire (1953)[91], Eq. (6.48) well estimates the wavelength of the antisymmetric waves growing on the observed film. However, the theory does not refer how rupture of the film takes place. Dombrowski and Johns (1963)[19] extends the theory for viscous liquid to explain the breakup of the sheet into ligaments and droplets.

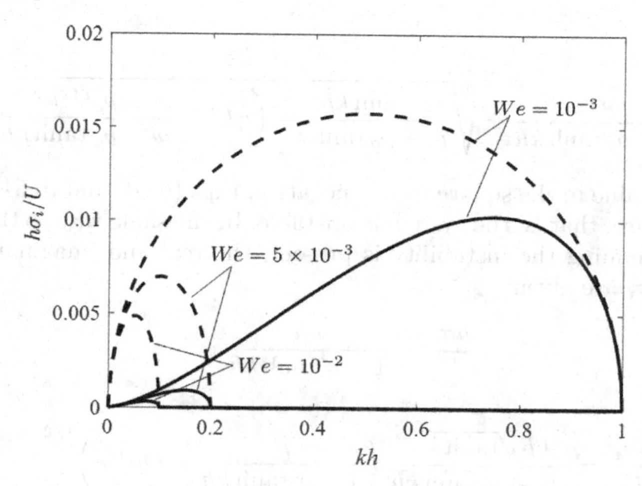

FIGURE 6.5
Dimensionless instability growth rate as a function of kh; solid line: symmetric wave, broken line: antisymmetric wave.

6.2.2 Symmetric waves

We similarly consider the stability of the symmetric waves on a two-dimensional liquid sheet of density ρ_l and thickness $2h$ moving with velocity U through air of density ρ_g (Fig. 6.4 top). The surface elevations of the symmetric waves, measured from the center of the sheet, are given as $\eta_{s+} = h - Aie^{i(kx-\sigma t)}$ for the top surface and $\eta_{s-} = -h + Aie^{i(kx-\sigma t)}$ for the bottom one (see Fig. 6.4 top). The velocity potential of the flow in the sheet is given as same as Eq. (6.35):

$$\phi(x,z,t) = \begin{cases} Ux + \phi_l(x,z,t) & (-h < z < h) \\ \phi_g(x,z,t) & (z < -h, z > h) \end{cases} \tag{6.49}$$

As noted in Section 4.4, the flow in symmetric waves on the sheet of thickness $2h$ is identical with one of the waves propagating on water depth h. Accordingly, the solution of two layer flow, progressive waves on constant depth in currents, already derived in Section 6.1, must be identical to the one we look for this problem. Therefore, we may use the velocity potential Eqs. (6.20), (6.21) and the dispersion relation Eq. (6.27) with $U_g = 0$, $U_l = U$, $h_l = h$ and $g = 0$:

$$\phi_l = -iA\frac{\sigma - Uk}{k}\frac{\cosh kz}{\sinh kh_l}\mathrm{e}^{i(kx-\sigma t)} \tag{6.50}$$

$$\phi_g = iA\frac{\sigma}{k}\mathrm{e}^{-kz}\mathrm{e}^{i(kx-\sigma t)} \tag{6.51}$$

where

$$\sigma = \frac{\rho_l U k}{\rho_l + \rho_g \tanh kh} \pm \sqrt{\frac{\tanh kh}{\rho_l + \rho_g \tanh kh}\left(\gamma k^3 - \frac{\rho_l \rho_g U^2 k^2}{\rho_l + \rho_g \tanh kh}\right)} \tag{6.52}$$

When the value in the square root is negative, Eqs. (6.50) and (6.51) exponentially increase, that is, the system is unstable. In the same way as the previous section, assuming the instability is present, the real and imaginary parts of $\sigma = \sigma_r + i\sigma_i$ are given

$$\frac{h\sigma_r}{U} = \frac{kh}{1 + \rho' \tanh kh} \tag{6.53}$$

$$\frac{h\sigma_i}{U} = \frac{kh\sqrt{\tanh kh}}{\sqrt{1 + \rho' \tanh kh}}\left(\frac{\rho'}{1 + \rho' \tanh kh} - khWe\right)^{1/2} \tag{6.54}$$

The shallow water approximation reduce Eq. (6.54) to

$$\frac{h\sigma_i}{U} = \frac{(kh)^{3/2}}{\sqrt{1 + \rho' kh}}\left(\frac{\rho'}{1 + \rho' kh} - khWe\right)^{1/2} \tag{6.55}$$

The growth rates for the symmetric waves for any We are smaller than the antisymmetric one (see Fig. 6.5). We also find that kh achieving the maximum growth rate for the symmetric waves is larger than the antisymmetric one.

6.3 Capillary instability on cylindrical jets

Consider inviscid liquid flow with constant axial velocity U of a cylindrical jet with the mean radius $\overline{a}$, having axial solid rotation with the angular velocity Ω. The base flow velocity in the jet, described in the cylindrical reference frame moving with the speed U, is given

$$\overline{\boldsymbol{u}} = (\overline{u_r}, \overline{u_\theta}, \overline{u_z}) = (0, \Omega r, 0) \tag{6.56}$$

The inviscid fluid motion in the jet is assumed to be governed by the Euler equation in the cylindrical coordinate (see Eqs. (1.82)–(1.84)):

$$\frac{\partial u_r}{\partial t} + u_r \frac{\partial u_r}{\partial r} + \frac{u_\theta}{r}\frac{\partial u_r}{\partial \theta} + u_z \frac{\partial u_r}{\partial z} - \frac{u_\theta^2}{r} = -\frac{1}{\rho}\frac{\partial p}{\partial r} \tag{6.57}$$

$$\frac{\partial u_\theta}{\partial t} + u_r \frac{\partial u_\theta}{\partial r} + \frac{u_\theta}{r}\frac{\partial u_\theta}{\partial \theta} + u_z \frac{\partial u_\theta}{\partial z} + \frac{u_r u_\theta}{r} = -\frac{1}{\rho r}\frac{\partial p}{\partial \theta} \tag{6.58}$$

$$\frac{\partial u_z}{\partial t} + u_r \frac{\partial u_z}{\partial r} + \frac{u_z}{r}\frac{\partial u_z}{\partial \theta} + u_z \frac{\partial u_z}{\partial z} = -\frac{1}{\rho}\frac{\partial p}{\partial z} \tag{6.59}$$

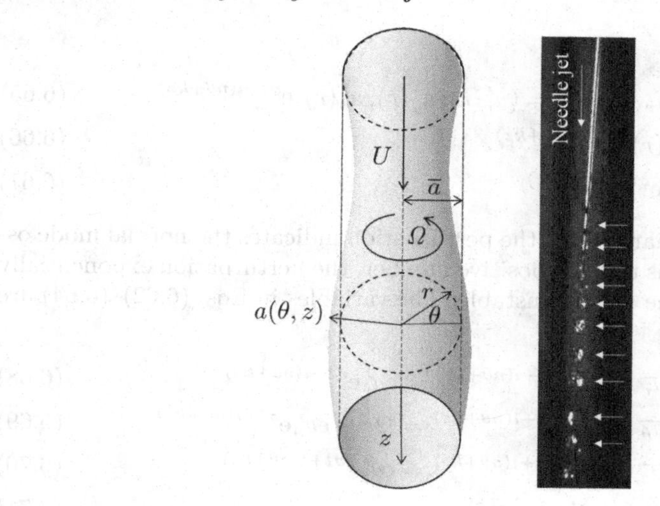

FIGURE 6.6
Flow in a rotating cylindrical jet (left), and breakup of a needle jet into drops at regular intervals (right).

and the continuity equation

$$\frac{1}{r}\frac{\partial (r u_r)}{\partial r} + \frac{1}{r}\frac{\partial u_\theta}{\partial \theta} + \frac{\partial u_z}{\partial z} = 0 \tag{6.60}$$

Substituting Eq. (6.56) into Eq. (6.57) integrated with the boundary condition that the capillary pressure $\overline{p} = \gamma/\overline{a}$ at $r = \overline{a}$ (see Eqs. (3.31) and (3.14)), the base flow pressure is given by

$$\overline{p} = \frac{\rho}{2}\Omega^2\left(r^2 - \overline{a}^2\right) + \frac{\gamma}{\overline{a}} \tag{6.61}$$

6.3.1 Solutions for perturbations

Assuming the variables described by the sum of the base flow and perturbation,

$$\boldsymbol{u}(r,\theta,z,t) = \overline{\boldsymbol{u}}(r) + \epsilon\hat{\boldsymbol{u}}(r,\theta,z,t) \tag{6.62}$$

$$p(r,\theta,z,t) = \overline{p}(r) + \epsilon\hat{p}(r,\theta,z,t) \tag{6.63}$$

$$a(r,\theta,z,t) = \overline{a} + \epsilon\hat{a}(\theta,z,t) \tag{6.64}$$

where the dimensionless parameter $\epsilon \ll 1$ and a is the free surface location (see Fig. 6.6) The perturbations are generally described by the exponential wave form:

$$\hat{\boldsymbol{u}} = (\hat{u_r}, \hat{u_\theta}, \hat{u_z}) = (u'_r(r), u'_\theta(r), u'_z(r))\mathrm{e}^{\sigma t + i(n\theta + kz)} \tag{6.65}$$

$$\hat{p} = p'(r)\mathrm{e}^{\sigma t + i(n\theta + kz)} \tag{6.66}$$

$$\hat{a} = a'\mathrm{e}^{\sigma t + i(n\theta + kz)} \tag{6.67}$$

If σ takes the imaginary value, the perturbation indicates the normal mode oscillation, while, if σ is the real positive number, the perturbation exponentially increases; that is, the flow is unstable. The variables in Eqs. (6.62)–(6.64) are specifically given by

$$u_r = \overline{u_r} + \epsilon u'_r \mathrm{e}^{\sigma t + i(n\theta + kz)} = \epsilon u'_r \mathrm{e}^{\sigma t + i(n\theta + kz)} \tag{6.68}$$

$$u_\theta = \overline{u_\theta} + \epsilon u'_\theta \mathrm{e}^{\sigma t + i(n\theta + kz)} = \Omega r + \epsilon u'_\theta \mathrm{e}^{\sigma t + i(n\theta + kz)} \tag{6.69}$$

$$u_z = \overline{u_z} + \epsilon u'_z \mathrm{e}^{\sigma t + i(n\theta + kz)} = \epsilon u'_z \mathrm{e}^{\sigma t + i(n\theta + kz)} \tag{6.70}$$

$$p = \overline{p} + \epsilon p' \mathrm{e}^{\sigma t + i(n\theta + kz)} \tag{6.71}$$

$$a = \overline{a} + \epsilon a' \mathrm{e}^{\sigma t + i(n\theta + kz)} \tag{6.72}$$

Substituting Eqs. (6.68)–(6.71) into Eqs. (6.57)–(6.60), the equation system of the perturbed variables at the order of ϵ is given by

$$(\sigma + in\Omega)\, u'_r - 2\Omega u'_\theta = -\frac{1}{\rho}\frac{\partial p'}{\partial r} \tag{6.73}$$

$$(\sigma + in\Omega)\, u'_\theta + 2\Omega u'_r = -\frac{in}{\rho r} p' \tag{6.74}$$

$$(\sigma + in\Omega)\, u'_z = -\frac{ik}{\rho} p' \tag{6.75}$$

$$\frac{1}{r}\frac{\partial (r u'_r)}{\partial r} + \frac{in}{r} u'_\theta + iku'_z = 0 \tag{6.76}$$

Eqs. (6.73)–(6.75) provide the perturbed velocities in terms of the perturbed pressure:

$$u'_r = -\frac{i}{\rho r}\frac{2n\Omega p' + (n\Omega - i\sigma)\, r\frac{\partial p'}{\partial r}}{\sigma^2 + 2in\Omega\sigma + (4 - n^2)\,\Omega^2} \tag{6.77}$$

$$u'_\theta = -\frac{1}{\rho r}\frac{(in\sigma - n^2\Omega)\, p' - 2r\Omega\frac{\partial p'}{\partial r}}{\sigma^2 + 2in\Omega\sigma + (4 - n^2)\,\Omega^2} \tag{6.78}$$

$$u'_z = -\frac{ikp'}{\rho\,(\sigma + in\Omega)} \tag{6.79}$$

The substitution of these velocities into the continuity equation, Eq. (6.76), gives the Bessel's differential equation:

$$\frac{dp'^2}{dr^2} + \frac{1}{r}\frac{dp'}{dr} - \frac{1}{(\sigma + in\Omega)^2}\left(\left(\frac{n^2}{r^2} + k^2\right)(\sigma + in\Omega)^2 + 4k^2\Omega^2\right)p' = 0 \tag{6.80}$$

The solution of Eq. (6.80) for the axisymmetric mode $n = 0$ is given by the modified Bessel function I_0 and K_0 (see Section A.8.2 in Appendix):

$$p' = C_1 I_0(\alpha r) + C_2 K_0(\alpha r) \tag{6.81}$$

where

$$\alpha = \frac{k\sqrt{\sigma^2 + 4\Omega^2}}{\sigma} \tag{6.82}$$

C_1 and C_2 are the constants. As $K_0(x)$ is singular at $x = 0$ (see Fig. 3.32 left), C_2 needs to be zero.

$$\therefore p' = C_1 I_0(\alpha r) \tag{6.83}$$

The substitution of p' into Eq. (6.77) gives

$$u_r' = -\frac{C_1\sigma}{\rho(\sigma^2 + 4\Omega)} \frac{dI_0(\alpha r)}{dr} \tag{6.84}$$

6.3.2 Linearized boundary conditions

The kinematic boundary condition at free surface ($r = a$) is given by Eq. (3.44):

$$u_r = \frac{\partial a}{\partial t} + \frac{u_\theta}{r}\frac{\partial a}{\partial \theta} + u_z \frac{\partial a}{\partial z} \quad (r = a) \tag{6.85}$$

This condition is linearized in the same way as Section 4.2. The Taylor approximation is used for the velocities:

$$u_r\,|_{r=a} \approx u_r\,|_{r=\overline{a}} + \epsilon\hat{a}\left.\frac{\partial u_r}{\partial r}\right|_{r=\overline{a}} + \cdots \tag{6.86}$$

$$u_\theta\,|_{r=a} \approx u_\theta\,|_{r=\overline{a}} + \epsilon\hat{a}\left.\frac{\partial u_\theta}{\partial r}\right|_{r=\overline{a}} + \cdots \tag{6.87}$$

$$u_z\,|_{r=a} \approx u_z\,|_{r=\overline{a}} + \epsilon\hat{a}\left.\frac{\partial u_z}{\partial r}\right|_{r=\overline{a}} + \cdots \tag{6.88}$$

The substitutions of Eqs. (6.86)–(6.88), (6.62), and (6.64) into the boundary condition Eq. (6.85) give

$$\overline{u_r} + \epsilon\hat{u}_r + \epsilon^2\hat{a}\frac{\partial \hat{u}_r}{\partial r} = \epsilon\frac{\partial \hat{a}}{\partial t} + \frac{\epsilon}{\overline{a}}\left(\overline{u_\theta} + \epsilon\hat{u}_\theta\right)\frac{\partial \hat{a}}{\partial \theta} + \epsilon\left(\overline{u_z} + \epsilon\hat{u}_z\right)\frac{\partial \hat{a}}{\partial z} \quad (r = \overline{a}) \tag{6.89}$$

where the base flow velocity $(\overline{u_r}, \overline{u_\theta}, \overline{u_z}) = (0, \Omega\overline{a}, 0)$ at $r = \overline{a}$. Considering only the first-order terms, the linearized kinematic boundary condition at $r = \overline{a}$ is given

$$\hat{u}_r = \frac{\partial \hat{a}}{\partial t} + \Omega\frac{\partial \hat{a}}{\partial \theta} \quad (r = \overline{a}) \tag{6.90}$$

The substitution of the exponential perturbations, Eqs. (6.65) and (6.67), into Eq. (6.90) gives

$$u'_r = (\sigma + in\Omega)\, a' \tag{6.91}$$

The dynamic boundary condition at the free surface ($r = a$), Eq. (3.51), of inviscid fluid is given

$$p = \gamma\kappa \approx \gamma \left(\frac{1}{a} - \frac{1}{a^2}\frac{\partial^2 a}{\partial \theta^2} - \frac{\partial^2 a}{\partial z^2} \right) \quad (r = a) \tag{6.92}$$

where the curvature κ is given by Eq. (3.25). Using the Taylor approximation, $p\mid_{r=a} \approx p\mid_{r=\overline{a}} + \epsilon\hat{a}(\partial p/\partial r)\mid_{r=\overline{a}}$, Eqs. (6.63) and (6.64) are substituted into Eq. (6.92):

$$\overline{p} + \epsilon\hat{p} + \epsilon\hat{a}\frac{\partial}{\partial r}(\overline{p} + \epsilon\hat{p}) = \gamma \left(\frac{\overline{a} - \epsilon\hat{a}}{\overline{a}^2 - \epsilon^2\hat{a}} - \frac{\epsilon}{(\overline{a} + \epsilon\hat{a})^2}\frac{\partial^2 \hat{a}}{\partial \theta^2} - \epsilon\frac{\partial^2 \hat{a}}{\partial z^2} \right) \quad (r = \overline{a}) \tag{6.93}$$

The linearized dynamic boundary condition for the order of ϵ at $r = \overline{a}$ is thus given by

$$\hat{p} + \hat{a}\frac{\partial \overline{p}}{\partial r} = -\gamma \left(\frac{\hat{a}}{\overline{a}^2} + \frac{1}{\overline{a}^2}\frac{\partial^2 \hat{a}}{\partial \theta^2} + \frac{\partial^2 \hat{a}}{\partial z^2} \right) \quad (r = \overline{a}) \tag{6.94}$$

The substitution of Eqs. (6.61), (6.66), and (6.67) into this boundary condition specifies the relation of the perturbed pressure and radius of the jet:

$$p' = a' \left(-\frac{\rho}{2}\Omega^2\overline{a} + \gamma \left(-\frac{1}{\overline{a}^2} + \frac{n^2}{\overline{a}^2} + k^2 \right) \right) \quad (r = \overline{a}) \tag{6.95}$$

6.3.3 Eigenvalue equation

The two boundary conditions defining u'_r and p', Eqs. (6.91) and (6.95), for the axisymmetric mode $n = 0$ must agree with the solutions of u'_r, Eq. (6.84), and p', Eq. (6.83), at $r = \overline{a}$:

$$\sigma a' = -\frac{C_1\sigma}{\rho(\sigma^2 + 4\Omega)}\frac{dI_0(\alpha\overline{a})}{dr}$$

$$a' \left(-\frac{\rho}{2}\Omega^2\overline{a} + \gamma \left(-\frac{1}{\overline{a}^2} + k^2 \right) \right) = C_1 I_0(\alpha\overline{a})$$

These equations give the eigenvalue equation for this system:

$$\sigma^2 + 4\Omega^2 = \frac{1}{\overline{a}^2} \left(\frac{1}{2}\Omega^2\overline{a}^3 + \frac{\gamma}{\rho}(1 - k^2\overline{a}^2) \right) \frac{dI_0(\alpha\overline{a})/dr}{I_0(\alpha\overline{a})} \tag{6.96}$$

From Eqs. (A.65) and (A.71) in Appendix, the derivative of the modified Bessel function, $dI_0(\alpha r)/dr = \alpha I_1(\alpha r)$. Eq. (6.96) may be expressed as

$$\sigma^2 + 4\Omega^2 = \frac{\alpha}{\overline{a}^2} \left(\frac{1}{2}\Omega^2\overline{a}^3 + \frac{\gamma}{\rho}(1 - k^2\overline{a}^2) \right) \frac{I_1(\alpha\overline{a})}{I_0(\alpha\overline{a})} \tag{6.97}$$

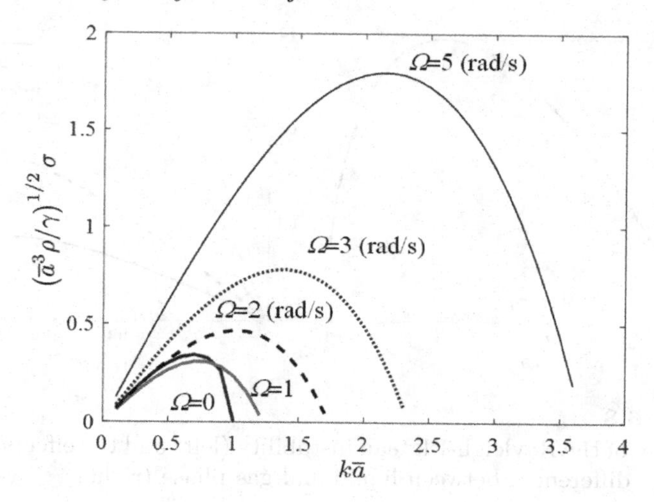

FIGURE 6.7
Dimensionless growth rate as a function of $k\overline{a}$.

Fig. 6.7 shows the dimensionless growth rate, depending on Ω, as a function of $k\overline{a}$. We find that the increase in axial rotation enhances instability growth at higher wave number. Accordingly, rapid breakup at shorter spacing can be expected with increase in Ω. A practical application of this finding to sea spray production by breaking waves is introduced in Section 8.2.2.

If $\Omega = 0$, Eq. (6.97) describes the well-known Rayleigh-Plateau instability, causing capillary breakup of a circular jet (see Fig. 6.6 right), whose growth rate is given by

$$\sigma^2 = \frac{\gamma k}{\rho \overline{a}^2}\left(1 - k^2\overline{a}^2\right)\frac{I_1(k\overline{a})}{I_0(k\overline{a})} \tag{6.98}$$

The breakup phenomena in the Rayleigh-Plateau mechanism has been applied to technology of an inkjet printer. Fig. 6.8 (left) shows the dimensionless growth rate of the Rayleigh-Plateau instability. The maximum growth rate $\sigma_{rp}\sqrt{\overline{a}^3\rho/\gamma} \approx 0.343$ is achieved at $k\overline{a} \approx 0.7$. The characteristic breakup time (in terms of capillary time) may be estimated by inverting the maximum growth rate:

$$\tau = \frac{1}{\sigma_{rp}} = 2.91\sqrt{\frac{\rho}{\gamma}\overline{a}^3} \tag{6.99}$$

The jet may breakup into drops with the spacing

$$L \approx U\tau = 2.91U\sqrt{\frac{\rho\overline{a}^3}{\gamma}} \tag{6.100}$$

In the gas-liquid two-layer system (liquid jet is surrounded by gas of density

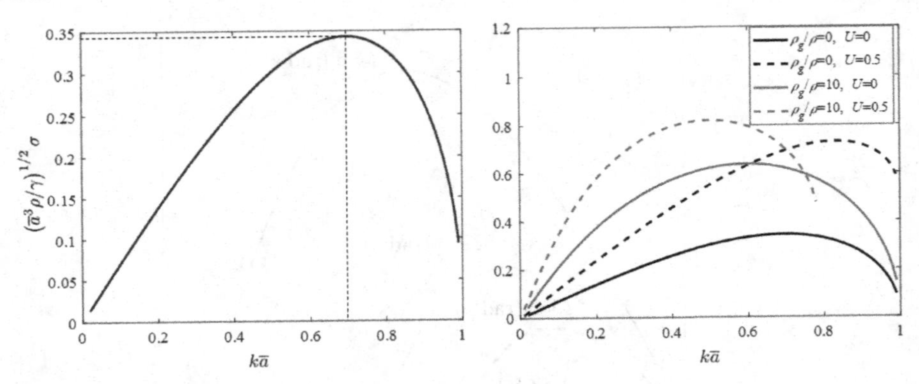

FIGURE 6.8
Growth rate of the Rayleigh-Plateau instability (left) and the effects of velocity and density differences between liquid and gas phase (right).

ρ_g), the dispersion relation is modified to

$$\sigma^2 = \frac{\gamma k}{\rho \overline{a}^2}\left(1 - k^2\overline{a}^2\right)\frac{I_1(k\overline{a})}{I_0(k\overline{a})}\left[1 + \frac{\rho_g}{\rho}\frac{K_0(k\overline{a})\,I_1(k\overline{a})}{K_1(k\overline{a})\,I_0(k\overline{a})}\right]^{-1} \tag{6.101}$$

which gives the maximum growth rate at lower wave number than that of free surface flow (see Fig. 6.8 right).

If there is the velocity difference between liquid and gas phases (the jet is ejected with velocity U in quiescent air), the Kelvin-Helmholtz instability affects the growth rate:

$$(\sigma - kU)^2 + \frac{\rho_g}{\rho}\sigma^2\frac{K_0(k\overline{a})\,I_1(k\overline{a})}{K_1(k\overline{a})\,I_0(k\overline{a})} = \frac{\gamma k}{\rho \overline{a}^2}\left(1 - k^2\overline{a}^2\right)\frac{I_1(k\overline{a})}{I_0(k\overline{a})} \tag{6.102}$$

Tomotika (1935) [100)] analytically estimated the viscous effect to the instability in the gas-liquid system. In the case of the viscosity ratio $\mu_l/\mu_g = 0.91$ (corresponding to the case of a cylindrical thread of lubricating oil surrounded by golden syrup, investigated in G.I. Taylor's experiments), analytical prediction $k\overline{a} = 0.568$ has satisfactory agrees with the experimental $k\overline{a} \approx 0.5$.

6.4 Crown splash

When a droplet impacts on a thin film, a circular wave, having discontinuous depth and velocity like as hydraulic jump, radically expands. A film-like jet is ejected obliquely upward at the discontinuous point to form a crown wall

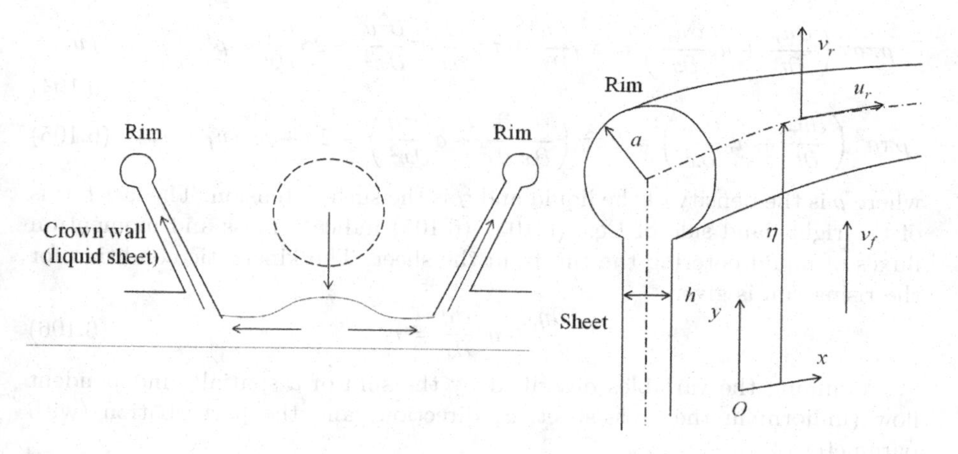

FIGURE 6.9
Illustration of the crown splash (left) and the parameters of motion of a rim bounding a sheet (right).

(see Fig. 6.9 left). The mechanism to produce the splashing jet from a liquid layer has been explained by Peregrine (1981)[80]. Since a tip of the crown wall has high curvature, surface tension works in the opposite direction to the projection of the wall, which moves the wall edge back (known as a retraction process), resulting in a formation of a blob-like rim with the circular axis along the edge (see Fig. 6.9 left). Since the rim radius consecutively increases owing the mass flux entering from the crown wall, the reduced surface tension destabilizes the rim along the axis, resulting in amplification of wave and finally breakup of the rim[2]. In this section, a linear stability analysis of the transverse disturbances of the rim, provided by Agbaglah et al.[2], is introduced.

6.4.1 Governing equation

Consider an edge of a planar liquid sheet of constant thickness h, rising with a constant velocity v_f, is bounded by a cylindrical rim, with radius $a(x,t)$, located at a vertical elevation of $\eta(x,t)$. The horizontal and vertical fluid velocities of the rim $(u_r(x,t),\ v_r(x,t))$, and constant vertical velocity in the sheet, v_f, are defined as shown in Fig. 6.9 (right). The mass and momentum conservation on the circular rim integrated over the cross section πa^2 is considered:

$$a\frac{\partial a}{\partial t}+\frac{a^2}{2}\frac{\partial u_r}{\partial x}+u_r a\frac{\partial a}{\partial x}=\frac{h}{2\pi}\left(v_f-v_r\right) \tag{6.103}$$

$$\rho\pi a^2\left(\frac{\partial u_r}{\partial t}+u_r\frac{\partial u_r}{\partial x}\right)=\pi\gamma\frac{\partial a}{\partial x}+\pi\gamma\frac{\partial}{\partial x}a^2\frac{\partial^2 a}{\partial x^2}+2\gamma\frac{\partial\eta}{\partial x}-\rho h\left(v_f-v_r\right)u_r \tag{6.104}$$

$$\rho\pi a^2\left(\frac{\partial v_r}{\partial t}+u_r\frac{\partial v_r}{\partial x}\right)=2\pi\gamma\left(\frac{\partial a}{\partial x}\frac{\partial\eta}{\partial x}+a\frac{\partial^2\eta}{\partial x^2}\right)-2\gamma+\rho h\left(v_f-v_r\right)^2 \tag{6.105}$$

where ρ is the density of the liquid and γ is the surface tension. The last terms of the right-hand side of Eqs. (6.103)–(6.105) indicate mass and momentum fluxes of liquid entering the rim from the sheet. The kinematic condition for the rising rim is given

$$\frac{\partial\eta}{\partial t}+u_r\frac{\partial\eta}{\partial x}=v_r \tag{6.106}$$

Assuming the variables described by the sum of a spatially independent flow (uniform in the cross-sheet, x, direction) and the perturbation (with parameter ϵ),

$$\eta(x,t)=\overline{\eta}(t)+\epsilon\hat{\eta}(x,t) \tag{6.107}$$

$$a(x,t)=\overline{a}(t)+\epsilon\hat{a}(x,t) \tag{6.108}$$

$$u_r(x,t)=\epsilon\hat{u_r}(x,t) \tag{6.109}$$

$$v_r(x,t)=\overline{v_r}(t)+\epsilon\hat{v_r}(x,t) \tag{6.110}$$

Substituting these variables into Eqs. (6.103)–(6.105), the equation systems for the base flow (zero-th order equations) and the first-order perturbations are given by:

<u>Base flow $O(\epsilon^0)$</u>

$$\overline{a}\frac{d\overline{a}}{dt}=\frac{h}{2\pi}\left(v_f-\overline{v_r}\right) \tag{6.111}$$

$$a^2\frac{d\overline{v_r}}{dt}=-\frac{2}{\pi}\frac{\gamma}{\rho}+\frac{h}{\pi}\left(v_f-\overline{v_r}\right)^2 \tag{6.112}$$

$$\frac{d\overline{\eta}}{dt}=\overline{v_r} \tag{6.113}$$

<u>First-order $O(\epsilon^1)$</u>

$$\overline{a}\frac{\partial\hat{a}}{\partial t}+\hat{a}\frac{\partial\overline{a}}{\partial t}+\frac{\overline{a}^2}{2}\frac{\partial\hat{u_r}}{\partial x}=-\frac{h}{2\pi}\hat{v_r} \tag{6.114}$$

$$\overline{a}^2\frac{\partial\hat{u_r}}{\partial t}=\frac{\gamma}{\rho}\frac{\partial\hat{a}}{\partial x}+\frac{\gamma}{\rho}\overline{a}^2\frac{\partial^3\hat{a}}{\partial x^3}+\frac{2\gamma}{\pi\rho}\frac{\partial\hat{\eta}}{\partial x}-\frac{h}{\pi}\left(v_f-\overline{v_r}\right)\hat{u_r} \tag{6.115}$$

$$\overline{a}^2\frac{\partial\hat{v_r}}{\partial t}+2\overline{a}\hat{a}\frac{\partial\overline{v_r}}{\partial t}=2\frac{\gamma}{\rho}\overline{a}\frac{\partial^2\hat{\eta}}{\partial x^2}-2\frac{h}{\pi}\left(v_f-\overline{v_r}\right)\hat{v_r} \tag{6.116}$$

$$\frac{\partial\hat{\eta}}{\partial t}=\hat{v_r} \tag{6.117}$$

We find that the base flow described by Eqs. (6.111)–(6.113) is time-dependent unlike the problems in the previous sections. Eq. (6.112) expresses that the mechanical balance of the surface tension and entering momentum flux contributes to the retraction motion of the rim. During the retraction process, the rim motion converges a quasi-steady state ($d\overline{v_r}/dt \to 0$) and the relative rim velocity achieves the so-called Taylor-Culick velocity [2]:

$$v_f - v_r = \sqrt{\frac{2\gamma}{\rho h}} \tag{6.118}$$

In this state, Eq. (6.111) states that the rim radius monotonically increases in time.

Following Agbaglah et al. (2013), [2] the base solutions at the initial state of the system are considered, assuming that base flow changes much more slowly than the perturbations. This, so-called, frozen approximation leads $\overline{a} \sim a_i$ and $d\overline{v_r}/dt \sim d\overline{v_{ri}}/dt$, where the subscript i indicates the initial state. Accordingly, the base state solutions Eqs. (6.111) and (6.112) are described by a parameter $\dot{v_{ri}} = d\overline{v_{ri}}/dt$:

$$\dot{a_i} = \frac{h}{2\pi a_i}(v_f - v_{ri}) \tag{6.119}$$

$$(v_f - v_{ri})^2 = \frac{\pi}{h}a_i^2\dot{v_{ri}} + \frac{2\gamma}{\rho h} \tag{6.120}$$

where $\dot{a_i}$ indicates the time derivative of a_i.

6.4.2 Solutions for perturbations

Introducing the exponential form of perturbation:

$$(\hat{a}, \hat{v_r}, \hat{u_r}, \hat{\eta}) = (a', v'_r, u'_r, \eta')e^{\sigma t + ikx} \tag{6.121}$$

The substitution into the first-order equations Eqs. (6.114)–(6.117) gives

$$a'(\sigma a_i + \dot{a_i}) + \frac{ik}{2}a_i^2 u'_r + \frac{h}{2\pi}v'_r = 0 \tag{6.122}$$

$$ik\left(k^2 a_i^2 - 1\right)a' + \left(\sigma a_i^2 + \frac{h}{\pi}(v_f - v_{ri})\right)u'_r - \frac{2ik\gamma}{\pi\rho}\eta' = 0 \tag{6.123}$$

$$2a_i\dot{v_{ri}}a' + \left(\sigma a_i^2 + \frac{2h}{\pi}(v_f - v_{ri})\right)v'_r + 2k^2 a_i\frac{\gamma}{\rho}\eta' = 0 \tag{6.124}$$

$$\sigma\eta' - v'_r = 0 \tag{6.125}$$

Writing this system in matrix form, as

$$\mathbf{MA} = 0 \tag{6.126}$$

where

$$\mathbf{M} = \begin{pmatrix} \sigma a_i + \dot{a_i} & \frac{ik}{2}a_i^2 & \frac{h}{2\pi} & 0 \\ ik\left(k^2 a_i^2 - 1\right) & \sigma a_i^2 + \frac{h}{\pi}(v_f - v_{ri}) & 0 & -\frac{2ik\gamma}{\pi\rho} \\ 2a_i\dot{v_{ri}} & 0 & \sigma a_i^2 + \frac{2h}{\pi}(v_f - v_{ri}) & \frac{2k^2\gamma}{\rho}a_i \\ 0 & 0 & -1 & \sigma \end{pmatrix}$$

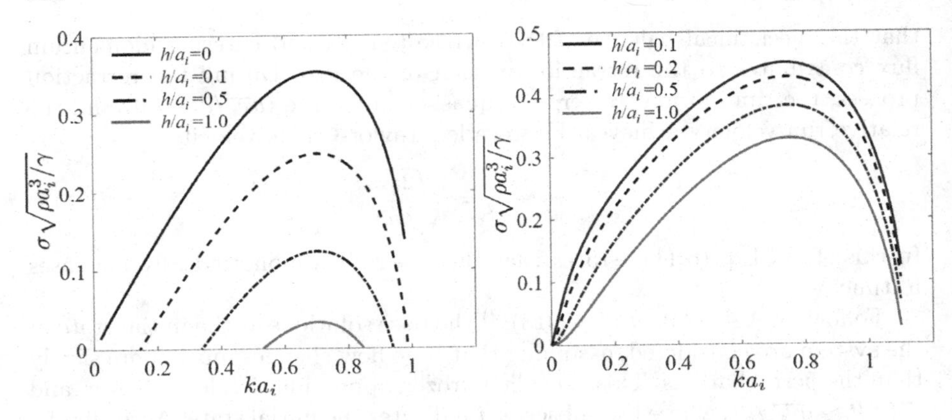

FIGURE 6.10
Dimensionless growth rate as a function of $a_i k$; dimensionless acceleration $\dot{v_{ri}}\sqrt{\rho a_i/\gamma}$=0 (left) and $\dot{v_{ri}}\sqrt{\rho a_i/\gamma} = -0.5$ (right).

and

$$\mathbf{A} = \begin{pmatrix} a' & u_r' & v_r' & \eta' \end{pmatrix}^T. \tag{6.127}$$

If the equation system has a non-trivial solution, the matrix $\mathbf{M}$ is singular. Therefore the determinant of $\mathbf{M}$ must be zero, which gives the eigenvalue equation of this system:

$$\begin{aligned} & a_i^4\sigma^4 + 7\dot{a}_i a_i^3\sigma^3 + \left\{ a_i^3 k^2 \left(2\frac{\gamma}{\rho} - \frac{1}{2} + \frac{a_i^2}{2}k^2 \right) + \frac{h}{\pi}a_i^2\dot{v_{ri}} + 4\frac{h}{\pi^2}\frac{\gamma}{\rho} + 6a_i^2\dot{a_i}^2 \right\}\sigma^2 \\ & + \left\{ 2a_i^2\dot{a}_i k^2 \left(4\frac{\gamma}{\rho} - 1 + a_i^2 k^2 \right) + 4\frac{h\gamma}{\pi^2\rho}\frac{\dot{a}_i}{a_i} \right\}\sigma \\ & + \frac{\gamma}{\rho}a_i^2 k^2 \left(4\frac{\dot{a_i}^2}{a_i} + 2\frac{\dot{v_{ri}}}{\pi} - k^2 + a_i^2 k^4 \right) = 0 \end{aligned} \tag{6.128}$$

where Eqs. (6.119) and (6.120) have been used to simplify some terms in the equation. Eq. (6.128) can be numerically solved. The perturbations grow in an unstable system when $\sigma > 0$.

Fig. 6.10 shows the dimensionless growth rates depending on the dimensionless sheet thickness h/a_i and rim acceleration $\dot{v_{ri}}\sqrt{\rho a_i/\gamma}$. We find from both cases of acceleration that the growth rate increases as the thickness decreases. In the quasi-steady case (Fig. 6.10 left), the maximum growth rate achieved at $ka_i \approx 0.7$ for any thickness. In particular, the growth rate curve for zero thickness (no sheet is attached with the rim) is identical to one of the Rayleigh-Plateau instabilities shown in Fig. 6.8 (left). Accordingly, the rim may be fragmented by the Rayleigh-Plateau mechanism in this case. We find the case of negative acceleration (Fig. 6.10 right) has higher growth rate than the case of zero acceleration. The growth rate increases in proportion to $\sqrt{ka_i}$ in a lower wave number range, indicating the Rayleigh-Taylor instability (Section 6.1.2) may affect the breakup process of the rim [2)]. In addition to this rim instability, if stability of the liquid sheet bounding the rim is considered (Section 6.2), the instability of the rim-sheet system grows at different wave numbers [121)], which provide statistical features of droplets fragmented through the rapture of the sheet and breakup of the rim [122)].

7

Ocean Waves

Ocean waves are generated and amplified through instability associated with wind shear and turbulence at an interface in a system of air and water flows, as noted in Section 6.1. Since wind velocity and direction locally follow an weather system above ocean, various wave components created in the systems over a large area are generally superposed, resulting in irregular wave field.

On the one hand, each wave component described by the linear wave theory, introduced in Section 4.3 and 4.6, can explain the major wave properties throughout the propagation including effects of bathymetry change and current. The velocities, pressure, and surface elevation of the progressive linear wave in the coordinate defined in Fig. 7.1 are described as

$$u = \frac{H}{2}\sigma\frac{\cosh k(h+z)}{\sinh kh}\cos(kx - \sigma t) \tag{7.1}$$

$$w = \frac{H}{2}\sigma\frac{\sinh k(h+z)}{\sinh kh}\sin(kx - \sigma t) \tag{7.2}$$

$$p = -\rho g z + \rho\frac{H}{2}\frac{\sigma^2}{k}\frac{\cosh k(h+z)}{\sinh kh}\cos(kx - \sigma t) \tag{7.3}$$

$$\eta = \frac{H}{2}\cos(kx - \sigma t) \tag{7.4}$$

where

$$\sigma^2 = gk\tanh kh \tag{7.5}$$

In this chapter, the wave height H is used instead of the amplitude A employed in Chapter 4.

Important features of ocean waves, including shoaling, refraction, evolution of wave groups, wave–current interaction, and irregular waves, are analytically addressed in this chapter. The method to deal with nonlinear waves, a well-known Stokes wave as an example, is also introduced in the last section of the chapter.

7.1 Properties of ocean waves

First, fundamental properties of wave propagation, mass, and energy transfer in waves are considered in this section.

DOI: 10.1201/9781003140160-7

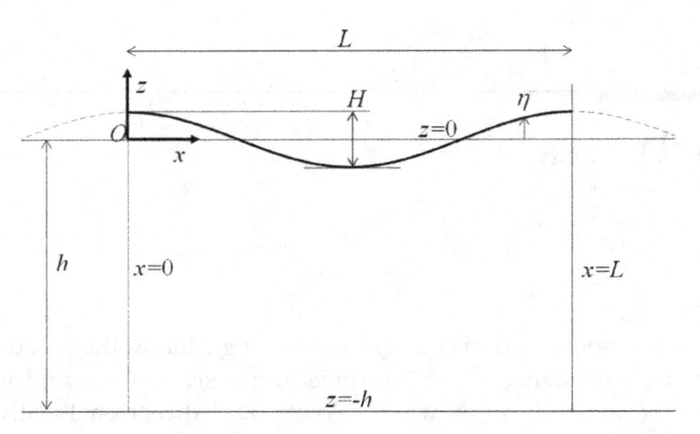

FIGURE 7.1
Linear wave propagating in constant depth.

7.1.1 Propagation of waves

A single progressive wave train in a planar wave field can be expressed in terms of oscillatory forms, $e^{i\theta}$, $\sin\theta$, or $\cos\theta$, where the phase is defined by

$$\theta(\boldsymbol{x}, t) = \boldsymbol{k} \cdot \boldsymbol{x} - \sigma t \tag{7.6}$$

as noted in Section 4.3 and 4.6. Accordingly, the frequency is defined in terms of θ by

$$\sigma = -\frac{\partial\theta}{\partial t} \tag{7.7}$$

Similarly, the wave number vector $\boldsymbol{k}$ is given by

$$\boldsymbol{k} = (k_x, k_y) = \left(\frac{\partial\theta}{\partial x}, \frac{\partial\theta}{\partial y}\right) = \boldsymbol{\nabla}\theta \quad \text{or} \quad k_i = \frac{\partial\theta}{\partial x_i} \tag{7.8}$$

where the horizontal differential operator $\boldsymbol{\nabla} = (\partial/\partial x, \partial/\partial y)$ is redefined in this section, and the index i=1, 2. $\boldsymbol{k}$ defines the direction of wave propagation along a wave ray. In another word, the wave ray is a curve which is tangent to the local $\boldsymbol{k}$ and is perpendicular to a line of constant phase (e.g. crest line).

Considering the trivial relation

$$\frac{\partial}{\partial t}(\boldsymbol{\nabla}\theta) - \boldsymbol{\nabla}\left(\frac{\partial\theta}{\partial t}\right) = 0 \quad \text{or} \quad \frac{\partial}{\partial t}\frac{\partial\theta}{\partial x_i} - \frac{\partial}{\partial x_i}\frac{\partial\theta}{\partial t} = 0 \tag{7.9}$$

Eqs. (7.7) and (7.8) give

$$\frac{\partial\boldsymbol{k}}{\partial t} + \boldsymbol{\nabla}\sigma = 0 \quad \text{or} \quad \frac{\partial k_i}{\partial t} + \frac{\partial\sigma}{\partial x_i} = 0 \tag{7.10}$$

This equation can be interpreted as conservation of the wave number; that is, if σ is spatially constant, the wave field never change in time. Consider another identity

$$\frac{\partial}{\partial y}\left(\frac{\partial \theta}{\partial x}\right)-\frac{\partial}{\partial x}\left(\frac{\partial \theta}{\partial y}\right)=0$$

With Eq. (7.8), this leads the irrotational condition on $\boldsymbol{k}$:

$$\boldsymbol{\nabla}\times\boldsymbol{k}=0 \quad \text{or} \quad \frac{\partial k_i}{\partial x_j}-\frac{\partial k_j}{\partial x_i}=0 \tag{7.11}$$

Assuming the dispersion relation is known and the frequency is expressed by

$$\sigma=\omega\left(\boldsymbol{k}\left(\boldsymbol{x},t\right),h\left(\boldsymbol{x}\right)\right) \tag{7.12}$$

Eq. (7.10) gives

$$\frac{\partial k_i}{\partial t}+\frac{\partial \omega}{\partial k_j}\frac{\partial k_j}{\partial x_i}=-\frac{\partial \omega}{\partial h}\frac{\partial h}{\partial x_i} \tag{7.13}$$

Introducing Eq. (7.11), this equation is modified to

$$\frac{\partial k_i}{\partial t}+c_{gj}\frac{\partial k_i}{\partial x_j}=-\frac{\partial \omega}{\partial h}\frac{\partial h}{\partial x_i} \tag{7.14}$$

where

$$c_{gj}=\frac{\partial \omega}{\partial k_j} \tag{7.15}$$

defines the propagation speed for k_i, termed the group velocity. Eq. (7.14) can be written in a characteristic form:

$$\frac{dk_i}{dt}=-\frac{\partial \omega}{\partial h}\frac{\partial h}{\partial x_i} \quad \text{on characteristic} \quad \frac{dx_i}{dt}=\frac{\partial \omega}{\partial k_i} \tag{7.16}$$

Considering the dispersion relation of progressive linear wave $\omega=\sqrt{gk\tanh kh}$, where $k=|\boldsymbol{k}|=\sqrt{k_ik_i}$, $\boldsymbol{k}$ of waves in deep water or in constant depth is constant on straight characteristics (see Fig. 7.2 left). Accordingly, the wave travels in the direction of $\boldsymbol{k}$ with the constant speed. When waves propagate on uneven depth, as $\partial h/\partial x_i \neq 0$, the values of k_i vary along the characteristics and thus the characteristics are no longer straight. For the simplest example of one-dimensional propagation

$$\frac{\partial \omega}{\partial h}\frac{\partial h}{\partial x}=\frac{g\sigma k^2}{\sinh 2kh}\frac{\partial h}{\partial x} \tag{7.17}$$

(see derivatives of hyperbolic functions in Section A.2 in Appendix). In this case, Eq. (7.16) becomes

$$\frac{dk}{dt}=-\frac{g\sigma k^2}{\sinh 2kh}\frac{\partial h}{\partial x} \tag{7.18}$$

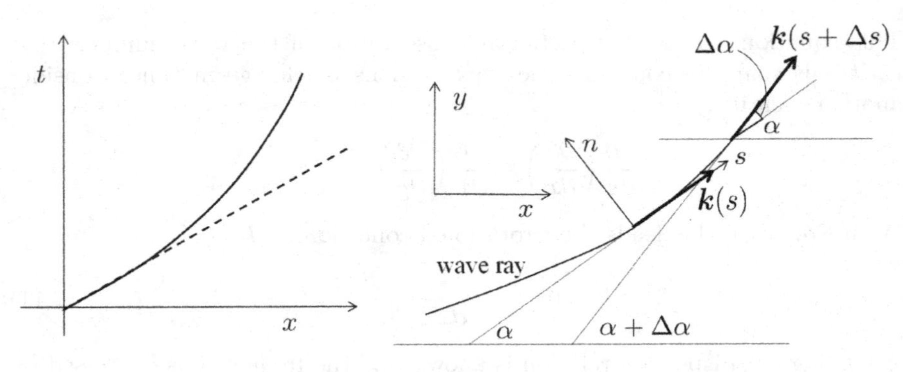

FIGURE 7.2
Characteristics of wave trains in deep water or constant depth (broken line) and in shoaling water (solid line); left. Illustration of the wave number of refracted waves on a curved wave ray; right.

Accordingly, when waves propagate toward shallower coast, as $\partial h/\partial x < 0$, the wave number increases (wavelength decreases) during the propagation (along the characteristics).

Consider propagation of σ, given by $\omega\left(\boldsymbol{k}\left(\boldsymbol{x},t\right),h\left(\boldsymbol{x},t\right)\right)$, Eq. (7.12), along the characteristics

$$\begin{aligned}\frac{d\sigma}{dt} &= \frac{\partial\sigma}{\partial t} + c_{gj}\frac{\partial\sigma}{\partial x_j} = \frac{\partial\omega}{\partial h}\frac{\partial h}{\partial t} + \frac{\partial k_j}{\partial t}\frac{\partial\omega}{\partial k_j} + c_{gj}\frac{\partial\omega}{\partial x_j} \\ &= \frac{\partial\omega}{\partial h}\frac{\partial h}{\partial t} - \frac{\partial\omega}{\partial k_j}\frac{\partial\omega}{\partial x_j} + c_{gj}\frac{\partial\omega}{\partial x_j} = \frac{\partial\omega}{\partial h}\frac{\partial h}{\partial t}\end{aligned} \tag{7.19}$$

Here Eqs. (7.10) and (7.15) have been used in the derivation. Eq. (7.19) indicates that if the dispersion relation is independent on time, the frequency is constant along each characteristic. For instance, the substitution of $\omega = \sqrt{gk}\tanh kh$ gives the last term of Eq. (7.19) as

$$\frac{\partial\omega}{\partial h}\frac{\partial h}{\partial t} = \frac{g\sigma k^2}{\sinh 2kh}\frac{\partial h}{\partial t} \tag{7.20}$$

Accordingly, the frequency never change during propagation unless sea floor temporally deforms.

$$\frac{d\sigma}{dt} = 0 \tag{7.21}$$

7.1.2 Refraction

When waves propagate shoaling water, refraction occurs owing to variations in water depth. Assuming a steady wave train, if the component of Eq. (7.14) in the direction of $\boldsymbol{k}$ is defined by k_s, the variation of k_s along the wave ray (s-axis) is given by

$$\frac{\partial k_s}{\partial s} = -\frac{1}{c_g}\frac{\partial \omega}{\partial h}\frac{\partial h}{\partial s} \tag{7.22}$$

Fig. 7.2 (right) illustrates the variation of the wave number on a wave ray curved through refraction. The variation of the normal component of $\boldsymbol{k}$ to the wave ray over Δs is given as $k_n(s+\Delta s) - k_n(s) = \Delta k_n = k_s \Delta\alpha$, where α is the direction of the ray. The derivative of k_n is thus given

$$\frac{\partial k_n}{\partial s} = \lim_{\Delta s \to 0}\frac{\Delta k_n}{\Delta s} = \lim_{\Delta s \to 0} k_s \frac{\Delta \alpha}{\Delta s} = k_s \frac{\partial \alpha}{\partial s} \tag{7.23}$$

According to the irrotational condition, Eq. (7.11), $\partial k_s/\partial n = \partial k_n/\partial s$. Therefore Eq. (7.23) gives

$$\frac{\partial k_s}{\partial n} = k_s \frac{\partial \alpha}{\partial s} \tag{7.24}$$

Since the normal component of Eq. (7.14), in steady state, is given by

$$\frac{\partial k_s}{\partial n} = -\frac{1}{c_g}\frac{\partial \omega}{\partial h}\frac{\partial h}{\partial n} \tag{7.25}$$

the direction of the wave ray varies with

$$\frac{\partial \alpha}{\partial s} = -\frac{1}{k_s c_g}\frac{\partial \omega}{\partial h}\frac{\partial h}{\partial n} \tag{7.26}$$

If the water depth is uniform in the direction perpendicular to the ray ($\partial h/\partial n = 0$), the wave train propagates on the straight ray. If there is normal gradient of depth ($\partial h/\partial n \neq 0$), as $\partial \alpha/\partial s \neq 0$, the refraction occurs and curves the ray.

7.1.3 Approximation of the dispersion relation

As noted, when a wave train with the frequency σ travels toward shallower coast, the wave number increases with decreasing water depth during propagation, which is defined by the dispersion relation, Eq. (7.5). Following a variation of the wave number, wave speed, defined by $c = \sigma/k$, decreases. The approximations of the dispersion and wave speed in deep and shallow water regimes are introduced by limits of hyperbolic functions (see Section 1.1.3).

According to the deep water approximation $\tanh kh \approx 1$, Eq. (1.36), the dispersion relation Eq. (7.5) is approximated as

$$\sigma^2 \approx g k_O \tag{7.27}$$

where $k_O (= 2\pi/L_O)$ is the wave number in deep water, L_O is the offshore wavelength. The wave speed in deep water (offshore wave speed) is thus given from Eq. (7.27) as

$$\begin{aligned} c_0 &= \frac{\sigma}{k_0} \\ &\approx \frac{g}{\sigma} = \frac{g}{2\pi}T \sim 1.56T \text{ (m/s)} \end{aligned} \tag{7.28}$$

The offshore wave velocity depends only on wave period or frequency.

In shallow water, the shallow water (long wave) approximation, $\tanh kh \approx kh$, Eq. (1.33), may be applied for Eq. (7.5):

$$\sigma^2 \approx ghk_l^2 \tag{7.29}$$

The wave speed in this regime is given

$$\begin{aligned} c_l &= \frac{\sigma}{k_l} \\ &\approx \sqrt{gh} \end{aligned} \tag{7.30}$$

We find c_l is dependent only on water depth and independent on wave period or frequency; that is, dispersion of waves is lost in this regime[1].

In the intermediate regime, as any hyperbolic approximations are unacceptable, Eq. (7.5) is used as is to define the wave speed:

$$\begin{aligned} c &= \frac{\sigma}{k} \\ &= \frac{g}{\sigma}\tanh kh \approx c_0 \tanh kh \end{aligned} \tag{7.31}$$

Fig. 7.3 compares the dimensionless frequency (left) and wave speed (right), estimated by Eq. (7.5), with the approximated values of plots. We find the deep and shallow water approximations that provide appropriate estimates of frequency and wave speed in each regime.

7.1.4 Mass transport

Consider displacement of a water particle (ξ, ζ) from arbitrary location (x, z) under progressive linear waves (Fig. 7.4 left), which is given as

$$\xi = \int u(x + \xi, z + \zeta)dt \tag{7.32}$$

$$\zeta = \int w(x + \xi, z + \zeta)dt \tag{7.33}$$

[1] Considering a wave group containing many wave components with different wavelengths in shallow water regime, the group never spread as all components propagate at the same speed of Eq. (7.30) in this regime (see Section 7.1.7).

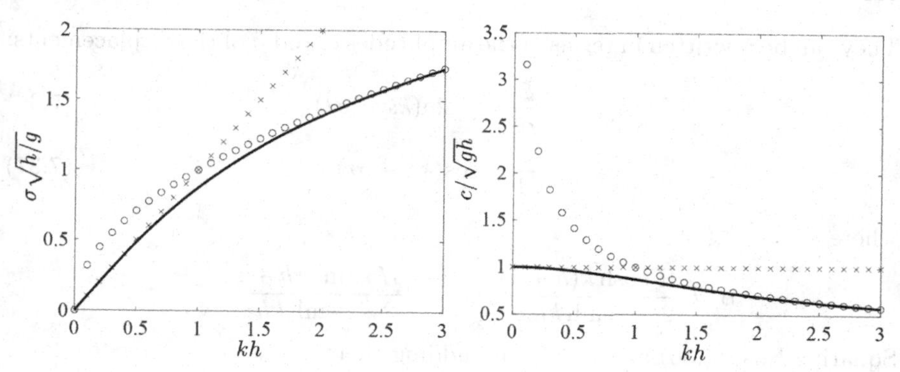

FIGURE 7.3
Dimensionless frequency (left) and wave speed (right) as a function of kh, estimated by the dispersion relation Eq. (7.5) (lines) and approximated values by Eqs. (7.27) (∘) and (7.29) (×).

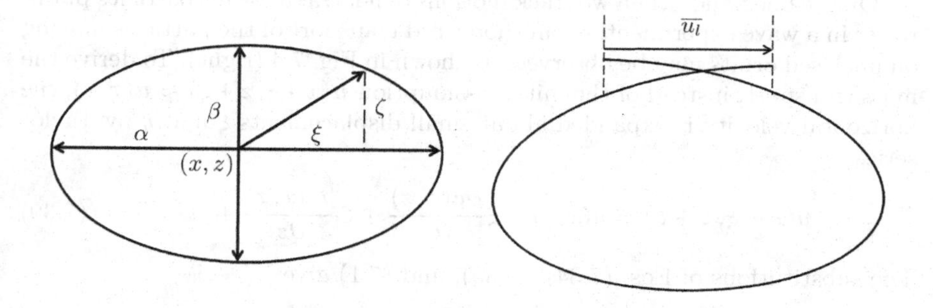

FIGURE 7.4
Elliptical closed trajectory of a fluid particle (left) and mass transport on an unclosed trajectory (right).

Assuming small displacement, we may approximated the velocity as $\boldsymbol{u}(x + \xi, z + \zeta) \approx \boldsymbol{u}(x, z)$, given by Eqs. (7.1) and (7.2), to substitute into Eqs. (7.32) and (7.33):

$$\xi \approx \int u(x, z)dt = -\frac{H}{2}\frac{\cosh k(h+z)}{\sinh kh}\sin(kx - \sigma t) \tag{7.34}$$

$$\zeta \approx \int w(x, z)dt = \frac{H}{2}\frac{\sinh k(h+z)}{\sinh kh}\cos(kx - \sigma t) \tag{7.35}$$

They can be rewritten in terms of the amplitudes α and β of the displacements:

$$\frac{\xi}{\alpha} = -\sin(kx - \sigma t) \tag{7.36}$$

$$\frac{\zeta}{\beta} = \cos(kx - \sigma t) \tag{7.37}$$

where

$$\alpha = \frac{H}{2}\frac{\cosh k(h+z)}{\sinh kh}, \quad \beta = \frac{H}{2}\frac{\sinh k(h+z)}{\sinh kh}$$

Squaring Eqs. (7.36) and (7.37) and adding them,

$$\left(\frac{\xi}{\alpha}\right)^2 + \left(\frac{\zeta}{\beta}\right)^2 = \sin^2(kx - \sigma t) + \cos^2(kx - \sigma t) = 1 \tag{7.38}$$

Accordingly, the particle locates on an ellipse with long axis α and short axis β. Because of the close trajectory, the particle returns to the same location after wave period T. Therefore there is no net mass transport on this assumption.

On the one hand, when we track motions of neutral buoyant particles put in water in a wave experiment, mean shoreward transport of the particles moving on unclosed orbits may be observed, as shown in Fig. 7.4 (right). To derive the mass transport, instead of the initial assumption $\boldsymbol{u}(x+\xi, z+\zeta) \approx \boldsymbol{u}(x, z)$, the horizontal velocity is expanded about small displacements ξ and η by Taylor series:

$$u(x+\xi, z+\zeta) = u(x,z) + \xi\frac{\partial u(x,z)}{\partial x} + \zeta\frac{\partial u(x,z)}{\partial z} + \cdots \tag{7.39}$$

The substitutions of Eqs. (7.34), (7.35), and (7.1) give

$$\begin{aligned} u(x+\xi, z+\zeta) &= \frac{H}{2}\sigma\frac{\cosh k(h+z)}{\sinh kh}\cos(kx - \sigma t) \\ &+ \frac{H^2}{4}\sigma k\frac{\cosh^2 k(h+z)}{\sinh^2 kh}\sin^2(kx - \sigma t) + \frac{H^2}{4}\sigma k\frac{\sinh^2 k(h+z)}{\sinh^2 kh}\cos^2(kx - \sigma t) + \cdots \end{aligned} \tag{7.40}$$

Taking time average of Eq. (7.40),

$$\begin{aligned} \overline{u_l} &= \frac{H^2}{4}\frac{\sigma k}{\sinh^2 kh}\left(\cosh^2 k(h+z)\overline{\sin^2(kx - \sigma t)} + \sinh^2 k(h+z)\overline{\cos^2(kx - \sigma t)}\right) \\ &= \frac{H^2}{8}\frac{\sigma k}{\sinh^2 kh}\cosh 2k\,(h+z) \end{aligned} \tag{7.41}$$

(see Section A.2 for transformation of hyperbolic functions in Appendix). If the total mean horizontal transport over finite depth is assumed to be zero, the mass transport velocity $\overline{u_m} = \overline{u_l} + U$ is integrated as

$$\int_{-h}^{0} \overline{u_m} dz = \int_{-h}^{0} \overline{u_l} dz + \int_{-h}^{0} U dz = 0 \tag{7.42}$$

where U is the constant describing the opposing uniform current. As U is readily determined by Eq. (7.42), the mass transport velocity can be specified

$$\overline{u_m} = \frac{H^2\sigma}{8\sinh^2 kh}\left(k\cosh 2k\,(h+z) - \frac{1}{2h}\sinh 2kh\right) \tag{7.43}$$

In particular, the deep water approximation of Eq. (7.41) is known as Stokes drift velocity:

$$\overline{u_s} = \frac{H^2}{4}\sigma k e^{2kz} \tag{7.44}$$

7.1.5 Wave energy

Potential energy of waves relative to a still water level is given by

$$E_p = \int_0^{\eta} \rho g z dz = \frac{1}{2}\rho g \eta^2 \tag{7.45}$$

Substituting the surface elevation, Eq. (7.4), and averaging over a wave period, the mean potential energy is given

$$\overline{E_p} = \frac{1}{2}\rho g \overline{\eta^2} \tag{7.46}$$

$$= \frac{\rho g}{16} H^2 \tag{7.47}$$

Kinematic energy q is determined by Eqs. (7.1), (7.2), and (7.5) as

$$\begin{aligned} q &= \frac{\rho}{2}\left(u^2 + w^2\right) \\ &= \frac{\rho}{2}\left(\frac{H\sigma}{2\sinh kh}\right)^2 \left(\cosh^2 k\,(h+z)\cos^2(kx - \sigma t) + \sinh^2 k\,(h+z)\sin^2(kx - \sigma t)\right) \\ &= \frac{\rho g}{8}\frac{H^2 k}{\sinh 2kh}\left(\cosh 2k\,(h+z) + \cos 2\,(kx - \sigma t)\right) \end{aligned} \tag{7.48}$$

(see transformations of hyperbolic functions in Section A.2). Assuming small amplitude waves, the integration of Eq. (7.48) over the depth gives

$$E_k = \int_{-h}^{\eta} q dz \approx \int_{-h}^{0} q dz = \frac{\rho g}{16}\frac{H^2}{\sinh 2kh}\left(\sinh 2kh + 2\frac{h}{k}\cos 2\,(kx - \sigma t)\right) \tag{7.49}$$

Averaging E_k over a wave period,

$$\overline{E_k} = \frac{\rho g}{16} H^2 \tag{7.50}$$

The total mean energy is thus given by

$$\overline{E} = \overline{E_p} + \overline{E_k} = \frac{\rho g}{8} H^2 \tag{7.51}$$

If surface elevation η is known, as $\overline{E_p}$ is given by Eqs. (7.46) or (7.47), the total mean energy of waves can be also estimated by

$$\overline{E} = 2\overline{E_p} \tag{7.52}$$

7.1.6 Energy flux

The wave energy transmitting through a cross section over depth under progressive waves is conserved along the direction of propagation. The horizontal transport of potential and kinematic energy, integrated over the depth, is given by

$$F_k = \int_{-h}^{\eta} \left(\rho g z + \frac{\rho}{2} \left(u^2 + w^2 \right) \right) u dz \tag{7.53}$$

The depth integration of the work done by the wave motion per unit time is estimated by

$$F_p = \int_{-h}^{\eta} p u dz \tag{7.54}$$

The energy flux, defined as the sum of F_k and F_p, is approximated for small amplitude waves as

$$F = F_k + F_p = \int_{-h}^{\eta} \left(\rho g z + \frac{\rho}{2} \left(u^2 + w^2 \right) \right) u dz + \int_{-h}^{\eta} p u dz \tag{7.55}$$

$$\approx \int_{-h}^{0} \left(\rho g z + p \right) u dz \tag{7.56}$$

Substituting Eqs. (7.1) and (7.3) and averaging over a wave period, the mean energy flux is defined as

$$\overline{F} \approx \frac{1}{T} \int_{t_0}^{t_0+T} \int_{-h}^{0} \rho g \sigma \left(\frac{H}{2} \right)^2 \frac{\cosh^2 k(h+z)}{\sinh kh \cosh kh} \cos^2 (kx - \sigma t) \, dz dt \tag{7.57}$$

Using addition theorem of hyperbolic function, Section A.2 in Appendix, and integrating the equation, $\overline{F}$ can be identified

$$\overline{F} = \frac{\rho g}{8} H^2 \frac{\sigma}{2k} \left(1 + \frac{2kh}{\sinh 2kh} \right) \tag{7.58}$$

Eq. (7.58) may be rewritten in terms of the total mean energy Eq. (7.51) as

$$\overline{F} = \overline{E} \frac{\sigma}{2k} \left(1 + \frac{2kh}{\sinh 2kh} \right) \tag{7.59}$$

$\overline{F}$ is conserved along the wave ray unless the energy is dissipated by wave breaking or friction on the floor, which is explained in later sections.

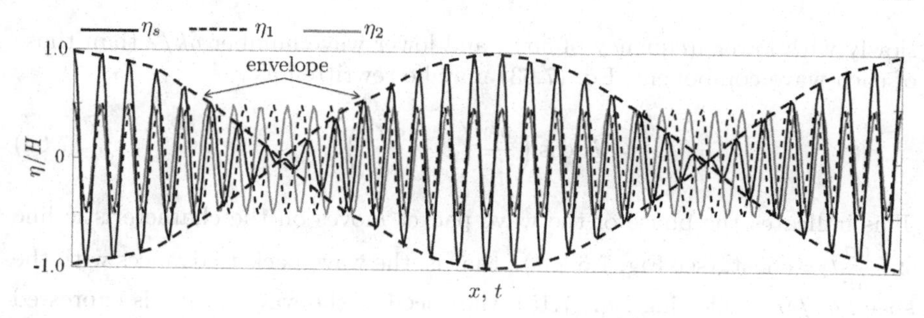

FIGURE 7.5
Wave packet composed of two wave trains.

7.1.7 Wave group

Any wave motion generally consists of a superposition of waves with various amplitudes and wavelengths. If we consider a disturbance containing infinitely many harmonic waves in a limited area on still water surface, the waves with long wavelength propagate faster than ones with shorter wavelengths (see Fig. 7.3 right), and the initial wave packet spread in time (the length of the wave packet temporally increases). The superposed waves may be sorted out during propagation into various groups of waves having approximately the same wavelength.

As the simplest case, consider two progressive waves with slightly different frequency $\delta\sigma$ and wave number δk, given by

$$\eta_1 = \frac{H}{2}\cos\left(\left(k - \delta k/2\right)x - \left(\sigma - \delta\sigma/2\right)t\right) \tag{7.60}$$

$$\eta_2 = \frac{H}{2}\cos\left(\left(k + \delta k/2\right)x - \left(\sigma + \delta\sigma/2\right)t\right) \tag{7.61}$$

The superposition of the two wave trains

$$\begin{aligned}
\eta_s &= \eta_1 + \eta_2 \\
&= \frac{H}{2}\left(\cos(kx - \sigma t)\cos\left(\delta\frac{k}{2}x - \delta\frac{\sigma}{2}t\right) + \sin(kx - \sigma t)\sin\left(\delta\frac{k}{2}x - \delta\frac{\sigma}{2}t\right)\right) \\
&\quad + \frac{H}{2}\left(\cos(kx - \sigma t)\cos\left(\delta\frac{k}{2}x - \delta\frac{\sigma}{2}t\right) - \sin(kx - \sigma t)\sin\left(\delta\frac{k}{2}x - \delta\frac{\sigma}{2}t\right)\right) \\
&= H\cos(kx - \sigma t)\cos\left(\frac{\delta k}{2}x - \frac{\delta\sigma}{2}t\right) \\
&= A\cos(kx - \sigma t)
\end{aligned} \tag{7.62}$$

where the time- and space-dependent amplitude of η_s

$$A = H\cos\left(\frac{\delta k}{2}x - \frac{\delta\sigma}{2}t\right) \tag{7.63}$$

Fig. 7.5 shows the surface elevations of the wave components η_1, η_2, and the superpositionη_s. We find A draws the envelope of the waves, which varies slowly with lower frequency of $\delta\sigma/2$ and lower wave number $\delta k/2$ than those of each wave component. Eq. (7.63) may be rewritten as

$$A = H\cos\left(\frac{\delta k}{2}\left(x - \frac{\delta\sigma}{\delta k}t\right)\right) \tag{7.64}$$

This indicates the phase of the wave packet moves on the characteristic line $x - \dfrac{\delta\sigma}{\delta k}t = \text{const.}$ (see Fig. 7.6 left) ; that is, the wave packet advances with the speed $\delta\sigma/\delta k$. Following Eq. (7.15), the speed of the wave packet is expressed as the group velocity:

$$c_g = \lim_{\delta k \to 0}\frac{\delta\sigma}{\delta k} = \frac{d\sigma}{dk} \tag{7.65}$$

Considering the dispersion relation $\sigma^2 = gk\tanh kh$, Eq. (7.5),

$$\begin{aligned}
2\sigma\frac{d\sigma}{dk} &= g\tanh kh + gkh\text{sech}^2 kh \\
\frac{d\sigma}{dk} &= \frac{g\sigma}{2}\frac{\tanh kh + kh\text{sech}^2 kh}{gk\tanh kh} = \frac{\sigma}{2k}\left(1 + \frac{kh}{\sinh kh\cosh kh}\right) \\
\therefore c_g &= \frac{c}{2}\left(1 + \frac{2kh}{\sinh 2kh}\right)
\end{aligned} \tag{7.66}$$

where the wave speed $c = \sigma/k$.

We may approximate c_g in shallow and deep water regimes. The shallow water approximation $\sinh 2kh \approx 2kh$ (see Fig. 1.2) gives

$$c_{gl} \approx c_l = \sqrt{gh} \tag{7.67}$$

Here the approximated phase speed should be referred to Section 7.1.3. We can confirm this property from the agreement of c/c_0 and c_g/c_0 in small kh in Fig. 7.6 (right). In deep water, since $2kh/\sinh 2kh \to 0$ at $2kh \to \infty$,

$$c_{g0} \approx \frac{c_0}{2} = \frac{\sigma}{k_0} = \frac{g}{2\sigma} \tag{7.68}$$

Since the group velocity is one-half of the phase speed in deep water (compare c/c_0 and c_g/c_0 in large kh in Fig. 7.6 right), the component waves pass through the wave packet as shown in Fig. 7.6 (left).

The group velocity in a gravity–capillary system is also derived by differentiating the dispersion, Eq. (4.51):

$$c_{gc} = \frac{c}{2}\left(\frac{1 + 3\frac{\gamma}{\rho g}k^2}{1 + \frac{\gamma}{\rho g}} + \frac{2kh}{\sinh 2kh}\right) \tag{7.69}$$

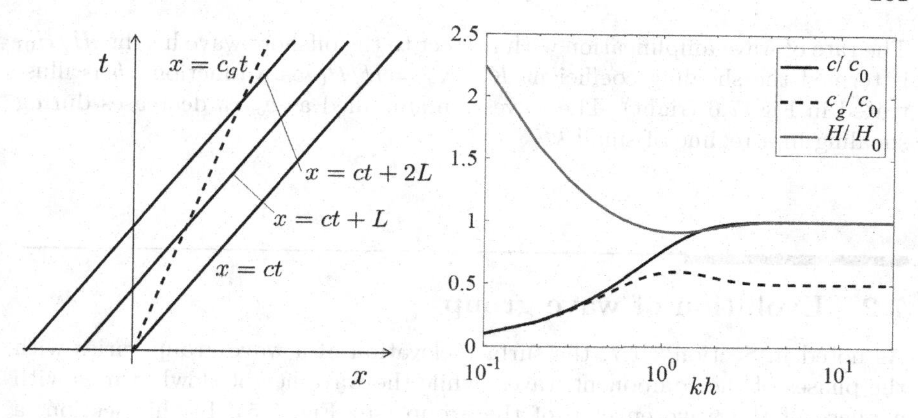

FIGURE 7.6
Characteristics of component wave crests (solid line) and wave group (broken line); left. The shoaling effects, H/H_0, c/c_0 and c_g/c_0, as a function of kh; right.

7.1.8 Shoaling

The mean energy flux of linear progressive waves, Eq. (7.59), can be expressed in terms of c_g by

$$\overline{F} = \overline{E} c_g \tag{7.70}$$

Assuming no energy dissipation through the propagation and steady wave filed, $\overline{F}$ is conserved along s axis on the wave ray:

$$\frac{d}{ds}\left(\overline{E} c_g\right) = 0 \tag{7.71}$$

The integration of Eq. (7.71) over the ray gives

$$\overline{E} c_g = \text{const.} \tag{7.72}$$

Accordingly, the energy fluxes at arbitrary two locations along the ray, say site 1 and 2, can be equated:

$$\overline{E} c_g \mid_1 = \overline{E} c_g \mid_2 \tag{7.73}$$

$$\frac{\rho g}{8} H_1^2 c_{g1} = \frac{\rho g}{8} H_2^2 c_{g2} \tag{7.74}$$

$$\therefore \frac{H_2}{H_1} = \sqrt{\frac{c_{g1}}{c_{g2}}} \tag{7.75}$$

If the phase velocity c_0 and wave height H_0 of deep water waves are known, the wave height at arbitrary location on the ray is given by

$$\frac{H}{H_0} = \sqrt{\frac{c_{g0}}{c_g}} = \sqrt{\frac{c_0}{2c_g}} = K_s \tag{7.76}$$

The rate of wave amplification with respect to the offshore wave height, H/H_0, is termed the shoaling coefficient K_s. $K_s = H/H_0$ as a function kh is illustrated in Fig. 7.6 (right). The waves are amplified as c_g/c_0 decreases during shoaling in a regime of small kh.

7.2 Evolution of wave group

As noted in Section 7.1.7, the surface elevation of a wave group varies with the phases of the component waves, while the wave height slowly varies with a phase of the wave envelop of the group (see Fig. 7.5). In this section, a general form of equation governing the group wave height is derived in a similar manner as Mei (2005).[60]

Assuming one-dimensional propagation in the x-direction, we introduce slow variables:

$$x_1 = \epsilon x, \quad \text{and } t_1 = \epsilon t \tag{7.77}$$

where $\epsilon \ll 1$ measures the ratio of two length and time scales. The wave amplitude must be a function of the slow variables:

$$\eta = A(x_1, t_1)\mathrm{e}^{i(kx-\sigma t)} \tag{7.78}$$

Similarly, the velocity potential takes the form

$$\phi = \varphi(x_1, z, t_1)\mathrm{e}^{i(kx-\sigma t)} \tag{7.79}$$

The amplitudes A and φ are expanded into a power series in ϵ:

$$A(x_1, t_1) = A_0(x_1, t_1) + \epsilon A_1(x_1, t_1) + \epsilon^2 A_2(x_1, t_1) + O(\epsilon^3) \tag{7.80}$$

$$\varphi(x_1, z, t_1) = \varphi_0(x_1, z, t_1) + \epsilon\varphi_1(x_1, z, t_1) + \epsilon^2\varphi_2(x_1, z, t_1) + O(\epsilon^3) \tag{7.81}$$

The perturbed forms of ϕ and η, ignoring perturbations of the second-order $O(\epsilon^2)$ or higher, are then given

$$\phi\left(x, x_1, z, t, t_1\right) = \left(\varphi_0\left(x_1, z, t_1\right) + \epsilon\varphi_1\left(x_1, z, t_1\right)\right)\mathrm{e}^{i(kx-\sigma t)} \tag{7.82}$$

$$\eta\left(x, x_1, t, t_1\right) = \left(A_0\left(x_1, t_1\right) + \epsilon A_1\left(x_1, t_1\right)\right)\mathrm{e}^{i(kx-\sigma t)} \tag{7.83}$$

The derivatives of $\phi = \phi\left(x, x_1, z, t, t_1\right)$ with respect to x and t are transformed as

$$\begin{aligned}
\frac{\partial\phi}{\partial x} &= \frac{\partial}{\partial x}\phi\left(x, x_1, z, t, t_1\right) + \frac{\partial x_1}{\partial x}\frac{\partial}{\partial x_1}\phi\left(x, x_1, z, t, t_1\right) \\
&= \frac{\partial}{\partial x}\phi\left(x, x_1, z, t, t_1\right) + \epsilon\frac{\partial}{\partial x_1}\phi\left(x, x_1, z, t, t_1\right) \\
&= \left(ik\left(\varphi_0 + \epsilon\varphi_1\right) + \epsilon\left(\frac{\partial\varphi_0}{\partial x_1} + \epsilon\frac{\partial\varphi_1}{\partial x_1}\right)\right)\mathrm{e}^{i(kx-\sigma t)}
\end{aligned} \tag{7.84}$$

$$
\begin{aligned}
\frac{\partial^2 \phi}{\partial x^2} &= \frac{\partial^2}{\partial x^2}\phi(x, x_1, z, t, t_1) + 2\epsilon \frac{\partial^2}{\partial x \partial x_1}\phi(x, x_1, z, t, t_1) + \epsilon^2 \frac{\partial^2}{\partial x_1^2}\phi(x, x_1, z, t, t_1) \\
&= \left(-k^2(\varphi_0 + \epsilon\varphi_1) + 2i\epsilon k\left(\frac{\partial \varphi_0}{\partial x_1} + \epsilon\frac{\partial \varphi_1}{\partial x_1}\right) + \epsilon^2\left(\frac{\partial^2 \varphi_0}{\partial x_1^2} + \epsilon\frac{\partial^2 \varphi_1}{\partial x_1^2}\right)\right) e^{i(kx-\sigma t)}
\end{aligned} \tag{7.85}
$$

$$
\begin{aligned}
\frac{\partial \phi}{\partial t} &= \frac{\partial}{\partial t}\phi(x, x_1, z, t, t_1) + \frac{\partial t_1}{\partial t}\frac{\partial}{\partial t_1}\phi(x, x_1, z, t, t_1) \\
&= \frac{\partial}{\partial t}\phi(x, x_1, z, t, t_1) + \epsilon\frac{\partial}{\partial t_1}\phi(x, x_1, z, t, t_1) \\
&= \left(-i\sigma(\varphi_0 + \epsilon\varphi_1) + \epsilon\left(\frac{\partial \varphi_0}{\partial t_1} + \epsilon\frac{\partial \varphi_1}{\partial t_1}\right)\right) e^{i(kx-\sigma t)}
\end{aligned} \tag{7.86}
$$

The Laplace equation is then expressed by

$$
\begin{aligned}
&-k^2(\varphi_0 + \epsilon\varphi_1) + 2i\epsilon k\left(\frac{\partial \varphi_0}{\partial x_1} + \epsilon\frac{\partial \varphi_1}{\partial x_1}\right) \\
&+ \epsilon^2\left(\frac{\partial^2 \varphi_0}{\partial x_1^2} + \epsilon\frac{\partial^2 \varphi_1}{\partial x_1^2}\right) + \frac{\partial^2 \varphi_0}{\partial z^2} + \epsilon\frac{\partial^2 \varphi_1}{\partial z^2} = 0
\end{aligned} \tag{7.87}
$$

The substitutions of Eq. (7.82) and Eq. (7.83) into the combined free-surface condition, Eq. (4.26), linearized kinematic condition, Eq. (4.20), and bottom boundary condition, Eq. (4.22), give

$$
\begin{aligned}
&-\sigma^2(\varphi_0 + \epsilon\varphi_1) - 2i\epsilon\sigma\left(\frac{\partial \varphi_0}{\partial t_1} + \epsilon\frac{\partial \varphi_1}{\partial t_1^2}\right) \\
&+ \epsilon^2\left(\frac{\partial^2 \varphi_0}{\partial t_1^2} + \epsilon\frac{\partial^2 \varphi_1}{\partial t_1^2}\right) + g\left(\frac{\partial \varphi_0}{\partial z} + \epsilon\frac{\partial \varphi_1}{\partial z}\right) = 0 \quad (z = 0)
\end{aligned} \tag{7.88}
$$

$$
-i\sigma(A_0 + \epsilon A_1) + \epsilon\left(\frac{\partial A_0}{\partial t_1} + \epsilon\frac{\partial A_1}{\partial t_1}\right) - \left(\frac{\partial \varphi_0}{\partial z} + \epsilon\frac{\partial \varphi_1}{\partial z}\right) = 0 \quad (z = 0) \tag{7.89}
$$

$$
\left\{ik(\varphi_0 + \epsilon\varphi_1) + \epsilon\left(\frac{\partial \varphi_0}{\partial x_1} + \epsilon\frac{\partial \varphi_1}{\partial x_1}\right)\right\}\epsilon\frac{\partial h}{\partial x_1} + \frac{\partial \varphi_0}{\partial z} + \epsilon\frac{\partial \varphi_1}{\partial z} = 0 \quad (z = -h) \tag{7.90}
$$

The equations at each order can be independently dealt with in the perturbation analysis. The zero-th order, $O(\epsilon^0)$, equations for Eqs. (7.87)–(7.90) are given by

$$
-k^2\varphi_0 + \frac{\partial^2 \varphi_0}{\partial z^2} = 0 \quad (-h < z < 0) \tag{7.91}
$$

$$
-\sigma^2\varphi_0 + g\frac{\partial \varphi_0}{\partial z} = 0 \quad (z = 0) \tag{7.92}
$$

$$
i\sigma A_0 + \frac{\partial \varphi_0}{\partial z} = 0 \quad (z = 0) \tag{7.93}
$$

$$\frac{\partial \varphi_0}{\partial z} = 0 \quad (z = -h) \tag{7.94}$$

The first-order, $O(\epsilon)$, equation system is:

$$-k^2\varphi_1 + \frac{\partial^2 \varphi_1}{\partial z^2} = -2ik\frac{\partial \varphi_0}{\partial x_1} \quad (-h < z < 0) \tag{7.95}$$

$$-\sigma^2\varphi_1 + g\frac{\partial \varphi_1}{\partial z} = 2i\sigma\frac{\partial \varphi_0}{\partial t_1} \quad (z = 0) \tag{7.96}$$

$$i\sigma A_1 + \frac{\partial \varphi_1}{\partial z} = \frac{\partial A_0}{\partial t_1} \quad (z = 0) \tag{7.97}$$

$$\frac{\partial \varphi_1}{\partial z} = -ik\varphi_0\frac{\partial h}{\partial x_1} \quad (z = -h) \tag{7.98}$$

The zero-th order equations can be solved in a similar way to the linear progressive waves in arbitrary depth (Section 4.3). However, in this section, we look for the solution for deep water waves ($h \to \infty$), for simplicity. A general solution of Eq. (7.91) takes the form

$$\varphi_0 = C_1\mathrm{e}^{kz} + C_2\mathrm{e}^{-kz} \tag{7.99}$$

As the deep water wave requires $\partial\varphi/\partial z = 0$ at $z \to -\infty$ (Eq. (7.94) for $h \to \infty$), C_2 must be zero. The substitution into Eq. (7.93) gives

$$\varphi_0 = -i\frac{\sigma}{k}A_0\mathrm{e}^{kz} \tag{7.100}$$

Substituting this into Eq. (7.92), as anticipated, the dispersion relation for deep water waves, Eq. (7.27), is derived:

$$\sigma^2 = gk \tag{7.101}$$

For the first-order system, the substitution of Eqs. (7.100) into (7.95) provides the first-order solution for Eq. (7.95) taking the form

$$\varphi_1 = C_3\mathrm{e}^{kz} + C_4\mathrm{e}^{-kz} - \frac{\partial A_0}{\partial x_1}\frac{\sigma}{2k^2}(2kz - 1)\,\mathrm{e}^{kz} \tag{7.102}$$

where C_4 is zero to satisfy $\partial\varphi_1/\partial z = 0$ when $h \to \infty$ (Eq. (7.98)). The substitution into Eq. (7.96) then gives

$$C_3\left(gk - \sigma^2\right) = 2\frac{\sigma^2}{k}\left(\frac{\partial A_0}{\partial t_1} + \left(\frac{\sigma}{4k} + \frac{g}{4\sigma}\right)\frac{\partial A_0}{\partial x_1}\right) \tag{7.103}$$

We find the dispersion relation Eq. (7.101) requires Eq. (7.103) to satisfy the condition:

$$\frac{\partial A_0}{\partial t_1} + \frac{\sigma}{2k}\frac{\partial A_0}{\partial x_1} = 0 \tag{7.104}$$

$$\therefore \frac{\partial A_0}{\partial t_1} + c_g \frac{\partial A_0}{\partial x_1} = 0 \tag{7.105}$$

where the group velocity of deep water waves $c_g = \sigma/2k$ (see Eq. (7.68)). The substitution of Eqs. (7.102) and (7.104) into the kinematic condition Eq. (7.97) gives the first-order solution:

$$\varphi_1 = -\left(i\frac{\sigma}{k}A_1 + \frac{\partial A_0}{\partial x_1}\frac{\sigma}{2k^2}(2kz-1)\right)e^{kz} \tag{7.106}$$

Eq. (7.105) can be rewritten as

$$\frac{\partial A_0^2}{\partial t} + c_g \frac{\partial A_0^2}{\partial x} = 0 \tag{7.107}$$

Here the subscript 1 has been omitted for simplicity. Eq. (7.107) describes the wave energy transfer with the group velocity c_g. Since $A_0^2 \partial c_g/\partial x = 0$ in deep water assumption used in this section, Eq. (7.107) may be written as

$$\frac{\partial A_0^2}{\partial t} + \frac{\partial c_g A_0^2}{\partial x} = 0 \tag{7.108}$$

It should be noted that this term $A_0^2 \partial c_g/\partial x$ has a roles in a shoaling process in shallower regime, as noted in Section 7.1.8. Accordingly, Eq. (7.108) is used as a general form of wave energy conservation. Eq. (7.108) may be expanded to a (x, y) planar field and rewritten in terms of the total mean wave energy $\overline{E}$ defined by Eq. (7.51):

$$\frac{\partial \overline{E}}{\partial t} + \frac{\partial c_{gj}\overline{E}}{\partial x_j} = 0 \tag{7.109}$$

$$\frac{\partial \overline{E}}{\partial t} + \nabla \cdot (\boldsymbol{c_g}\overline{E}) = 0 \tag{7.110}$$

This equation, known as an energy balance equation, can be reduced to Eq. (7.71) if steady wave field is assumed.

Eq. (7.108) divided by σ can be modified as

$$\frac{\partial}{\partial t}\left(\frac{A_0^2}{\sigma}\right) + \frac{A_0^2}{\sigma^2}\frac{\partial \sigma}{\partial t} + \frac{\partial}{\partial x}\left(c_g \frac{A_0^2}{\sigma}\right) + \frac{A_0^2}{\sigma^2}c_g\frac{\partial \sigma}{\partial x} = 0 \tag{7.111}$$

Assuming σ is expressed by the dispersion relation ω (see Section 7.1.1)

$$\sigma = \omega\left(k\left(x,t\right), h(x)\right) \tag{7.112}$$

Introducing Eqs. (7.10) and (7.15), the time derivative of σ can be expressed by

$$\frac{\partial \sigma}{\partial t} = \frac{\partial \omega}{\partial k}\frac{\partial k}{\partial t} = -c_g \frac{\partial \omega}{\partial x} \tag{7.113}$$

Eq. (7.111) is thus rewritten by

$$\frac{\partial}{\partial t}\left(\frac{A_0^2}{\omega}\right) + \frac{\partial}{\partial x}\left(c_g \frac{A_0^2}{\omega}\right) = 0 \tag{7.114}$$

This equation is known as an action balance equation that describes conservation of the wave action A_0^2/ω. The general form of this equation in a planar field, introducing $\overline{E}$, is given by

$$\frac{\partial}{\partial t}\left(\frac{\overline{E}}{\omega}\right) + \nabla \cdot \left(\boldsymbol{c_g} \frac{\overline{E}}{\omega}\right) = 0 \tag{7.115}$$

The wave action equation has been used for prediction of irregular waves represented by directional wave spectrum $S(\omega, \alpha)$, where α is the wave direction. Introducing the wave action spectrum $N(\omega, \alpha) = S(\omega, \alpha)/\omega$, Eq. (7.115) is replaced by

$$\frac{\partial N}{\partial t} + \nabla \cdot (\boldsymbol{c_g} N) = 0 \tag{7.116}$$

The application of this equation is introduced in Section 7.4.

7.3 Wave–current interactions

When ocean waves propagate through a region with a current, the wave properties, such as wavelength, height, velocity, and direction, may change as a result of a modification of the dispersion relation by the current. As the simplest example, consider a progressive wave train with the phase speed c_r in quiescent water. The waves in a following current (velocity $U > 0$) can move faster than those in quiescent water, while the propagation in the opposing one $U < 0$ becomes slower. Accordingly, the absolute phase velocity in the current field c is given by

$$c = c_r + U \tag{7.117}$$

Multiplying by k, the so-called Doppler relation is obtained as

$$\sigma = \sigma_r + Uk \tag{7.118}$$

where σ_r is the relative frequency in the frame moving with U.

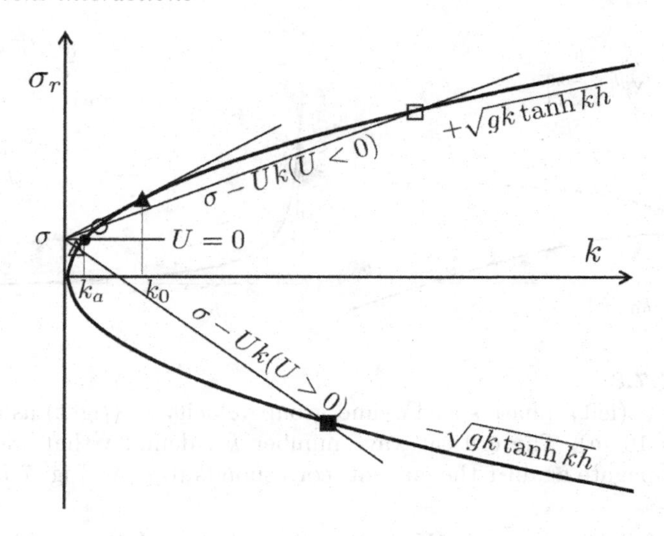

FIGURE 7.7
Graphical illustration of the solutions of Eq. (7.120). Curved lines indicate $\pm\sqrt{gk\tanh kh}$, and their tangential angle indicates the group velocity. Slope angles of straight lines indicate the current velocity. ◯, ▲, and □ indicate the solutions for $U<0$, △, and ■ indicate the ones for $U>0$, and • for $U=0$.

On the one hand, Section 6.1 provides the dispersion relation of free surface waves in uniform current, Eq. (6.24). Neglecting the capillary effect, the dispersion is expressed by

$$(\sigma - Uk)^2 = gk\tanh kh \tag{7.119}$$

$$\therefore \sigma_r = \sigma - Uk = \pm\sqrt{gk\tanh kh} \tag{7.120}$$

Eq. (7.120) may have multiple solutions for k for given σ, depth, and current velocity. Fig. 7.7 graphically indicates the solutions when waves and currents run parallel. In the case of $U>0$, waves move downstream in a following current, while, for $U<0$, waves may move upstream in an opposing current. If no current $U=0$, Eq. (7.120) agrees the normal dispersion relation, Eq. (7.5), and thus a unique solution σ at $k=k_a$ is obtained (see • in Fig. 7.7).

In $U<0$, there are one or two solutions given at wave numbers higher than k_a (see ◯, ▲, and □), indicating that the wavelength of the waves is reduced by the opposing current. In particular, when tangential angle of $\sqrt{gk\tanh kh}$, describing a group velocity $d\sigma/dk = c_g$, agrees with the slope angle of $|U|k$, only one solution (▲) is given at k_0. In this case of $c_g = |U|$, the wave energy cannot transfer upstream and is stopped by the current. Accordingly, the energy of longer waves with a solution in $k<k_0$ (◯) can propagate against the current, as $c_g > |U|$, while the energy of shorter waves in $k>k_0$ (see □) is swept downstream.

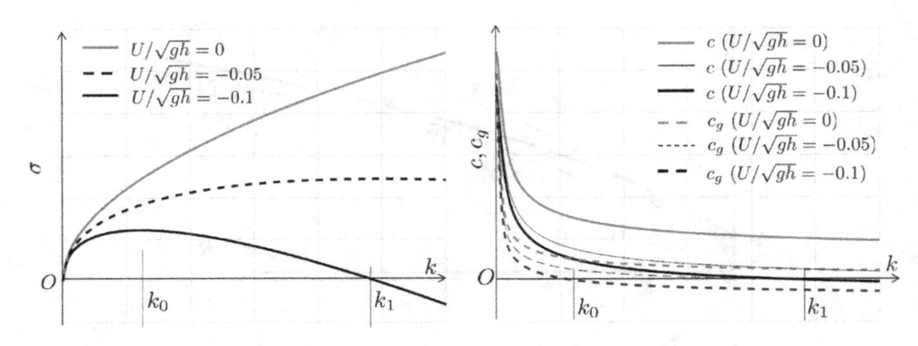

FIGURE 7.8
Frequency σ (left), phase speed c, and group velocity c_g (right) as a function of k ($h = 10$ m). The critical wave number k_0, defining that wave energy cannot propagate against the current, corresponds to k_0 in Fig. 7.7.

In the following current ($U > 0$), waves of the solutions with a smaller wave number than k_a (△) propagate faster than the current. The energy of shorter waves with another solution (■) is also swept downstream as $C_g < U$.

Fig. 7.8 shows the typical variations in $\sigma (= Uk + \sqrt{gk \tanh kh})$, phase velocity $c = \sigma/k$, and group velocity $d\sigma/dk = c_g$ as a function of k in opposing currents ($U < 0$). σ, c, and c_g all decrease with the opposing velocity. we find that c_g for $U/\sqrt{gh} = -0.1$ becomes zero at k_0 where the tangential angle of σ is zero ($d\sigma/dk = 0$). This point corresponds to ▲ in Fig. 7.7, which is a boundary of the responses whether the wave energy can propagate in the wave direction or carry downstream by the current. We also find $c = 0$ for $U/\sqrt{gh} = -0.1$ at $k = k_1$ where the wave phase cannot propagate against the current. From Eq. (7.120), the limit of wave propagation is given by

$$c_{cr} = U + \sqrt{\frac{g}{k}\mathrm{tanh}kh} = 0 \qquad (U < 0) \tag{7.121}$$

In shallow water, $\mathrm{tanh}kh \approx kh$ (Eq. (1.33)). Introducing critical opposing current velocity U_c and substituting $U = -U_c$ into Eq. (7.121), a critical Froude number is defined as

$$F_{rc} = \frac{U_c}{\sqrt{gh_c}} = 1 \tag{7.122}$$

where h_c is the critical depth. The Froude number $F_r = U_-/\sqrt{gh}$ (where U_- is the opposing current velocity) defines subcritical flow where shallow water waves can move upstream against the current when $F_r < F_{cr}$, and supercritical flow ($F_r > F_{cr}$) where waves are carried downstream.

7.3.1 Refraction owing to currents

Refraction of waves may occur as a result of planar variations in current. We consider the current-induced refraction in the same way as in Section 7.1.2.

Assuming the frequency is given as a function of $\boldsymbol{k}$ and $\boldsymbol{U}$, i.e. $\sigma = \omega(\boldsymbol{k}, \boldsymbol{U})$ (see Eq. (7.120)), Eqs. (7.10) and (7.11) give

$$\begin{aligned}\frac{\partial k_i}{\partial t} + \frac{\partial \sigma}{\partial x_i} &= \frac{\partial k_i}{\partial t} + \frac{\partial \omega}{\partial k_j}\frac{\partial k_j}{\partial x_i} + \frac{\partial \omega}{\partial U_j}\frac{\partial U_j}{\partial x_i} \\ &= \frac{\partial k_i}{\partial t} + \left(c_{gj} + \frac{k_j}{k}U_j\right)\frac{\partial k_i}{\partial x_j} + k\frac{\partial U_j}{\partial x_i} = 0 \end{aligned} \tag{7.123}$$

where c_{gj} is the group velocity without current. Considering k_s in s direction on the wave ray in steady wave state, Eq. (7.123) gives

$$\frac{\partial k_s}{\partial s} = -\left(c_g + U_s\right)^{-1} k\frac{\partial U_s}{\partial s} \tag{7.124}$$

where U_s is the s component of the current velocity. This equation indicates that the variation in the current velocity along the ray changes in wave number (or wavelength) and thus wave speed in the propagation.

The direction of the ray is given from Eq. (7.24) as

$$\frac{\partial \alpha}{\partial s} = -\left(c_g + U_s\right)^{-1} \frac{\partial U_s}{\partial n} \tag{7.125}$$

indicating that the wave angle changes where the current velocity is uneven in the normal to the ray.

Including the depth-induced refraction Eqs. (7.22) and (7.26), the wave propagation in shoaling water with currents is described by

$$\frac{\partial k_s}{\partial s} = -\left(c_g + U_s\right)^{-1}\left(\frac{\partial \omega}{\partial h}\frac{\partial h}{\partial s} + k\frac{\partial U_s}{\partial s}\right) \tag{7.126}$$

$$\frac{\partial \alpha}{\partial s} = -\left(c_g + U_s\right)^{-1}\left(\frac{1}{k}\frac{\partial \omega}{\partial h}\frac{\partial h}{\partial n} + \frac{\partial U_s}{\partial n}\right) \tag{7.127}$$

7.3.2 Wave height variation in a current field

As noted, the energy of waves propagates with the velocity $\boldsymbol{U} + \boldsymbol{c_g}$. The action balance equation, Eq. (7.115), is thus modified to be general form including the effect of current:

$$\frac{\partial}{\partial t}\left(\frac{\overline{E}}{\sigma}\right) + \boldsymbol{\nabla}\cdot\left(\left(\boldsymbol{U} + \boldsymbol{c_g}\right)\frac{\overline{E}}{\sigma}\right) = 0 \tag{7.128}$$

where $\overline{E}$ is the mean total energy (see Section 7.1.5). In the steady wave state, the energy along the ray in the current parallel to the ray can be expressed by

$$\frac{\partial}{\partial s}\left(\left(U + c_g\right)\frac{\overline{E}}{\sigma}\right) = 0 \tag{7.129}$$

We consider the simplest example that deep water waves in quiet water, with the phase velocity c_0 and wave height H_0, enter a region with the current along

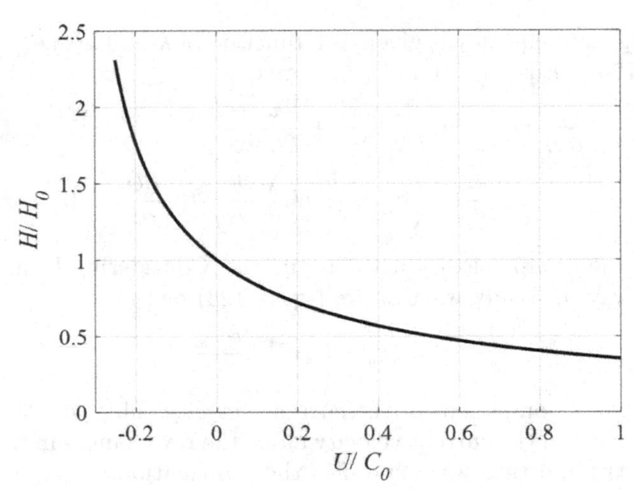

FIGURE 7.9
Amplification of deep water waves as a function of U/c_0.

the ray. Considering $c_g = c/2$ and $\sigma = g/c$ in deep water, the integration of Eq. (7.129) gives

$$H^2\left(U + \frac{c}{2}\right)c = \frac{1}{2}H_0^2 c_0^2 \tag{7.130}$$

The wave amplification due to current is thus given by

$$\frac{H}{H_0} = \frac{c_0}{(2U + c)\,c} = \left(\frac{c}{c_0}\left(\frac{c}{c_0} + 2\frac{U}{c_0}\right)\right)^{-1/2} \tag{7.131}$$

Fig. 7.9 shows the wave amplification H/H_0 as a function of U/c_0. We find the wave height increases in opposing current ($U < 0$) and decreases in following one ($U > 0$), as anticipated.

7.4 Irregular waves

Irregular wave field, observed in open ocean, consists of many wave components with different wave periods, wave heights, and wave directions (see Fig. 7.10). The irregular surface elevation may be expressed by the superposition of N waves coming from angle α:

$$\eta = \sum_{n=1}^{N} \frac{H_n}{2} \cos\left(k_n \cos\alpha x + k_n \sin\alpha y - \sigma_n t + \delta_n\right) \tag{7.132}$$

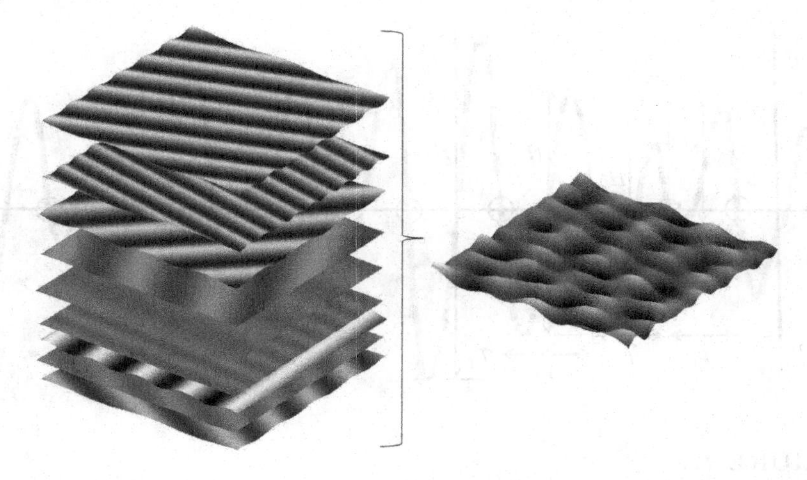

FIGURE 7.10
Multi-directional waves with wide-ranging wave numbers are superposed in real sea.

where δ is the phase lag. The resulting complex surface forms cannot be directly dealt with by harmonic wave theories studied in the previous chapters. There are two major approaches to quantify properties of irregular waves; statistical approach to define statistical parameters characterizing the wave records, and spectrum approach to decompose the irregular waves into linear harmonic wave components, which is introduced in this section.

7.4.1 Statistical representation of wave height

Consider a time record of surface elevation of irregular waves shown in Fig. 7.11. An individual wave on the record is defined by a so-called zero-up-crossing method. In this method, when each time that negative surface elevation intersects the mean level upward is marked on the record (see circles in Fig. 7.11), the individual waves are identified between the adjacent marks of the zero-up-crossings and numbered as $i = 1, 2, \ldots, N$. The wave heights and periods for every waves are measured from the record and indicated as H_i and T_i for the i-th wave. The N sets of the wave height and period are used to statistically define representative wave parameters.

The mean wave height and period are defined in the usual way by

$$\overline{H} = \frac{1}{N}\sum_{i=1}^{N} H_i \tag{7.133}$$

$$\overline{T} = \frac{1}{N}\sum_{i=1}^{N} T_i \tag{7.134}$$

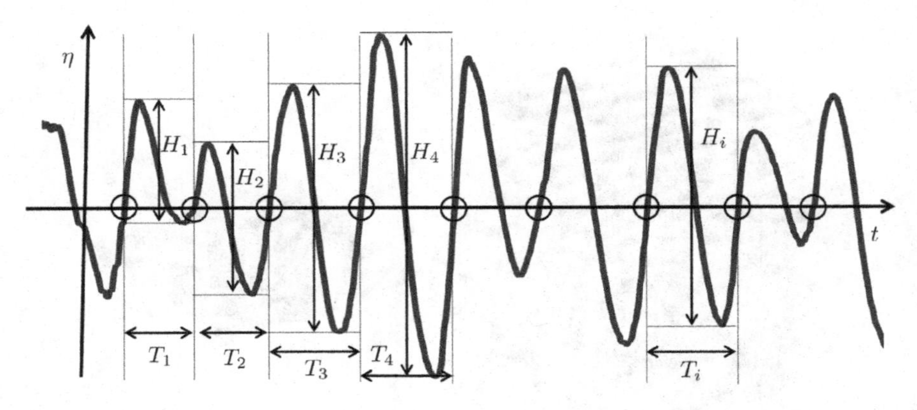

FIGURE 7.11
Numbered irregular waves identified by the zero-up-crossing method.

The root-mean-square wave height is given by

$$H_{rms} = \sqrt{\frac{1}{N} \int_{i=1}^{N} H_i^2} \tag{7.135}$$

For another important definition, termed a significant wave, all N waves are sorted by the wave height and rearranged in descending order. For example, if the largest wave height is H_4 in Fig. 7.11, the second largest is H_3, and the lowest is H_1, the orders of wave height $H(n)$ and period $T(n)$, $n = 1, \ldots, N$, are given by

$$\{H(1), H(2), \ldots, H(N)\} = \{H_4, H_3, \ldots, H_1\} \tag{7.136}$$

$$\{T(1), T(2), \ldots, T(N)\} = \{T_4, T_3, \ldots, T_1\} \tag{7.137}$$

The significant wave height, denoted as either H_s or $H_{1/3}$, is defined as the average of the highest $N/3$ waves:

$$H_{1/3} = \frac{1}{N/3} \sum_{n=1}^{N/3} H(n) \tag{7.138}$$

The significant wave period is also estimated by the identical definition:

$$T_{1/3} = \frac{1}{N/3} \sum_{n=1}^{N/3} T(n) \tag{7.139}$$

Similarly, the general forms defined by the average of the highest pN waves

($p \leq 1$) can be defined as

$$H_p = \frac{1}{pN} \sum_{n=1}^{pN} H(n) \tag{7.140}$$

$$T_p = \frac{1}{pN} \sum_{n=1}^{pN} T(n) \tag{7.141}$$

Statistical relations between these representative wave heights are given when a specific probability distribution of wave height is assumed. As an example, Rayleigh distribution gives the relation, $H_{1/10} = 1.80H_{rms} = 1.27H_{1/3} = 2.03\overline{H}$. Further analysis of wave statistics should be referred to another specialized textbook[117].

7.4.2 Wave spectrum

Assuming a time record of surface elevation, $\eta(t)$, during a period $\mathcal{T}$ is known, $\eta(t)$ can be expressed by Fourier series:

$$\eta(t) = \sum_{n=-N/2}^{N/2} \mathcal{F}(n)e^{in\omega t} \tag{7.142}$$

where $\omega = 2\pi\mathcal{T}/N$, N is the number of the data. The Fourier coefficient for the nth harmonics is given by

$$\mathcal{F}(n) = \frac{1}{\mathcal{T}} \int_t^{t+\mathcal{T}} \eta(t)e^{-in\omega t}dt \tag{7.143}$$

Considering Perseval's theorem

$$\overline{\eta^2} = \frac{1}{\mathcal{T}} \int_t^{t+\mathcal{T}} \eta^2(t)dt = \sum_{n=-N/2}^{N/2} |\mathcal{F}(n)|^2 \tag{7.144}$$

Eqs. (7.46) and (7.52) provide the relation between the mean total wave energy $\overline{E}$ and $\mathcal{F}$;

$$2\overline{E_p} = \overline{E} = \rho g\overline{\eta^2} \tag{7.145}$$

$$\therefore \frac{\overline{E}}{\rho g} = \sum_{n=-N/2}^{N/2} |\mathcal{F}(n)|^2 \tag{7.146}$$

Accordingly, the square of $\mathcal{F}(n)$ describes the nth harmonic component of the mean wave energy, which defines wave energy spectrum:

$$S(n) = |\mathcal{F}(n)|^2 \tag{7.147}$$

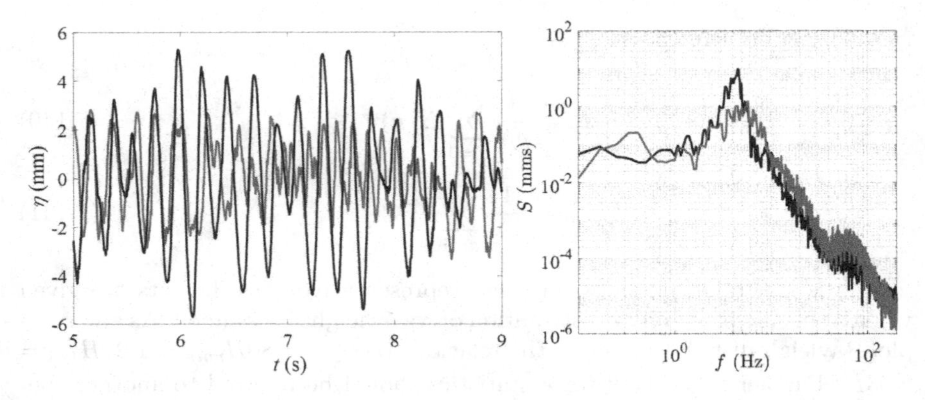

FIGURE 7.12
Time record of surface elevation (left) and its spectrum (right) of irregular waves at an early stage of generation by steady wind, observed in a wind wave experimental tank; fetch $X = 40$ cm (gray line) and 160 cm (black line). The corresponding wave forms are shown in Fig. 3.19.

In a general case of irregular waves coming from various directions, the wave energy spectrum with respect to a wave direction (α), termed directional spectrum, is introduced. Eq. (7.142) is then rewritten as

$$\eta = \sum_{n=-N/2}^{N/2} \mathcal{F}(n,\ \alpha) e^{in\omega t} d\alpha \tag{7.148}$$

The spectrum is thus redefined as $S(n, \alpha) = |\mathcal{F}(n, \alpha)|^2$. In general, the directional spectrum $S(n, \alpha)$ is modeled as the product of the frequency spectrum and the so-called spreading function $D(n, \alpha)$:

$$S(n, \alpha) = S(n)\, D(n, \alpha) \tag{7.149}$$

Specific spreading functions and their properties should be referred to Goda (1996)[116].

The spectrum form generally depends on the distance over water that the wind blows, termed fetch. Fig. 7.12 shows the typical surface elevations and their spectra at an early stage of wave generation by experimental wind. The initial wave forms are known to be very irregular as capillary and gravity waves with frequencies over a wide range, induced by wind, are superposed[33] (see the corresponding wave forms in Fig. 3.19). While overall growth of the waves owing to wind can be found from a comparison of the time records of η between fetch $X = 40$ cm (gray line) and 160 cm (black line), the spectrum can quantitatively describe the growth of wave energy at specific dominant frequencies depending on the fetch.

According to Mitsuyasu et al. (1974)[32], dimensionless wave height $\hat{H}$ and period $\hat{T}$ has the empirical relation with the dimensionless fetch $\hat{X} = gX/U_{10}^2$:

$$\hat{H} = 2.15 \times 10^{-3} \hat{X}^{0.504} \tag{7.150}$$

$$\hat{T} = 5.07 \times 10^{-2} \hat{X}^{0.330} \tag{7.151}$$

where $\hat{H} = gH_{1/3}/U_{10}^2$, $\hat{T} = gT_{1/3}/2\pi U_{10}$, U_{10} is the wind velocity at 10 m above the mean sea surface (generally used as a reference wind velocity). Similarly, Hasselmann et al. (1973)[42] proposed the fetch-dependent wave properties for $\hat{X} \leq 10^{-4}$:

$$\frac{g^2 \overline{E}}{U_{10}^4 (\rho g)} = 1.6 \times 10^{-7} \hat{X} \tag{7.152}$$

$$\frac{U_{10} f_p}{g} = 3.5 \hat{X}^{-0.33} \tag{7.153}$$

where $\overline{E}$ is the mean total wave energy and f_p is the frequency of the spectrum peak. Since $\overline{E} \sim \hat{H}^2/8$ (see Eq. (7.51)) and $f_p \approx 1.1 T_{1/3}$ (see Goda (1988)[26]), both sets of the relations, Eqs. (7.150)–(7.153), provide $\hat{H} \propto \hat{X}^{1/2}$ and $\hat{T} \propto \hat{X}^{1/3}$, which leads the relation $\hat{H}^2 \propto \hat{T}^3$. In fact, Toba (1972)[119] proposed the slope

$$H^* = 0.062 T^{*3/2} \tag{7.154}$$

where $H^* = gH_{1/3}/u_*^2$, $T^* = gT_{1/3}/u_*$, and u_* is the friction velocity, Eq. (2.30) (see Section 2.1.4). These statistical correlations indicate that the wave height and period cannot be independently defined in the equilibrium state; that is, they are uniquely determined by the wind parameters.

7.4.3 Standard wave spectra

There are three major standard forms of wave spectra, assuming the wind-wave equilibrium state (constant wind blowing for very long time), commonly used for simulating irregular waves; spectra of Pierson-Moskowitz, JONSWAP, and Bretschneider-Mitsuyasu. These model spectra have a common function form in the frequency f; $S(f) \propto f^{-5} \exp(-f^{-4})$, describing the exponential increase of S in lower frequencies in proportion to $\exp(-f^{-4})$, and the spectrum decay following f^{-5} in higher frequency range. The specific functions of these spectra are listed below.

7.4.3.1 Pierson-Moskowitz spectrum

Pierson and Moskowitz (1964)[82] proposed a simple form of wave spectrum for fully developed wind seas (see Fig. 7.13 left):

$$S(f) = \alpha g^2 (2\pi)^{-4} f^{-5} \exp\left(-\frac{5}{4}\left(\frac{f}{f_p}\right)^{-4}\right) \tag{7.155}$$

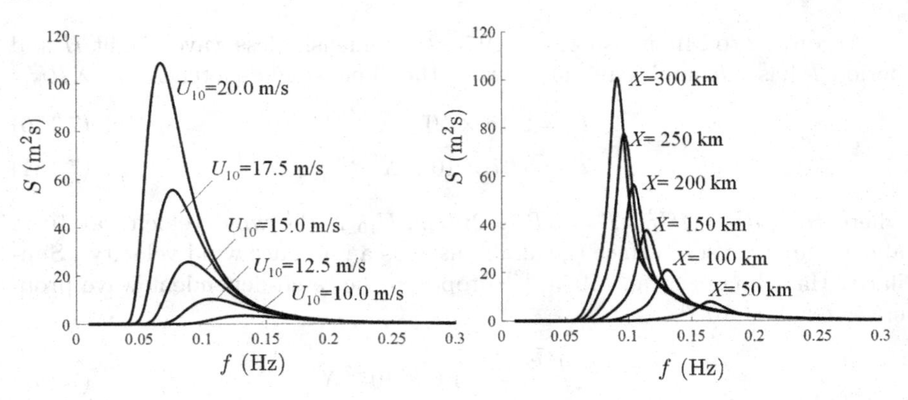

FIGURE 7.13
Typical forms of Pierson-Moskowitz spectrum for different wind velocities (left), and JONSWAP spectrum for different fetches (right).

where $\alpha = 8.1 \times 10^{-3}$, the peak frequency $f_p = 0.879 g\pi U_{19.5}/2\pi$, the wind velocity at an elevation of 19.5 m above the mean surface level $U_{19.5} \approx 1.026 U_{10}$.

7.4.3.2 JONSWAP (Joint North Sea Wave Observation Project) spectrum

Hasselmann et al. (1973) [42] considered effects of nonlinear wave–wave interactions, arising through long-distance propagation of nonlinear waves, to steepen the spectrum on the basis of data collected in the Joint North Sea Wave Observation Project. The JONSWAP spectrum takes the form of Pierson-Moskowitz spectrum multiplied by an peak enhancement factor γ^δ, resulting in a steep spectrum form (Fig. 7.13 right):

$$S(f) = \alpha g^2 (2\pi)^{-4} f^{-5} \exp\left(-\frac{5}{4}\left(\frac{f}{f_p}\right)^{-4}\right)\gamma^\delta \tag{7.156}$$

where $\gamma = 3.3$, $\alpha = 7.6 \times 10^{-2} \hat{X}^{-0.22}$, $f_p = 3.5 F^{-0.33} g/U_{10}$, dimensionless fetch $\hat{X} = gX/U_{10}^2$ (X; fetch), $\delta = \exp\left(-\frac{1}{2C^2}\left(f/f_p - 1\right)^2\right)$, f_p is the peak frequency of the spectrum. C is given by

$$C = \begin{cases} 0.07 & (f \leq f_p) \\ 0.09 & (f > f_p) \end{cases} \tag{7.157}$$

7.4.3.3 Bretschneider-Mitsuyasu spectrum

Bretschneider (1968) [10] proposed the spectrum for the fully developed wind waves in terms of the significant wave height and period. Mitsuyasu (1970) [67] adjusted the coefficients of this spectrum, known as Bretschneider-Mitsuyasu spectrum:

$$S(f) = C_1 H_{1/3} T_{1/3}^{-4} f^{-5} \exp\left(-C_2 \left(T_{1/3} f\right)^{-4}\right) \tag{7.158}$$

where $C_1 = 0.257$ and $C_2 = 1.03$. Goda (1988)[26] reexamined the relation $T_{1/3}$ and f_p and modified the coefficients $C_1 = 0.205$ and $C_2 = 0.75$

7.4.4 Prediction of wave spectrum

While the action balance equation, Eqs. (7.116) and (7.128), describes the conservation of wave action spectrum $N(\sigma, \alpha) = S(\sigma, \alpha)/\sigma$, the effects of energy transfer from wind to grow waves and energy dissipation to wave decay as an extra source term St to the equation for predicting the spectrum N:

$$\frac{\partial N}{\partial t} + \boldsymbol{\nabla} \cdot \left(\left(\boldsymbol{c_g} + \boldsymbol{U}\right) N\right) = \frac{S_t}{\sigma} \tag{7.159}$$

A variety of wave prediction models, based on Eq. (7.159), has been provided and available to compute wave spectrum. The most models employ the total source S_t as the sum of the contributions of wave growth by the wind (S_w), nonlinear transfer of wave energy (S_{nl}), wave decay due to white-capping ($S_{ds,w}$), bottom friction ($S_{ds,b}$), and depth-induced wave breaking ($S_{ds,br}$); $S_t = S_w + S_{nl} + S_{ds,w} + S_{ds,b} + S_{ds,br}$. The specific source models should be referred to manuals of the public models, such as WAM, SWAN, and WAVE-WATCH.

7.5 Stokes wave

Wave properties defined by a linear wave theory, introduced in Chapter 4, 6, and extended in previous sections of this chapter, describe most wave states observed in ocean and coastal areas. On the one hand, when wave amplitude is high enough to violate the small amplitude assumption, nonlinear effects need to be considered. In this section, the second-order perturbation method is introduced as an extension of linear theory to weakly nonlinear theory, to derive the well-known Stokes wave solution.

7.5.1 Perturbation method

The perturbation method has been introduced to linearize boundary conditions at free surface and to derive the first-order solution of the Laplace equation in Chapter 4 and 6. In this section, the second-order perturbations are considered to obtain the wave solution with weal nonlinearity.

We assume a two-dimensional irrotational flow in the constant depth, h, on (x, z) domain where the origin of vertical z-axis is defined at a still water level, as shown in Fig. 7.1. The surface elevation η and velocity potential ϕ are expanded into a power series in small quantity ϵ:

$$\eta(x,t) = \epsilon\eta_1(x,t) + \epsilon^2\eta_2(x,t) + O(\epsilon^3) \tag{7.160}$$

$$\phi(x,z,t) = \epsilon\phi_1(x,z,t) + \epsilon^2\phi_2(x,z,t) + O(\epsilon^3) \tag{7.161}$$

The substitution of Eq. (7.161) into the Laplace equation Eq. (4.1) gives

$$\frac{\partial^2\phi}{\partial x^2} + \frac{\partial^2\phi}{\partial z^2} = \epsilon\left(\frac{\partial^2\phi_1}{\partial x^2} + \frac{\partial^2\phi_1}{\partial z^2}\right) + \epsilon^2\left(\frac{\partial^2\phi_2}{\partial x^2} + \frac{\partial^2\phi_2}{\partial z^2}\right) + O(\epsilon^3) = 0 \tag{7.162}$$

The perturbed form of the impermeable bottom condition, Eq. (3.38), is given

$$\frac{\partial\phi}{\partial z} = \epsilon\frac{\partial\phi_1}{\partial z} + \epsilon^2\frac{\partial\phi_2}{\partial z} + O(\epsilon^3) = 0 \quad (z = -h) \tag{7.163}$$

The substitutions of Eqs. (7.160) and (7.161) into the kinematic boundary condition at free surface ($z = \eta$), Eq. (3.41), give

$$\begin{aligned}
&\frac{\partial\phi}{\partial z} - \frac{\partial\eta}{\partial t} - \frac{\partial\phi}{\partial x}\frac{\partial\eta}{\partial x} \\
&= \left(\epsilon\frac{\partial\phi_1}{\partial z} + \epsilon^2\frac{\partial\phi_2}{\partial z} + O(\epsilon^3)\right) - \left(\epsilon\frac{\partial\eta_1}{\partial t} + \epsilon^2\frac{\partial\eta_2}{\partial t} + O(\epsilon^3)\right) \\
&- \left(\epsilon\frac{\partial\phi_1}{\partial x} + \epsilon^2\frac{\partial\phi_2}{\partial x} + O(\epsilon^3)\right)\left(\epsilon\frac{\partial\eta_1}{\partial x} + \epsilon^2\frac{\partial\eta_2}{\partial x} + O(\epsilon^3)\right) = 0 \quad (z = \eta)
\end{aligned} \tag{7.164}$$

This equation is rearranged in order of ϵ as

$$\epsilon\left(\frac{\partial\phi_1}{\partial z} - \frac{\partial\eta_1}{\partial t}\right) + \epsilon^2\left(\frac{\partial\phi_2}{\partial z} - \frac{\partial\eta_2}{\partial t} - \frac{\partial\phi_1}{\partial x}\frac{\partial\eta_1}{\partial x}\right) + O(\epsilon^3) = 0 \quad (z = \eta) \tag{7.165}$$

Similarly, the perturbed form of the dynamic boundary condition at $z = \eta$, Eq. (3.57), is given

$$\begin{aligned}
&gz + \epsilon\frac{\partial\phi_1}{\partial t} + \epsilon^2\frac{\partial\phi_2}{\partial t} + \frac{\epsilon^2}{2}\left(\left(\frac{\partial\phi_1}{\partial x}\right)^2 + \left(\frac{\partial\phi_1}{\partial z}\right)^2\right) + O(\epsilon^3) \\
&= \epsilon C_1 + \epsilon^2 C_2 + O(\epsilon^3) \quad (z = \eta)
\end{aligned} \tag{7.166}$$

The quantities at the surface location $z = \eta$ are expanded in the Taylor series about $z = 0$ (see Section 4.2):

$$\begin{aligned}
\phi_1\big|_{z=\eta} &= \phi_1\big|_{z=0} + \left(\epsilon\eta_1 + \epsilon^2\eta_2 + O(\epsilon^3)\right)\left.\frac{\partial\phi_1}{\partial z}\right|_{z=0} \\
&+ \frac{\left(\epsilon\eta_1 + \epsilon^2\eta_2 + O(\epsilon^3)\right)^2}{2}\left.\frac{\partial^2\phi_1}{\partial z^2}\right|_{z=0} + O(\epsilon^3)
\end{aligned} \tag{7.167}$$

$$\phi_2|_{z=\eta} = \phi_2|_{z=0} + \left(\epsilon\eta_1 + \epsilon^2\eta_2 + O(\epsilon^3)\right) \left.\frac{\partial \phi_2}{\partial z}\right|_{z=0}$$
$$+ \frac{\left(\epsilon\eta_1 + \epsilon^2\eta_2 + O(\epsilon^3)\right)^2}{2} \left.\frac{\partial^2 \phi_2}{\partial z^2}\right|_{z=0} + O(\epsilon^3) \tag{7.168}$$
$$gz\,|_{z=\eta} = gz\,|_{z=0} + \left(\epsilon\eta_1 + \epsilon^2\eta_2 + O(\epsilon^3)\right) g \tag{7.169}$$

Substituting them into Eqs. (7.165) and (7.166) and neglecting the perturbations of the third order or higher, the boundary conditions approximated at $z = 0$ are given as

$$\epsilon\left(\frac{\partial \phi_1}{\partial z} - \frac{\partial \eta_1}{\partial t}\right) + \epsilon^2\left(\frac{\partial \phi_2}{\partial z} - \frac{\partial \eta_2}{\partial t} + \eta_1\frac{\partial^2 \phi_1}{\partial z^2} - \frac{\partial \phi_1}{\partial x}\frac{\partial \eta_1}{\partial x}\right) = 0 \quad (z = 0) \tag{7.170}$$

$$\epsilon\left(\frac{\partial \phi_1}{\partial t} + g\eta_1\right) + \epsilon^2\left(\frac{\partial \phi_2}{\partial t} + g\eta_2 + \eta_1\frac{\partial^2 \phi_1}{\partial t \partial z} + \frac{1}{2}\left(\left(\frac{\partial \phi_1}{\partial x}\right)^2 + \left(\frac{\partial \phi_1}{\partial z}\right)^2\right)\right)$$
$$= \epsilon C_1 + \epsilon^2 C_2 \quad (z = 0) \tag{7.171}$$

Each order of the terms in Eqs. (7.162), (7.163), (7.170), (7.171), and (7.163) is independently dealt with to identically satisfy the equation. The system of equations of the first-order $O(\epsilon)$ is thus given

$$\frac{\partial^2 \phi_1}{\partial x^2} + \frac{\partial^2 \phi_1}{\partial z^2} = 0 \quad (-h < z < 0) \tag{7.172}$$
$$\frac{\partial \phi_1}{\partial z} = \frac{\partial \eta_1}{\partial t} \quad (z = 0) \tag{7.173}$$
$$\frac{\partial \phi_1}{\partial t} + g\eta_1 = C_1 \quad (z = 0) \tag{7.174}$$
$$\frac{\partial \phi_1}{\partial z} = 0 \quad (z = -h) \tag{7.175}$$

The equation system of the second-order $O(\epsilon)$:

$$\frac{\partial^2 \phi_2}{\partial x^2} + \frac{\partial^2 \phi_2}{\partial z^2} = 0 \quad (-h < z < 0) \tag{7.176}$$
$$\frac{\partial \phi_2}{\partial z} - \frac{\partial \eta_2}{\partial t} + \eta_1\frac{\partial^2 \phi_1}{\partial z^2} - \frac{\partial \phi_1}{\partial x}\frac{\partial \eta_1}{\partial x} \quad (z = 0) \tag{7.177}$$
$$\frac{\partial \phi_2}{\partial t} + g\eta_2 + \eta_1\frac{\partial^2 \phi_1}{\partial t \partial z} + \frac{1}{2}\left(\left(\frac{\partial \phi_1}{\partial x}\right)^2 + \left(\frac{\partial \phi_1}{\partial z}\right)^2\right) = C_2 \quad (z = 0) \tag{7.178}$$
$$\frac{\partial \phi_2}{\partial z} = 0 \quad (z = -h) \tag{7.179}$$

We find the first-order system of equations Eqs. (7.172)–(7.175) is identical with the one of the linearized equations, Eqs. (4.19)–(4.23), introduced in

Section 4.2. Accordingly, they give the progressive linear wave solution, Eq. (4.54):

$$\phi_1 = -i\frac{H}{2}\frac{\sigma_1}{k_1}\frac{\cosh k_1 (h+z)}{\sinh k_1 h} e^{i(k_1 x - \sigma_1 t)} \tag{7.180}$$

$$\eta_1 = \frac{H}{2} e^{i(k_1 x - \sigma_1 t)} \tag{7.181}$$

where $\sigma_1^2 = gk_1 \tanh k_1 h$, k_1 and σ_1 indicate the wave number and the angular frequency of the first-order wave, respectively.

7.5.2 Second-order solutions

We next consider the progressive wave form of the solution for Eq. (7.176) (see Section 4.3 for deriving a general solution of the Laplace equation):

$$\phi_2 = \left(C_1 e^{k_2 z} + C_2 e^{-k_2 z}\right) i e^{i(k_2 x - \sigma_2 t)} \tag{7.182}$$

where k_2 and σ_2 are the second-order wave number and frequency, respectively. As the bottom condition, Eq. (7.179), determines $C_1 = C_2 e^{2k_2 h}$, Eq. (7.182) is rewritten as

$$\phi_2 = C_1' \cosh k_2 (h+z) \, i e^{i(k_2 x - \sigma_2 t)} \tag{7.183}$$

The substitution of the first-order solution Eqs. (7.180) and (7.181) into Eq. (7.178) gives

$$\frac{\partial \phi_2}{\partial t} + g\eta_2 - \left(\frac{H}{2}\right)^2 \sigma_1^2 \left(1 - \frac{1}{2\sinh^2 k_1 h}\right) e^{2i(k_1 x - \sigma_1 t)} = C_2 \tag{7.184}$$

(see transformations of hyperbolic functions Section A.2 in Appendix). We find that ϕ_2 of Eq. (7.183) must have the harmonic oscillation form consistent with $e^{2i(k_1 x - \sigma_1 t)}$ to satisfy Eq. (7.184); that is, $k_2 = 2k_1$ and $\sigma_2 = 2\sigma_1$. Eq. (7.182) thus takes the form:

$$\phi_2 = C_1' \cosh 2k_1 (h+z) \, i e^{2i(k_1 x - \sigma_1 t)} \tag{7.185}$$

Substituting into Eq. (7.184) and assuming the mean surface level is zero; $\overline{\eta_2} = 0$,

$$\eta_2 = \frac{1}{g}\left(-2C_1'\sigma_1 \cosh 2k_1 h + \frac{H^2}{4}\sigma_1^2 \left(1 - \frac{1}{2\sinh^2 k_1 h}\right)\right) e^{2i(k_1 x - \sigma_1 t)} \tag{7.186}$$

The substitution of the first order ϕ_1, η_1, the second-order ϕ_2, Eq. (7.185), and η_2, Eq. (7.186), into the kinematic boundary condition Eq. (7.177) gives

$$k_1 C_1' \sinh 2k_1 h - \frac{\sigma_1}{g}\left(2C_1'\sigma_1 \cosh 2k_1 h - \frac{H^2}{4}\sigma_1^2\left(1 - \frac{1}{2\sinh^2 k_1 h}\right)\right)$$

$$- \frac{H^2}{4}\sigma_1 k_1 \tanh k_1 h = 0$$

$$\therefore C_1' = -\frac{3}{32}\frac{H^2 \sigma_1}{\sinh^4 k_1 h} \tag{7.187}$$

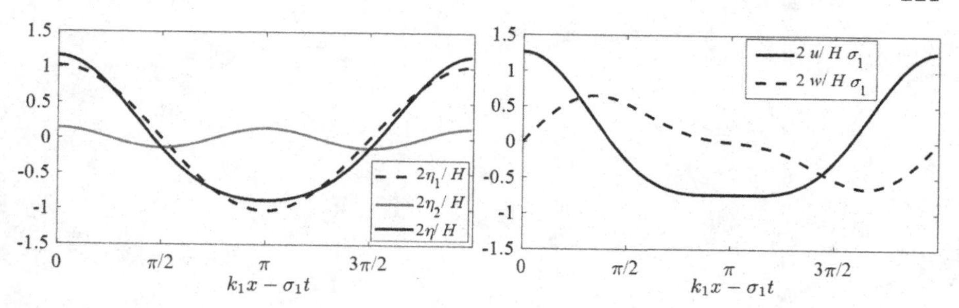

FIGURE 7.14
Surface elevation (left) and velocity (right) of the second-order Stokes wave at $z = 0$ ($T = 8$ s, $h = 5$ m).

The second-order solutions of the velocity potential and surface elevation are finally determined:

$$\phi_2 = -\frac{3}{32}H^2\sigma_1\frac{\cosh 2k_1(h+z)}{\sinh^4 k_1h}ie^{2i(k_1x-\sigma_1t)} \tag{7.188}$$

$$\eta_2 = \frac{H^2}{16}k_1\frac{\cosh 2k_1h+2}{\sinh^2 k_1h\tanh k_1h}e^{2i(k_1x-\sigma_1t)} \tag{7.189}$$

The trigonometric form of the solution is given by

$$Re[\phi_2] = \frac{3}{32}H^2\sigma_1\frac{\cosh 2k_1(h+z)}{\sinh^4 k_1h}\sin 2(k_1x-\sigma_1t) \tag{7.190}$$

$$Re[\eta_2] = \frac{H^2}{16}k_1\frac{\cosh 2k_1h+2}{\sinh^2 k_1h\tanh k_1h}\cos 2(k_1x-\sigma_1t) \tag{7.191}$$

The second-order Stokes waves are then described by

$$\begin{aligned}\phi = &\frac{H}{2}\frac{\sigma_1}{k_1}\frac{\cosh k_1(h+z)}{\sinh k_1h}\sin(k_1x-\sigma_1t)\\ &+\frac{3}{32}H^2\sigma_1\frac{\cosh 2k_1(h+z)}{\sinh^4 k_1h}\sin 2(k_1x-\sigma_1t)\end{aligned} \tag{7.192}$$

$$\eta = \frac{H}{2}\cos(k_1x-\sigma_1t)+\frac{H^2}{16}k_1\frac{\cosh 2k_1h+2}{\sinh^2 k_1h\tanh k_1h}\cos 2(k_1x-\sigma_1t) \tag{7.193}$$

Fig. 7.14 illustrates the Stokes wave form and velocity at $z = 0$. Because of the second harmonic component, the Stokes wave has the steeper wave crest, with higher horizontal velocity, and flatter trough than those of the linear waves.

8

Breaking Wave Dynamics

Ocean wave motion, well approximated by irrotational and inviscid flow, rapidly changes into rotational turbulent flow when the waves break. Formations of the organized vortices, generated under breaking waves, cause various mechanical processes; local deformation and mixing of sea surfaces, entrapment of air bubbles, sediment suspension, which supports biological activities in a marine ecosystem through the gas and heat transport and the material convection. The entrained air bubbles contribute to disturb the flow, dissipate wave energy, and generate sea spray during degassing process. The mechanisms to induce the variety of breaking processes and the consequences are considered in this chapter. Research progress about air–sea interactions relevant to momentum and gas transfers through marine boundary layers is also introduced at the end of the chapter.

8.1 Shoaling wave breaking

When ocean waves generated in deep water arrives a shoaling water, the wave height (H) increases and the wave speed (c) decreases during the shoaling process, as noted in Section 7.1.8 (see Fig. 7.6). As the fluid velocity (u) is in proportion to the wave height, Eq. (7.1), the relative fluid velocity u/c in shoaling waves rapidly increases in propagation toward shore. When u/c exceeds the unity (fluid motion becomes faster than the wave), the flow at the wave crest precedes the wave and projects to splash on the forward water surface to initiate breaking processes. In this section, fundamental properties of breaking types and breaker limits, depending on depth and bed slope, are introduced.

8.1.1 Breaker type

Typical behaviors of broken waves in the surf zone have been visually characterized as four breaker types; spilling, plunging, collapsing, and surging breakers (see Fig. 8.1). Major two types, spilling and plunging breakers, are caused by the different mechanisms of wave breaking; the wave steepness (H/L) exceeds the highest limit, such as Eq. (8.2), for the former, and the crest velocity

DOI: 10.1201/9781003140160-8

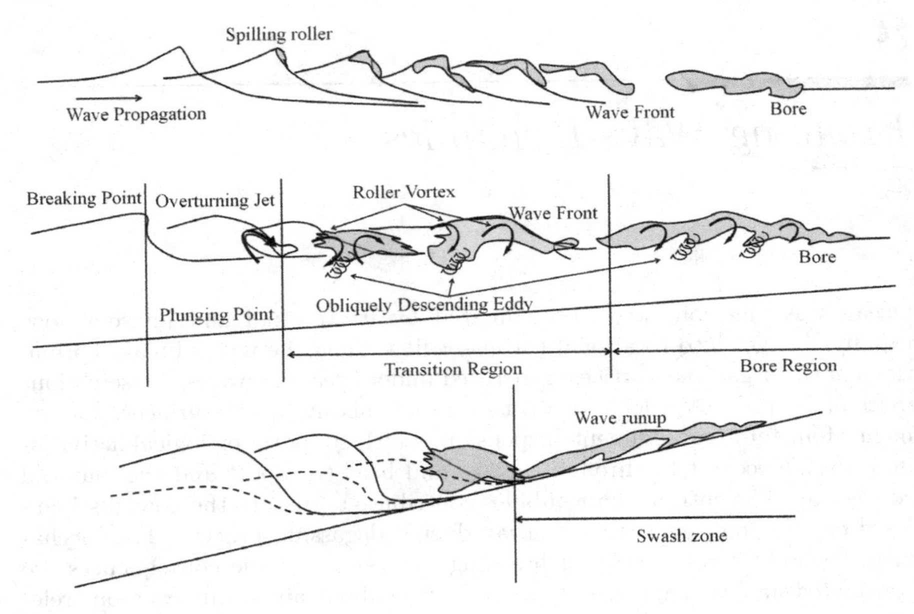

FIGURE 8.1
Types of breakers; spilling breaker (top), plunging breaker (middle), collapsing (solid wave form), and surging (broken wave form) breakers (bottom).

precedes the wave speed for the latter. The specific features of the breaking processes at these four types are given below.

Spilling breaker
The spilling breaker occurs when the wave crest of steep symmetric wave form becomes unstable and spills down on the forward wave slope, resulting in a formation of a roller vortex at the front face of the crest (see Fig. 8.1 top). This breaker occurs in deeper depth and on milder bottom slope than the other types.

Plunging breaker
In the rapid shoaling on steep bottom slope, the wave form becomes unsymmetric with sheer wave face and mild rear wave slope. The crest overturns to plunge onto the forward water surface (see Fig. 8.1 middle). The rotational flow in the jet around an air tube (barrel) creates a primary roller vortex, while the secondary jets are repeatedly ejected forward to splash, resulting in an array of roller vortices in the transition region. The splashing waves culminate in turbulent bores propagating toward the shore in a bore region until the energy is sufficiently dissipated.

Collapsing breaker
While the amplified wave crest does not overturn, the bottom face steepens and then collapses near a shore, resulting in turbulent water spreading on a swash. This type is observed when long period waves arrive on a steep shore.

Surging breaker

A toe of asymmetric smooth surface of long period swells, surging on a shore, advances up a swash zone with minor breaking.

8.1.1.1 Surf similarity parameter

Breaker type is dependent on the bottom slope $m = \tan\theta$ and the deep-water steepness H_0/L_0. Battjes (1974)[9] proposed a surf similarity parameter to quantify features of the breakers:

$$\xi_0 = \frac{m}{\sqrt{H_0/L_0}} \tag{8.1}$$

where H_0 and L_0 are the offshore wave height and wavelength, respectively. H_0/L_0 are termed offshore wave steepness. When $\xi_0 \leq 0.5$, the spilling breaker can occur, while the plunging breaker may be observed in the range of $0.5 < \xi_0 \leq 3.3$. The surging or collapsing breakers are expected in $\xi_0 > 3.3$. More specific identification of the breaker type by ξ_0 is introduced in the next section (Fig. 8.3).

8.1.2 Breaker limits

Since wave breaking is caused by strong nonlinear unsteady fluid motion, it is difficult to theoretically describe and generalize the breaker limits for multiple mechanisms of breaking onsets noted in Section 8.1.1. As a reference of the spilling breaker onset, the analytical limit of steepness of a deep water wave[63] has been considered:

$$\left(\frac{H_0}{L_0}\right)_b = 0.142 \tag{8.2}$$

Miche (1944)[62] extends Eq. (8.2) to a shallower regime:

$$\frac{H_b}{L_b} = 0.142 \tanh k_b h_b \tag{8.3}$$

where the subscript b indicates the quantity at the breaking location. McCowan (1894)[59] estimated the highest limit of a solitary wave in finite water depth:

$$\frac{H_b}{h_b} = 0.78 \tag{8.4}$$

Here the relative breaking wave height H_b/h_b is termed the breaker index commonly indicated by γ_b.

Numbers of experimental investigations have been performed to identify the parameters generalizing γ_b. Goda (1974)[27] compiled the previous experimental results to estimate an empirical breaker index:

$$\frac{H_b}{h_b} = 0.17\frac{L_0}{h_b}\left(1 - \exp\left[-1.5\frac{\pi h}{L_0}\left(1 + 15\tanh^{4/3}\theta\right)\right]\right) \tag{8.5}$$

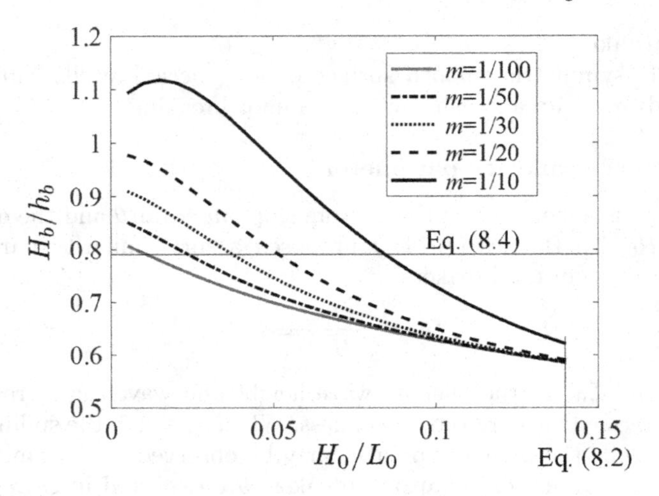

FIGURE 8.2
Breaker index H_b/h_b as a function of the offshore wave steepness H_0/L_0, proposed by Camenen and Larson (2007)[13].

Camenen and Larson (2007)[13] also reexamined large amount of published experimental data and established new formula of the optimal breaker index:

$$\frac{H_b}{h_b} = 0.284 \left(\frac{H_0}{L_0}\right)^{-1/2} \tanh\left(f^* \pi \left(\frac{H_0}{L_0}\right)^{1/2}\right) \tag{8.6}$$

$$f^* = 0.87 + \left(0.32 + 14\frac{H_0}{L_0}\right) \sin\left[\frac{\pi}{2}\left(\frac{m}{m_{max}}\right)^n\right] \tag{8.7}$$

where $m = \tan\theta$ is bottom slope, $m_{max} = 0.10 + 1.6H_0/L_0$. The exponent $n = 1 + 14H_0/L_0$ when $m \leq m_{max}$ or $n = -(1 + 20H_0/L_0)$ when $m > m_{max}$. Fig. 8.2 illustrates the breaker index proposed by Camenen and Larson (2007)[13].

Camenen and Larson (2007)[13] related the breaker types with the surf similarity parameter ξ_0, Eq. (8.1), and H_b/h_b given by Eq. (8.6), which provides the quantitative definition of the boundaries of breaker regimes:

$$\begin{array}{lll} \text{spilling/plunging} & \xi_0 = 1.5 - 1.25H_b/h_b & (0.26 < \xi_0 < 0.6) \quad (8.8) \\ \text{plunging/collapsing} & \xi_0 = 4.4 - 2.5H_b/h_b & (0.8 < \xi_0 < 2.9) \quad (8.9) \\ \text{collapsing/surging} & \xi_0 = 1.7 + 3.3H_b/h_b & (3.6 < \xi_0 < 6) \quad (8.10) \end{array}$$

Fig. 8.3 shows the breaker types defined on a $\xi_0 - \gamma_b$ plane. The lowest limit of γ_b can be estimated by Eq. (8.6) with the steepness limit $H_0/L_0 = 0.142$ by Eq. (8.2).

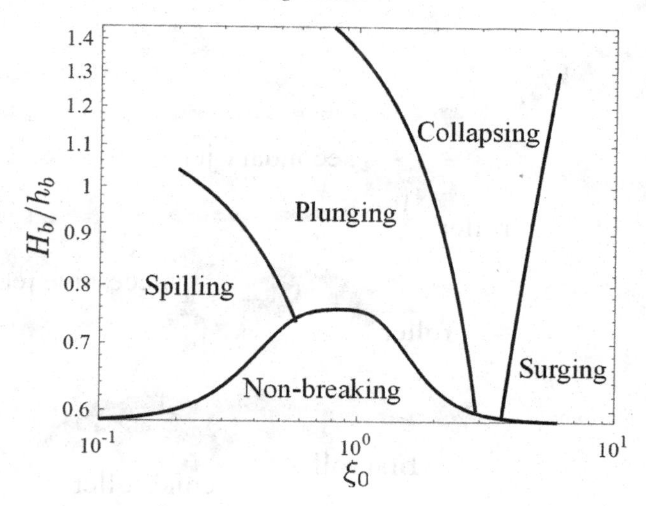

FIGURE 8.3
Breaker types separated on a $\xi_0 - H_b/h_b$ plane.

8.2 Turbulence and vortices in breaking waves

Ocean waves are well described by horizontal-vertical two-dimensional potential flow along the propagation. Wave breaking changes the flow state to three-dimensional rotational fluid motion. Vortices and turbulence, produced by wave splashes, are developed during the wave propagation in the surf zone. While air bubbles are entrained in the waves, sediments on the surf zone beach are lifted and suspended by the effects of turbulence, resulting in a formation of complex air- and sediment-laden turbulent flows. The dynamics of fluid motion governed by the organized vortex structures in breaking waves is introduced in this section.

8.2.1 Vortex structures

As noted in Section 8.1.1, in the plunging breaker, a shoaling wave crest overturns and splashes onto the forward wave surface. The rotational fluid motion within the jet surrounding the air tube evolves into a primary roller vortex at an early stage of wave breaking (see Fig. 8.4). Secondary jets are sequentially ejected to produce another roller vortex in the repeating splashing process, resulting in an array of horizontal roller vortices in the transition region. The energetic splashing waves then culminate in turbulent bores propagating toward the shore in a bore region (see Fig. 8.1).

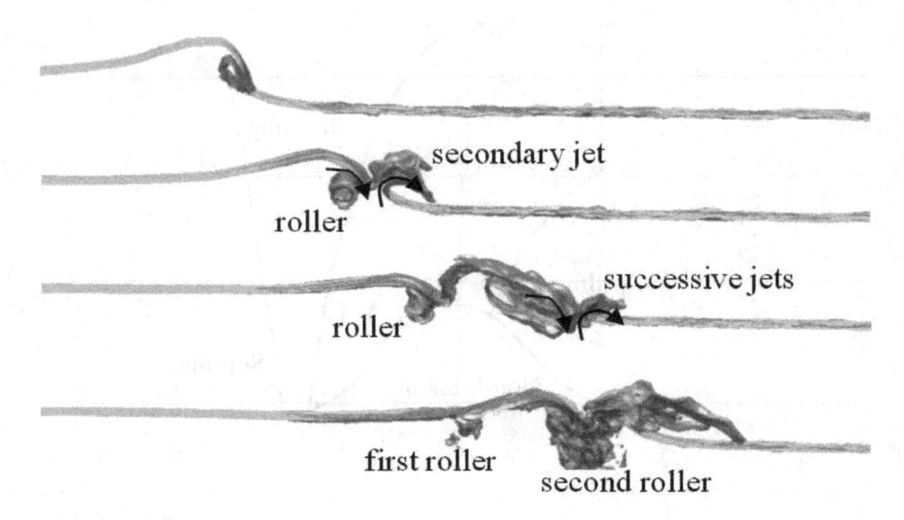

FIGURE 8.4
Computed surface forms of the plunging wave in a splash-up process.

Nadaoka et al.[73)] observed the formation of three-dimensional vortex structures involving obliquely descending eddies (ODEs) that develop along the oblique principal axis of fluid strain, behind the rollers (see Fig. 8.17). Watanabe et al.[123)] identified the mechanism that forms the ODEs during the wave-breaking process. During the first splash, the rotational flow in the secondary jet occurs in front of the roller vortex, resulted from the rotational overturning flow; that is, two roller vortices with the identical rotation adjacently aligns (see Fig. 8.5 left), indicating the flow is governed by corotating vortex pair introduced in Section 3.4.3.2. The streamlines around the corotating vortices intersect at a stagnation point between the pair (see

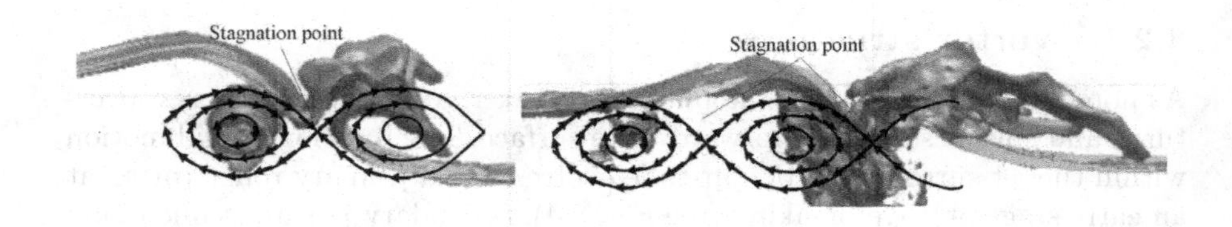

FIGURE 8.5
Streamlines and stagnation point at the first wave plunging (left) and the second one (right).

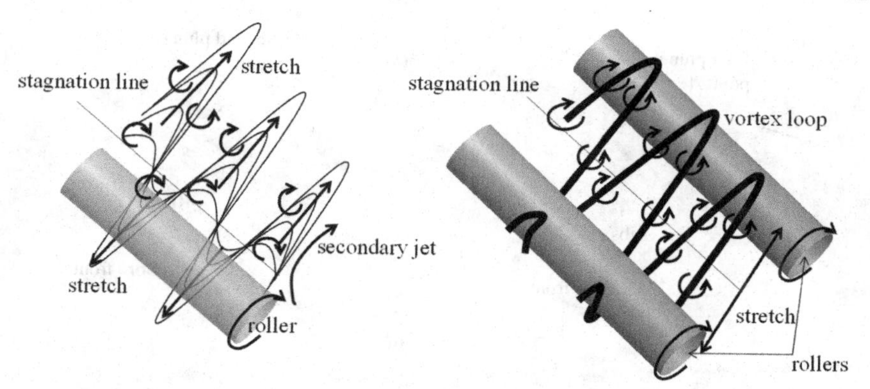

FIGURE 8.6
Schematic illustrations of the evolution of stretched vortex filament initially located at the stagnation line at the first plunging (left) and the vortex loop evolving into the rib vortices during the oblique stretch between the top of the forward roller to the bottom of the rear roller.

Fig. 3.15 and also Fig. 8.5 left). Fluid at the stagnation point is highly stretched in the oblique direction of the principal axis of the fluid strain (see oppositely oriented arrows at the stagnation point, indicating fluid stretch, in Fig. 8.5 left).

Consider a straight vortex filament, extending parallel to the primary roller, located near the stagnation point (see Fig. 8.6 left). If small disturbance of the orientation of the filament exists, as shown in Fig. 8.6 (left), the portions of the filament offshore side of the stagnation point are stretched in obliquely downward toward the bottom of the rear roller, while the shoreward portion of the filament is stretched obliquely upward along the axis of the secondary jet. As the amplitude of the deformation of the filament increases with the stretch, the initial straight filament evolves into wavy form, termed a vortex loop (Fig. 8.6 left). Since the filament has axial rotation, the stretched filament at the both sides of the bend has the opposite vorticity, that is, a pair of counter-rotating vortices (see Section 3.4.3.1) in the direction of stretch are produced on the oppositely oriented axis of the vortex loop. When the secondary jet splashes to produce new roller vortex (see Fig. 8.5 right), the loop is longitudinally stretched on the shear layer extending from the top of the preceding roller to the bottom of the one behind it, resulting in a formation of a rib structure, while the bends of the loop are wrapped by the both rollers (see Fig. 8.6 right).

Fig. 8.7 shows the evolution of the vortex structure during the splash cycle. We find the vortex loops evolving into the rib vortices are produced at successive splashes (*a* and *b*). The ODEs are visualized, as in Fig. 8.17, when air bubbles are entrapped within these rib vortices obliquely stretched

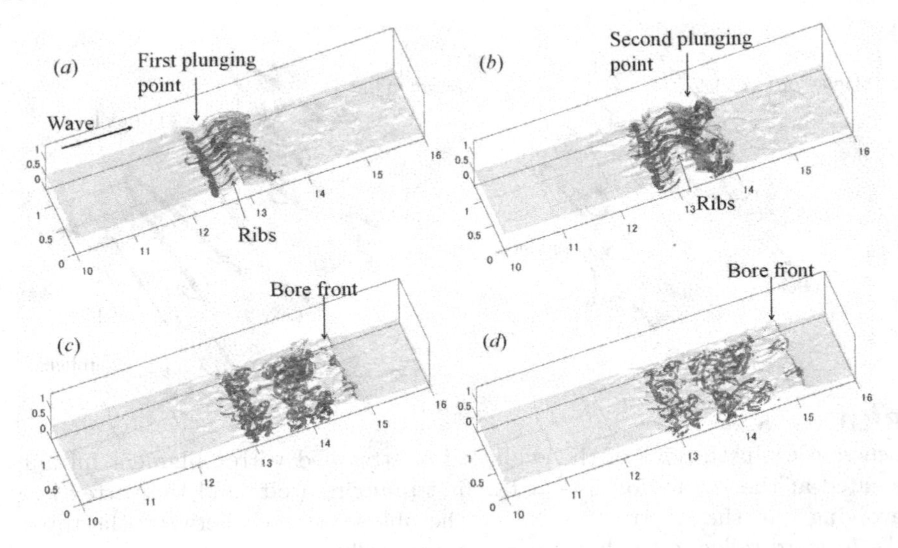

FIGURE 8.7
Computed vortex cores, indicating the organized vortex structure, during the breaking process.

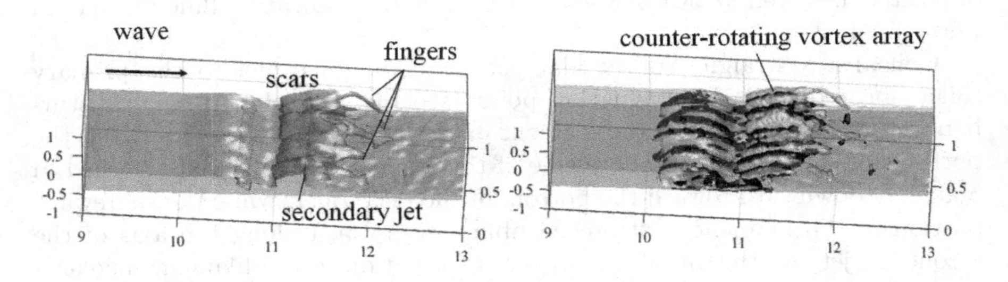

FIGURE 8.8
Computed free surface (left) and iso-surface of streamwise vorticity (right); positive and negative vorticity are indicated by white and black, respectively.

between the rollers[123]. When the vortex stretch is reduced by attenuating rollers, the rib vortices interact with the adjacent ones and twist each other until they are fully dissipated (c and d).

Fig. 8.8 shows the computed streamwise component of the vorticity (rotation with the axis in the direction of propagation) beneath the breaking wave surface. We find multiple pairs of the counter-rotating vortices are aligned

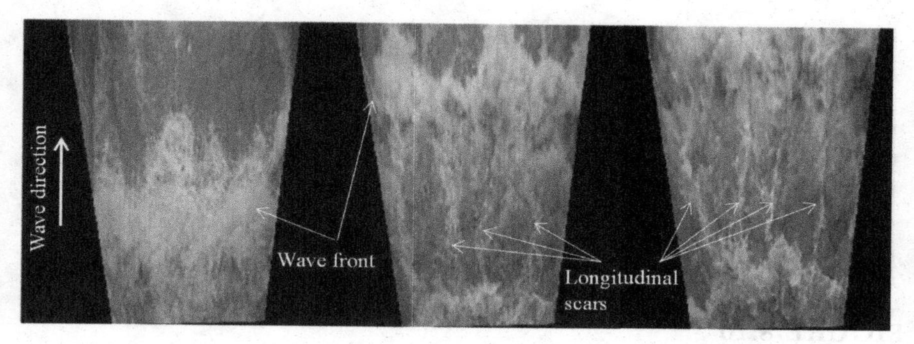

FIGURE 8.9
Sequence of photographs of experimental breaking wave taken from top.

to the direction of wave crest, which resulted from a lateral array of the rib vortices.

8.2.2 Surface–vortex interactions

As introduced in Section 3.5.2, when counter-rotating vortices placed beneath a free surface, the surface is entrained into the inner water mass and thus depressed, forming scars (see Fig. 3.20). Watanabe et al. (2005)[123)] found that the longitudinal extensions of scars are formed on the rear surface of breaking waves and they locate above the rib vortices with counter-rotating vorticity. Accordingly, the counter-rotating streamwise vortex pairs produced at the wave plunging (Fig. 8.8 right), entrain the surface into the bulk water and create the scars between the pairs laterally aligned to the direction of wave crest (see Fig. 8.8 left).

Fig. 8.9 shows the sequence of photographs of the experimental breaking wave taken from top. The extensions of longitudinal aerated scars are observed behind the wave front, which is a signature of the organized rib vortices located beneath the wave surface.

Watanabe and Mori (2008)[124)] observed variations of the temperature distributions on breaking wave surfaces by infrared measurements. In the experiment, as the water surface was initially heated, the bulk interior water was colder than the surface. They found longitudinal linear extensions of cold water emerges on the rear surface of breaking waves at a regular lateral interval in the direction of wave crest (see Fig. 8.10), indicating analogy with evolution of the scars owing to the subsurface counter-rotating vortices developed behind the wave front (see Fig. 8.8). As noted in Section 3.4.3.1, a pair of counter-rotating vortices induce the flow between them (see Fig. 3.13). If multiple pairs of the vortices are horizontally placed, upward and downward flows are alternatively induced between adjacent vortices (see Fig. 3.14). If we

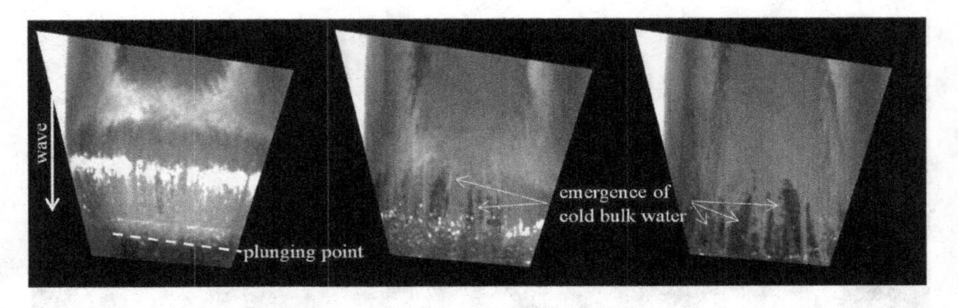

FIGURE 8.10
Sequence of infrared images of the breaking wave (surface temperature of the wave surface).

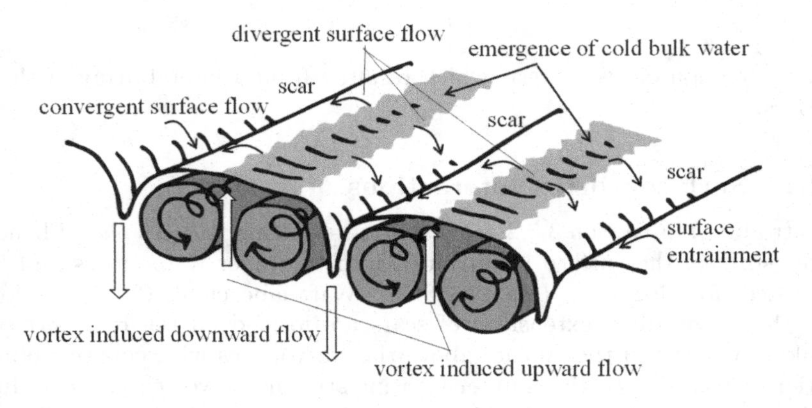

FIGURE 8.11
Schematic illustration of the surface renewal by the flows induced by multiple pairs of counter-rotating vortices produced in breaking waves

consider the pairs are located close to the surface, the upward induced flow transport bulk water mass to the surface, creating a blob-like surface form in the diverging surface flow (see Fig. 3.20 left), while the downward induced flow results in the converging surface flow to transport the surface between the vortices into the bulk, developing the scar (see Fig. 3.20 right). Fig. 8.11 illustrates the mechanism of the surface replacement owing to pairs of counter-rotating vortices close to the rear surface of breaking waves, which explains the temperature variations on the breaking wave surfaces shown in Fig. 8.10. Since the upward induced flows carry cold bulk water to the surface, the cold water appears on the surface along the longitudinal rib vortices beneath the surface, resulting in linear patterns of low surface temperatures shown in Fig. 8.10. The heated surface is brought to the scars by the convergent surface flows and transported into depths by the vortex-induced downward flows. In

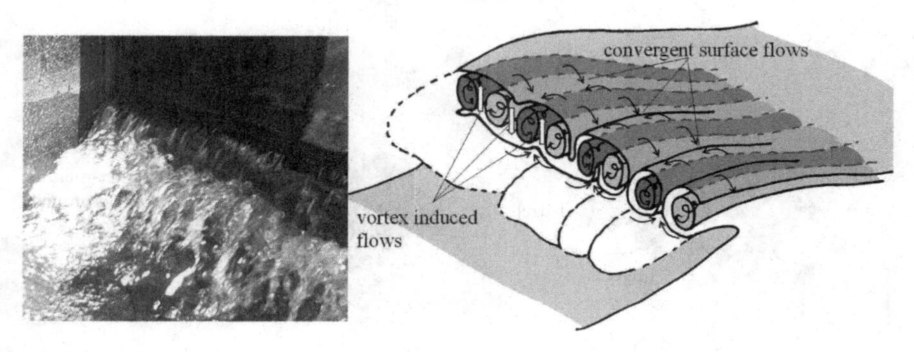

FIGURE 8.12
Photograph of finger jets of experimental plunging wave (left) and schematic illustration of the mechanism to produce the finger jets.

this way, the rib vortices contribute to renew the wave surface. During this process, the ocean surface heated by solar radiation is entrained into the bulk water to rapidly warm up the surface layer of ocean in addition to heat conduction across the sea surface, while the bulk water is exposed on the surface to be heated. Accordingly, the organized rib structure of vortices produced in breaking waves has a role to exchange heat between atmosphere and ocean.

Even if a planar form of overturning jet of plunging wave splashes on the surface (like as Fig. 8.4 top), a laterally uniform, planar shape of the secondary jet is never produced. As shown in Fig. 8.12 (left), typically finger-shaped jets are often observed. Saruwatari et al. (2009)[1)] explained the mechanism to produce the finger jets via the scarifying process of the secondary jet. As observed in Fig. 8.8, the multiple pairs of streamwise counter-rotating vortices create longitudinally extending scars laterally aligned at regular spacing on the surface of the main body (or neck) of secondary jet. While the finger jets extend from the scarified surface of main body, the lateral spacing between the fingers correlates with one of the scars on the surface of main body. Accordingly, the scars developing on the secondary jet laterally fragment the jet into the fingers. As illustrated in Fig. 8.12 (right), since the counter-rotating vortices within the jet are intensified by strong stretch along the projecting axis, the scars formed on the both of top and bottom surfaces of the jet are deepen by the intensified convergent surface flows, and finally penetrate the jet to separate the surface.

Fig. 8.13 shows the computed finger jets produced by the plunging breaker simulating the experimental breaking wave of Fig. 8.12 (left). We find the organized vortex structure is also disintegrated and collapsed by breakup of the main body of secondary jet into fingers. The breakup of finger jets into drops has been observed in Saruwatari et al.[1)].

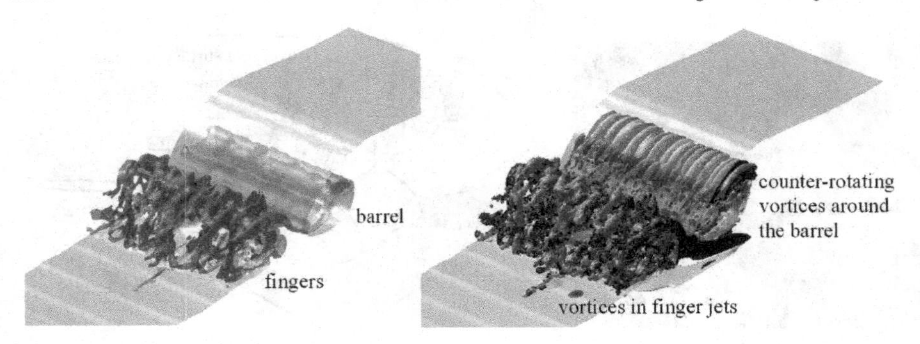

FIGURE 8.13
Computed surface of finger jets of plunging wave (left) and iso-surface of the streamwise component of vorticity (right).

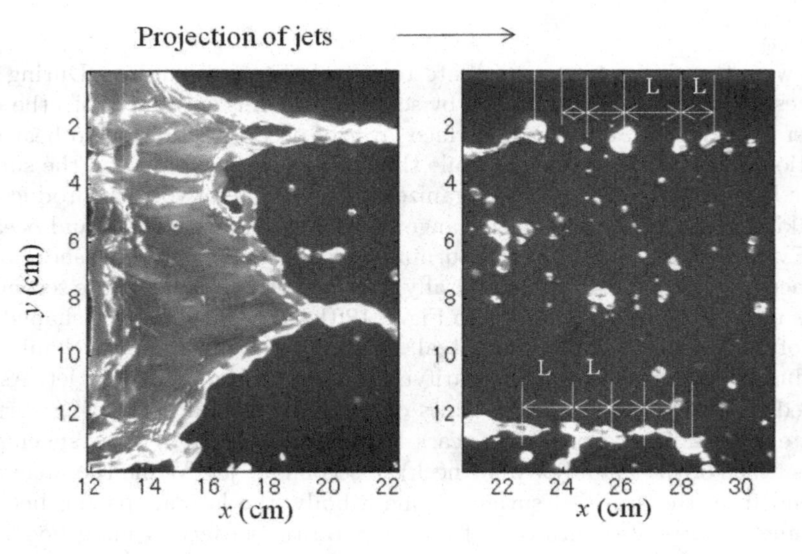

FIGURE 8.14
Example of visualized droplets fragmented from finger jets.

Fig. 8.14 shows the experimental droplets fragmented from the finger jet. While it is well known that the fragment of a cylindrical jet occurs through Rayleigh-Plateau mechanism, described by Eq. (6.98), the experimental ratio of the jet radius a and the unstable wavelength L, $a/L = 0.25$–0.31, takes much larger value than the one estimated by Rayleigh-Plateau mechanism, $a/L = 0.11$. As introduced in Section 6.3, the axial rotation of the cylindrical jet significantly increases a/L (see Fig. 6.8), that is, number of droplets increases with the axial rotation for the specific radius of jet. As already noted, the finger jet has axial rotation (see Figs. 8.8 and 8.13), the jet may breakup at shorter interval than that estimated by theory of Rayleigh-Plateau instability. Accordingly, the rib vortices entrapped in the fingers contributes increasing number of sea spray fragmented from the finger jets via rotational enhancement of capillary instability.

8.2.3 Vortex-induced suspension of sediment

Sediment transport is one of the most important issues of coastal engineering in terms of shoreline control. While bed load transport is often considered for predicting coastal process, suspended load owing to turbulent diffusion in breaking waves, which is major mechanism of the sediment transport in the surf zone, has not been understood. The previous models poorly predicted the sediment concentration in the surf zone.

While a well-known Rause's concentration of suspended sediment[84), which is similar to the one introduced in Section 2.2.5.2, may be used for open channel flows, the exponential vertical distribution, also introduced in Section 2.2.5.1, has been observed in the surf zone[76)]. The general form of the exponential mean concentration is given by

$$\overline{C} = C_0 \exp\left[-\frac{w_t}{D_T}(z - z_0)\right] \tag{8.11}$$

where C_0 is the concentration at the reference level z_0, w_t is the terminal velocity of a sediment particle, and D_T is the turbulent diffusion coefficient ($=\nu_T/S_{cT}$; ν_T is eddy viscosity and S_{cT} is the turbulent Schmidt number, introduced in Section 2.2.3).

Van Rijn[85)] proposed the three-layer model of the sediment concentration.

$$D_T = 0.004 d_* U_\delta \delta_s, \tag{8.12}$$

where the particle parameter $d_* = d_{50}\left(\left(s-1\right)g/\nu^2\right)^{1/3}$, U_δ is the peak orbital velocity, s is the specific gravity, ν is the kinematic viscosity, the thickness of the sediment mixing layer $\delta_s = 0.3h\left(H/h\right)^{0.5}$, h is the water depth, and H is the wave height. Larson and Kraus[51)] suggested the mixing coefficient

$$D_T = k_d h\left(\epsilon/\rho\right)^{1/3}, \tag{8.13}$$

where the energy dissipation rate $\epsilon \approx \partial\overline{F}/\partial x$, $\overline{F}$ is the mean energy flux, Eq. (7.59), and the dimensionless empirical coefficient $k_d \approx 0.03$. Goda[28)] integrated the past observation datasets of the mean sediment concentration for the empirical prediction by

$$\overline{C} = C_0 \exp\left[-\frac{6}{h}\epsilon^{-0.2}(z - z_0)\right] \tag{8.14}$$

However Wang[106)] found underpredictions of the experimental vertical profiles of the sediment concentration by any conventional models.

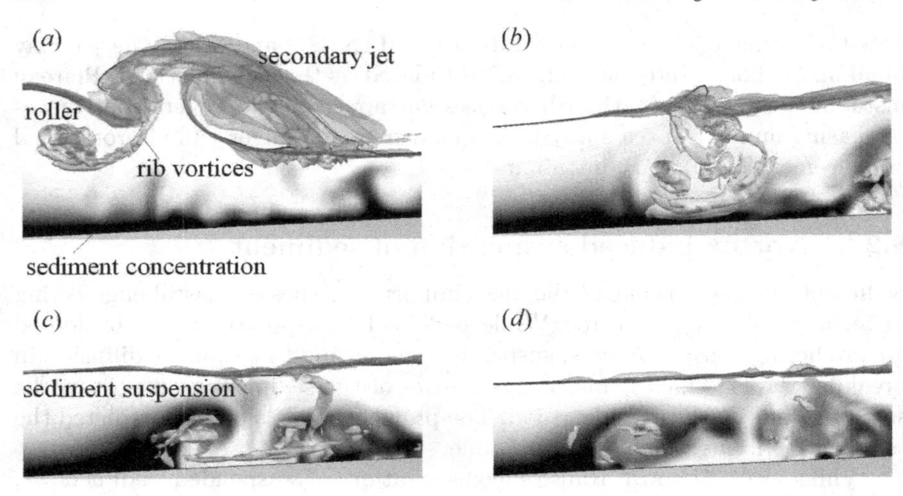

FIGURE 8.15
Cross-sectional concentration of sediment (denoted by a gray-scale) and vortex cores (white surface).

Otsuka et al. (2017)[39)] found a major mechanism of the sediment suspension owing to the counter-rotating rib vortices from physical and numerical wave experiments. Fig. 8.15 shows the cross-sectional distributions of sediment concentration and vortices produced in the plunging waves. When the rib vortices, developed on the shear layer between the bottom of the roller vortex and the preceding secondary jet at (*a*), approach the sea bed (see Fig. 8.15*b*), the sediment at the rear part of the vortices is carried upward to increase the concentration over the depth (Fig. 8.15*c* and *d*).

Fig. 8.16 illustrates the mechanism of the sediment suspension owing to the streamwise counter-rotating vortices. When the array of the counter-rotating vortices, as shown in Fig. 8.8, approach the sediment bed, convergent and divergent flows are alternatively induced on the bed. The sediment is swept away in the divergent flows, causing local erosion, while the convergent flow locally accumulates the sediment between the pairs. Since upward and downward flows are alternatively induced between the multiple pairs of counter-rotating vortices horizontally placed, as noted in Section 3.4.3.1 (see also Fig. 8.16), the accumulated sediment is carried by the upward induced flow and vertically transported through the pairs. Otsuka et al. (2017)[39)] modeled the vortex-induced suspension of sediment by introducing the velocity induced by two-point vortices, Eqs. (3.121) and (3.122). The upward velocity at the center of the counter-rotating point vortices horizontally placed with distance λ is given

$$w_c = \frac{2\mathit{\Gamma}}{\pi} \frac{\lambda}{\lambda^2 + 4\left(z - z_v\right)^2} \tag{8.15}$$

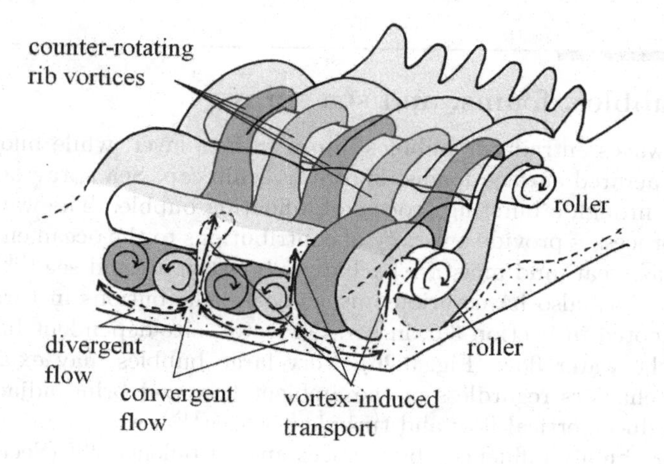

FIGURE 8.16
Schematic illustration of the mechanism of vortex-induced suspension of sediment.

where Γ is the circulation, and z_v is the vertical level of center of the vortices. Assuming Rankine vortices with vorticity ω (see Section 3.3.3), the circulation at the edge of the vortex core is given by $\Gamma = 2\pi r_c^2\omega = 2\pi r_c v_c$, which is used as the representative circulation of the point vortices, where r_c is the core radius and the tangential velocity $v_c = r_c\omega$. If two model parameters λ and v_c are determined by experimental parameterization of the transverse flows induced by the counter-rotating vortices[40], the temporal mean of the upward induced carrier velocity, $\overline{w_c}$, per unit width may be approximately given by

$$\overline{w_c} \approx \frac{w_c}{2}(-0.77F_r + 0.66) \tag{8.16}$$

where w_c at the bed ($z = -h$) is given by

$$w_c \approx 4r_c\sqrt{gh}F_r \frac{B\,(-0.64F_r + 0.61)}{B^2\,(-0.64F_r + 0.61)^2 + 4\,(-h - z_v)^2} \tag{8.17}$$

Here B is the reference transverse length (width of a flume), the parameterized Froude number $F_r = 0.47\xi_0 + 0.09$, and ξ_0 is the surf similarity parameter, Eq. (8.1). The mean concentration including the effect of upward carrier flow induced by the vortices is given as

$$\overline{C} = C_0 \exp\left[-\frac{w_t - \overline{w_c}}{D_T}(z - z_0)\right]. \tag{8.18}$$

where D_T is estimated by Eq. (8.13). According to Otsuka et al.[39], Eq. (8.18) reasonably predicts the mean suspension concentration under violent breakers.

8.3 Bubbles, foams, and sea sprays

Breaking waves entrain air bubbles into a surface layer, while buoyant bubbles form aerated surface foams, known as whitecap. Sea spray aerosols are produced through a bursting process of a floating bubble. The wave-induced aeration processes provide a variety of contributions to the ocean environment through gas, heat, and moisture exchanges between air and sea[41;61].

The bubbles also have various mechanical contributions in turbulent sea flows. As noted in Section 3.6, in aerated water, size-dependent bubble drag modifies the water flow (Fig. 3.26). Very large bubbles may exhibit active buoyant behaviors regardless of the ambient flow and bring adjacent water mass to induce vertical flow and turbulent wakes[118], while motions of small bubbles are highly influenced by vortices and turbulence[110] (Section 3.6.3). According to Lamarre and Melville[48], half of breaking wave energy was dissipated via bubble entrainment.

In this section, a series of the aeration processes induced by wave breaking are introduced.

8.3.1 Generation and entrainment of bubbles

According to Kiger and Duncan (2012)[43], mechanisms of air entrainment in a plunging breaker are (i) the overturning jet drags air into the water around the first plunging point, (ii) entrapment by subsequent impacts of secondary jet, (iii) entrainment between the rear side of the secondary jet and the upper surface of the overturning jet, (iv) entrainment over the splash-up process in the transition region, and (v) entrainment at the leading edge of the turbulent bore in the bore region. Deane and Stokes (2002)[25] defined two phases in the lifetime of wave-generated bubbles. The first phase, called acoustically active phase (as newly created bubbles emit pulses of sounds), occurs as bubbles are entrained and fragmented under splashes. Bubbles from 2 mm down to at least 0.1 mm radius, created by the impacts of jets, persist through the entire acoustically active phase. Larger bubbles (2 mm to ≥ 10 mm) are produced by collapsing cavity (barrel) in shorter period during the first phase. In the later, second phase, called acoustically quiescent phase (bubble creation emitting sound has been ceased), the bubble plume evolves in turbulent diffusion, advection, buoyant degassing, and dissolution.

When the entrained air encounters the vortices intensified in the organized structure (Section 8.2.1), since water pressure locally depresses in any vortex, as noted in Section 3.3.3, the air is entrapped within the cores of vortices for forming aerated vortex filaments[57] (see Fig. 8.17 left). When buoyancy of an air bubble, bubble drag in the vortex flow, and pressure gradient in the vortex core are in balance, the bubbles are trapped and spirally rotating within the vortex[123]. Nadaoka et al.[73] observed the spiral motion of bubbles trapped in the rib vortices and called it an ODE (Fig. 8.17 right). In this case, the

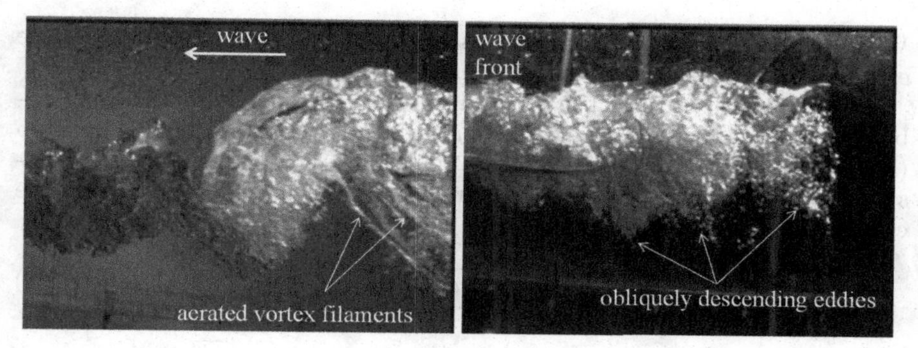

FIGURE 8.17
Side view of a breaking wave with aerated vortex filaments extending from the roller vortex obliquely downward (left) and three-quarter view of a breaking wave creating obliquely descending eddies (right).

bubbles entrapped in the vortices remain in sea water for a long time, against buoyant degassing. As residence time of air within these vortices increases, gas dissolution into seawater enhances[118].

8.3.2 Bubble size distributions

The aeration effects to physical and chemical environment strongly depend on the bubble size distribution. There have been many acoustic investigations measuring the resonant frequencies of bubble sounds to identify the bubble sizes[52]. Vagle and Farmer[104] used a multi-frequency acoustical backscatter technique to estimate bubble size distributions at different depths off the coast of California. They found a logarithmic spectrum slope for bubble sizes larger than the peak radius ($a \approx$ 20–30 μm); it varied from less than a^{-4} close to the ocean surface to a^{-7} at the bottom of the bubble cloud. Wu (1989)[114] considered a diffusion process of the mean concentration of bubbles near the sea surface and proposed the exponential vertical distribution of bubble population N as a solution form of the turbulent diffusion for immiscible particles (see Section 2.2.5.1):

$$\frac{N(z,a)}{N_0} = f(a) \exp\left[\frac{z}{z_b}\right] \tag{8.19}$$

where N_0 is the bubble population at the sea surface, and $f(a)$ is the frequency of occurrence of bubble having radius a; e.g. $f(a) \sim a^{-4}$ for close to the sea surface[104]. z_b is the characteristic length of the vertical entrainment of bubbles, suggested as[114]

$$z_b = 0.4 \quad (U_{10} < 7 \text{ m/s}) \tag{8.20}$$

$$z_b = 0.4 + 0.12(U_{10} - 7) \quad (U_{10} > 7 \text{ m/s}) \tag{8.21}$$

When N_0 is empirically determined as $N_0 = 57U_{10}^{3.5}$, Eq. (8.19) provides the estimate of the bubble population at arbitrary level z.

The bubble size scaling has been considered under an assumption that turbulent pressure fluctuations are the dominant mechanism for bubble breakup (Hinze 1955[34], Garrett and Farmer 2000[24]). Assuming the dynamic pressure fluctuations, exceeding surface tension, are responsible for breakup, the criteria may be given by the turbulent Weber number

$$We = \frac{\rho}{\gamma}\overline{u'^2}a \tag{8.22}$$

where ρ is the density of fluid, γ is the surface tension, $\overline{u'^2}$ is the average value of the squares of velocity differences over the scale of the bubble. Assuming homogeneous turbulence, according to Kolmogorov's second similarity hypothesis (see Section 2.1), the turbulence pattern is solely determined by energy dissipation ϵ (which has dimension of $[L^2T^{-3}]$, where L and T are representative length and time, respectively).

$$\overline{u'^2} \sim (\epsilon a)^{2/3} \tag{8.23}$$

Eq. (8.22) is thus expressed by

$$We = \frac{\rho}{\gamma}\epsilon^{2/3}a^{5/3} \tag{8.24}$$

When exceeding the critical value, breakup occurs for bubbles with radius larger than the Hinze scale:

$$a_H = We_c^{3/5}\left(\frac{\gamma}{\rho}\right)^{3/5}\epsilon^{-2/5} \tag{8.25}$$

where We_c is the critical Weber number. Deane and Stokes (2002)[25] used $We_c = 4.7$ and $\epsilon = 13\text{Wkh}^{-1}$ for $a_H = 1.0$ mm which defines the smallest radius of fragmented bubbles in their experiments.

Garrett and Farmer (2000)[24] proposed a rational spectrum slope of the bubble size. Number of bubbles, $N(a)$, fragmented by turbulence from the entrained air with discharge, Q (with the dimension $[L^3T^{-1}]$), is given by

$$N(a) \sim Q\epsilon^{-1/3}a^{-10/3} \tag{8.26}$$

The Hinze scale, a_H, cutoffs the size spectrum of Eq. (8.26). While Deane and Stokes (2002)[25] validate Eq. (8.26) for $a > a_H$ in experiments, they found the size spectrum for smaller bubbles than a_H follows another power-law scaling:

$$N(a) \sim Q\left(\frac{\gamma}{\rho}\right)^{-3/2}V^2a^{-3/2} \tag{8.27}$$

where V is the jet velocity of breaking waves.

Yu et al. (2019)[128] provide the analytical description of the bubble size scaling. Considering a spherical bubble of radius a at a depth $2a\alpha$ under free surface, where the constant α is of order 1, the energy required to entrain a bubble is given by

$$E_b(a) = \frac{4}{3}\pi g a^3 (2a\alpha) + 4\pi\frac{\gamma}{\rho}a^2 \tag{8.28}$$

where the first and second terms of right-hand side represent potential and surface energies, respectively. As the total number of bubbles of radius a entrained across a surface area A_h is estimated to be $N(a)A_h da$, the total energy required to entrain is given by

$$N(a)A_h da E_b(a) = N(a)A_h da\left(\frac{8}{3}\pi g a^4\alpha + 4\pi\frac{\gamma}{\rho}a^2\right) \tag{8.29}$$

On the one hand, assuming the turbulence with the length scale of k^{-1} is responsible for the entrainment of bubbles with radius $a \sim k^{-1}$, the turbulent energy spectrum density $E(k)dk$ may be given by

$$E(k)dk \sim E(a^{-1})da^{-1} = E(a^{-1})a^{-2}da \tag{8.30}$$

The turbulent energy within a volume $A_h a$ below the free surface supplies for managing entrainment:

$$E(a^{-1})a^{-2}da A_h a \sim N(a)A_h da\left(\frac{8}{3}\pi g a^4\alpha + 4\pi\frac{\gamma}{\rho}a^2\right) \tag{8.31}$$

Assuming homogeneous turbulence, the Kolmogorov's spectrum, Eq. (2.9), may be introduced:

$$E(a^{-1}) \propto \epsilon^{2/3}a^{5/3} \tag{8.32}$$

The substitution of Eq. (8.32) into Eq. (8.31) gives the general description of the bubble size spectrum:

$$N(a) \propto \left(\frac{2}{3}g a\alpha + \frac{\gamma}{\rho a}\right)^{-1}\epsilon^{2/3}a^{-7/3} \tag{8.33}$$

For large bubble, as $2ga\alpha/3 \gg \gamma/\rho a$, neglecting the surface tension, $N(a)$ is governed by the gravity (buoyancy):

$$N_g(a) \propto g^{-1}\epsilon^{2/3}a^{-10/3} \tag{8.34}$$

The spectrum slope of $a^{-10/3}$ is coincident with Eq. (8.26) for larger bubbles. Another limit for small bubbles, which the surface tension governs $N(a)$, is given

$$N_s(a) \propto \left(\frac{\gamma}{\rho}\right)^{-1}\epsilon^{2/3}a^{-4/3} \tag{8.35}$$

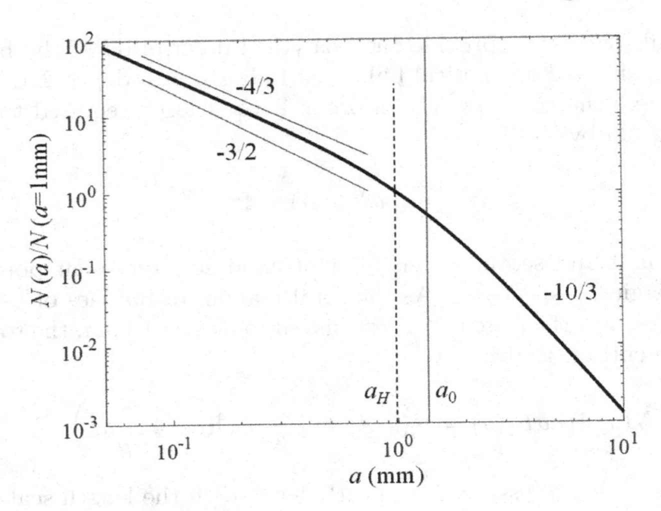

FIGURE 8.18
Bubble size spectrum slopes of $-10/3$ (Eq. (8.26) and Eq. (8.34)), $-3/2$ (Eq. (8.27)) and $-4/3$ (Eq. (8.35)), together with the general spectrum form given by Eq. (8.33) (solid line, $\alpha = 6$).

The boundary between the different spectrum slopes of Eqs. (8.34) and (8.35) is determined by equating the both effects, $2ga\alpha/3 = \gamma/\rho a$;

$$a_0 = \sqrt{\frac{3}{2}\frac{\gamma}{\alpha\rho g}} \tag{8.36}$$

We find a_0 is determined regardless of ϵ, unlike the Hinze scale a_H, and is comparable to the capillary length $l_c = \sqrt{\gamma/\rho g}$, i.e. ratio of surface tension and gravity (see Section 3.6.4).

Fig. 8.18 illustrates the bubble size distribution of Eq. (8.33) together with the power-scaling by Eq. (8.27). The $a^{-4/3}$ slope derived by Yu et al. (2019) [128] is slightly milder than $a^{-3/2}$ observed by Deane and Stokes (2003) [25]. There is also a difference between Eqs. (8.36) and (8.25). It should be noted that the spectrum slopes and their boundary radius for wave-generated bubbles may also depend on the phase, depth, and scales of wave breaking, while the general form of Eq. (8.33) is available under the assumption of steady and equilibrium state. Deane and Stokes [25] observed the variable spectrum slopes and Hinze scales according to the plume age. The initial slopes $a^{-1.8}$ ($a < a_H$) and $a^{-4.9}$ ($a > a_H$) become steeper with increasing time and decreasing void fraction because large bubbles, with buoyancy that dominates over drag, rapidly ascend and disappear through buoyant degassing, while smaller bubbles are involved in turbulence and are passively transported at depth [25;61].

8.3.3 Gas transfer from bubbles

While parameterizations of gas diffusion process across sea surface have been conventionally used for predicting the air-sea gas transfer, introduced in Section 8.4.2, the conventional gas transfer models may underestimate the observed gas transfer during periods of bubble penetration [18]. Gas in a bubble entrained in seawater is dissolved across the interface over the lifetime of the bubbles. As breaking waves create huge number of bubbles, gas transfer is significantly enhanced through large total area of bubble interfaces. In particular, dissolution of oxygen and carbon dioxide directly supports biological activities in a marine ecosystem.

The gas dissolution from a boundary (interface) is described by a diffusion process of the concentration at the boundary C_a to the bulk concentration C_0 far from the boundary, which is mathematically expressed in terms of a mass transfer velocity (or specifically a gas transfer velocity in this problem) k_D, as introduced in Section 2.2.2 (see Eq. (2.75)). Accordingly, the gas diffusion flux, j, is given

$$j = k_D(C_a - C_0) \tag{8.37}$$

Assuming a static spherical bubble of diameter d in quiescent water, the diffusion flux at the bubble interface is expressed by Eq. (2.79). The gas transfer velocity is therefore given as

$$k_D = 2\frac{D}{d} \tag{8.38}$$

where D is the diffusion coefficient. As noted in Section 2.2.2.2, the Higbie penetration model, describing a surface renewal, gives Eq. (2.83):

$$k_D = 1.13Re^{1/2}Sc^{1/2}\frac{D}{d} \tag{8.39}$$

where Re is Reynolds number and Sc is Schmidt number. Levich [54] theoretically derived the mass transfer from a bubble for creeping flow:

$$k_D = 0.65Re^{1/2}Sc^{1/2}\frac{D}{d} \tag{8.40}$$

Frössling's equation [22] for a rigid sphere is given by:

$$k_D = \left(2 + 0.6Re^{1/2}Sc^{1/3}\right)\frac{D}{d} \tag{8.41}$$

While these models assume the bubble surface is rigid, termed a freezing bubble (see Section 3.6.2), often used for a small bubble, when a large bubble with deformation, shown in Fig. 3.25, rises at faster terminal velocity (see Fig. 3.26 right), concentration boundary layer formed over the deformed bubble surface may have trailed along the lee side of the bubble and been shed by vortex wakes in the turbulent flow [105].

Niida and Watanabe (2018) [118] measured the concentration of oxygen dissolved from bubble plumes (bubbles are successively emitted from needle fixed

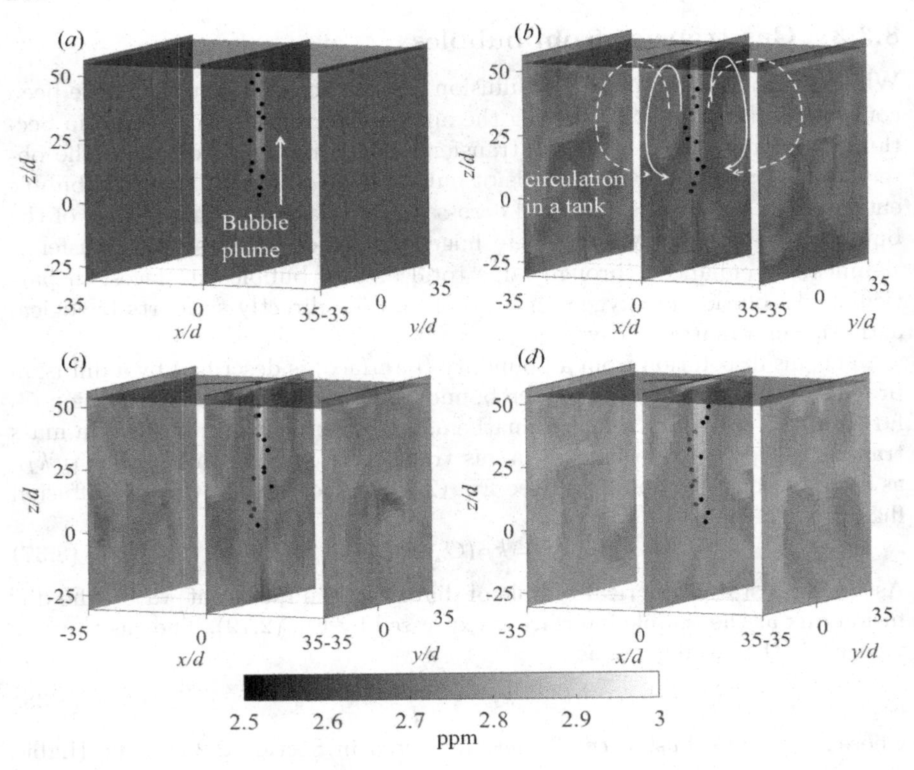

FIGURE 8.19
Oxygen concentrations of water where bubbles (black dots) are successively released from the level $z/d = 0$ in a rectangular water tank; (*a*) time from the bubble emission $t = 1$ s, (*b*) $t = 60$ s, (*c*) $t = 120$ s, and (*d*) $t = 180$ s. The bubble plume induces circulation of turbulent fluid flows in a tank, which spreads dissolved oxygen along the axis of the bubble plume to the whole fluid region in the tank.

on a bottom of tank) by Laser Induced Fluorescence (LIF) technique, and empirically determined the transfer velocity for deformed bubbles with turbulent behaviors, given as

$$k_D = \left(\alpha \mathrm{Re}^{0.5} \mathrm{Sc}^{0.5}\right) \frac{D}{d}, \tag{8.42}$$

where the size-dependent constant α =1.13 for $d < 3.6$ mm, and $\alpha = 2.21$ for $d \geq 3.6$ mm. They introduced Eq. (8.42) into a large eddy simulation of bubble-laden turbulent flows as a source model of oxygen (see Fig. 8.19).

8.3.4 Whitecapping

As noted in Section 3.6.4, when ascending air bubbles arrive clean water surface, they coalesce with the surface, that is, bubbles disappear on the surface. Since surfactant prevents coalescence, bubbles in seawater containing a variety of surfactants, can persist on the surface for a while. Floating bubbles can aggregate on curved surfaces to form bubble clusters (Fig. 3.29).

The visible appearance of aerated water and clustered foam is known as a whitecap. When wind velocity exceeds 4 m/s, we may observe an aerated spilling wave crest, termed active whitecap, which decays into the mature whitecap or hazy form patch [68] (Fig. 3.33). Whitecap fraction (or coverage), estimated from visual images of sea surfaces, measures the extents influenced by wave-induced turbulence and aerated water, which has been used as a parameter of gas exchange (Section 8.4.2) and sea spray production (Section 8.3.5).

There have been various ocean observations to parameterize the whitecap fraction W in a power law of wind velocity. Monahan and Muircheartaigh (1980) [69] proposed the two optimal power law expressions of W:

$$W = 2.95 \times 10^{-6} U_{10}^{3.52} \tag{8.43}$$

$$W = 3.84 \times 10^{-6} U_{10}^{3.41} \tag{8.44}$$

where U_{10} is the wind velocity at 10 m above the mean sea surface. Wu (1988) [113] proposed a similar bulk formula:

$$W = 2 \times 10^{-6} U_{10}^{3.75} \tag{8.45}$$

Arsher et al. (2002) [6] proposed the form of cubic law:

$$W = 3.7 \times 10^{-6} \left(U_{10} - 1.2\right)^3 \tag{8.46}$$

Fig. 8.20 shows W predicted by Eqs. (8.43)–(8.46).

Toba and Koga (1986) [98] considered turbulent boundary layer formed over progressive waves and introduced a so-called wind-wave Reynolds number:

$$R_B = u_*^2 / \nu \sigma_p \tag{8.47}$$

where ν is the kinematic viscosity of air, u_* is the friction velocity (see Section 2.1.4) and σ_p is the spectral peak frequency of the wind waves (see Section 7.4.2). The representative length of R_B, $L_s = u_* T_{1/3} \approx 2\pi u_* / \sigma_p$, corresponds to the drift distance of water particle on wave surface owing to wind stress over one wave period, where $T_{1/3}$ is the significant wave period (see Section 7.4.1). Zao and Toba (2001) [129] found R_B well predicts W at higher correlation than the other parameters, such as wave period, wind speed, and friction velocity:

$$W = 3.88 \times 10^{-7} R_B^{1.09} \tag{8.48}$$

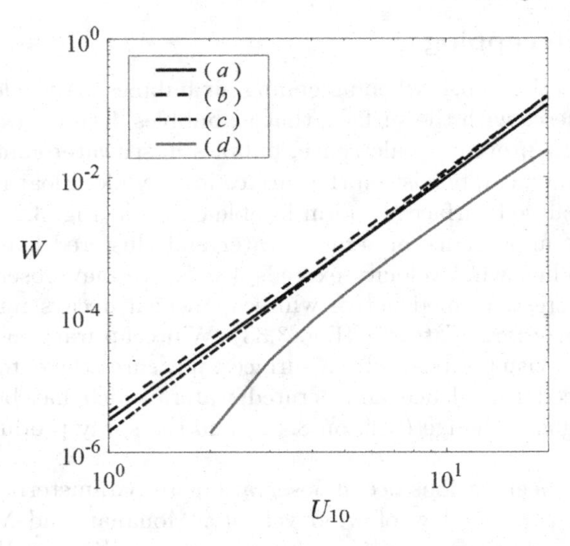

FIGURE 8.20
Whitecap fractions as a function of U_{10}; (*a*) Eq. (8.43), (*b*) Eq. (8.44), (*c*) Eq. (8.45), and (*d*) Eq. (8.46).

If wave height is used as the length scale of R_B, $R_H = u_* H_{1/3}/\nu$ can be defined, which is also used to parameterize W:

$$W = 4.02 \times 10^{-7} R_H^{0.96} \tag{8.49}$$

The wave age, defined as the ratio of the phase velocity of waves at the spectral peak frequency, c_p, and the friction velocity u_*, has been used as a parameter for W in a fetch-limited ocean near coast where waves are generally young. Lafon et al. (2003)[47] used data measured at a site of 60 km offshore in the Gulf of Lion (north-west Mediterranean) and parameterized W as

$$W = 2.0 \times 10^4 \left(\frac{c_p}{u_*}\right)^{-4.9} \tag{8.50}$$

Callaghan et al. (2008)[12] analyzed data observed at a site where is 3 km away from the south shore of Martha's Vineyard, USA, and proposed

$$W = 3.11 \times 10^{-4} \left(\frac{c_p}{U_{10}}\right)^{-4.93} \tag{8.51}$$

$$W = 1.81 \times 10^3 \left(\frac{c_p}{u_*}\right)^{-4.63} \tag{8.52}$$

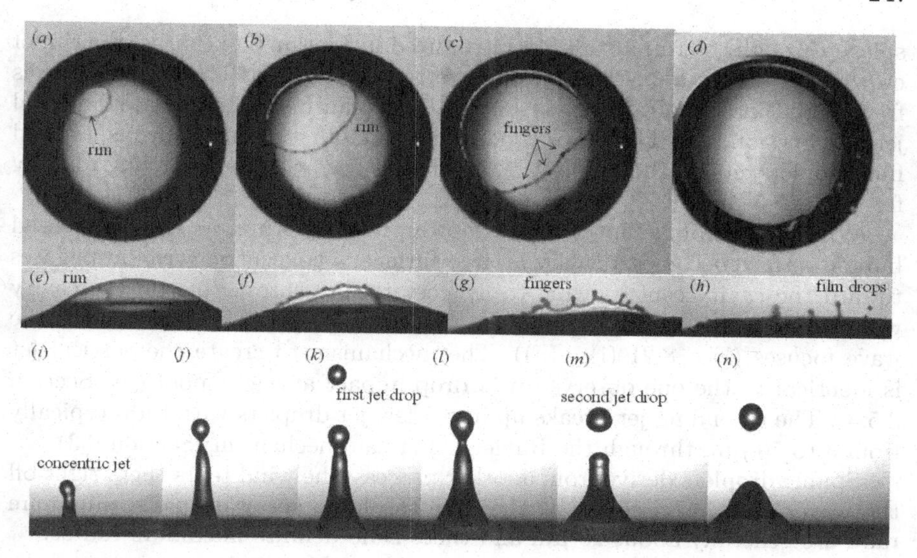

FIGURE 8.21
Consecutive backlit images of a bursting bubble: (a) - (d) plan view, (e) - (h) side view, and an ascending concentric jet that is fragmented into jet drops: (i) - (n).

8.3.5 Sea spray and marine aerosol

Sea spray has important roles for meteorological and environmental processes, including the exchanges of gases, moisture and heat, transport of chemical materials[78)], and the dynamics of the marine atmospheric boundary layer[46)]. The contributions of sea sprays to particular atmospheric processes depend on the size distributions of the spray. This in turn defines the total surface area, where mass exchanges occur via dissolution and evaporation, and the concentration and residence time in the atmosphere, which modify momentum transport between air and water as additional drag forces.

Sea spray droplets come, basically, in several varieties: film droplets, jet droplets, and spume droplets, in addition to drop fragments of finger jets caused by the subsurface vortices of breaking waves noted in Section 8.2.2. Film and jet droplets derive from one process – air bubbles bursting at the sea surface. As noted in Section 3.6.4.2, when a bubble arrives the surface, the top part bubble, known as a bubble cap, appears above the ambient meniscus surface. The difference in capillary pressure between the film of the bubble cap and the meniscus surface generates the flow from the former to the later at the edge of the cap, leading to the formation of a local pinching separation of the surface to create a hole of the film[55)] (see Fig. 8.21 (a) - (h)). While the capillary pressure radically acting on the rim of the hole outward spreads out the opening, the growing rim becomes unstable and undulate along the rim to form fingers. The evolution of this bursting process is similar to that of crown splash through the rim instability introduced in Section 6.4; that is, the rim of crown wall becomes unstable through retraction process. The fingered rim is fragmented into small drops via Rayleigh-Plateau instability of a cylindrical jet (see Section 6.3). The drops produced in the bursting mechanisms, termed film droplets, create the smallest range of sea spray with radii ranging roughly from 0.5 to 5 μm.

When the bubble cap disappears owing the bursting, as the submerged bubble surface becomes a cavity on free surface, a concentric wave propagates toward the cavity center owing to hydrostatic pressure acting on the cavity wall, resulting in a formation of the ascending jet at the center where the wave focuses (Fig. 8.21 (i) - (n)). The mechanism to create the vertical jet is identical to the one observed in a drop impact at water pool (see Section 3.5.3). The ascending jet breaks up into a few jet droplets with radii typically from 3 to 50 μm, through the Rayleigh-Plateau mechanism (Section 6.3).

Spume droplets derive from another process; the wind tears them right off the wave crests. Consequently, they are the large spray droplets; minimum radii are generally about 20 μm and there is no definite maximum radius.

Wu et al. (1984) [115)] measured sea spray within the lowest meter of the atmospheric surface layer with an optical instrument supported on a raft. The measured diameter d of droplets follow a two-segment equilibrium spectrum; the division of two size regions within the production zone (the ejection range of jet drops z <20 cm) and that outside. For 150 μm $< d <$ 400 μm ($z <$ 20 cm) or 50 μm $< d <$ 200 μm ($z >$ 20 cm), the frequency of occurrence empirically follows

$$f \sim d^{-2.8} \tag{8.53}$$

When $d >$ 400 μm ($z <$ 20 cm) or $d >$ 200 μm ($z >$ 20 cm)), a steeper power-law scaling was observed:

$$f \sim d^{-8} \tag{8.54}$$

Wu (1989) [114)] empirically estimated the number of film drops produced by a bubble of diameter d in the range 4 mm $< d <$ 10 mm as

$$n_f(d) = 1.07d^{2.15} \tag{8.55}$$

The number of jet drops was exponentially approximated:

$$n_j(d) = 7\mathrm{e}^{-d/3} \tag{8.56}$$

Assuming bubble population is given by Eq. (8.19) with $f(d) \sim d^{-4}$, the population at the sea surface $z = 0$ is estimated to be $N_0(d) \sim 57U_{10} - 3.5d^{-4}$. The total number of drops produced by film bursting (n_{f0}) and jet breakup (n_{j0}) is given by $n_f N_0$ and $n_j N_0$, respectively:

$$n_{f0} \sim d^{-1.85} \tag{8.57}$$

$$n_{j0} \sim \mathrm{e}^{-d/3} d^{-4} \tag{8.58}$$

FIGURE 8.22
Violent wave overtopping observed when shoaling waves impact at a sea wall.

Marine aerosol particulate flux has been generally expressed by a spray generation function (SGF); number of drops with radius r produced per square meter of surface per second per micrometer increment in r. Smith et al. (1993)[89] approximate the observed flux as the sum of two log-normal distributions of the form:

$$\frac{dF}{dr_{80}} = \sum_{i=1}^{2} A_i \exp\left[-f_i\left(\ln\frac{r_{80}}{r_i}\right)^2\right] \quad (1 \le r_{80} \le 25\ \mu\text{m}) \tag{8.59}$$

where r_{80} is the drop radius reduced to a relative humidity of 80%, $f_1 = 3.1$, $f_2 = 3.3$, $r_1 = 2.1\ \mu$m, $r_2 = 9.2\ \mu$m, A_i is determined by the wind speed 14 m above the mean sea surface; $\log(A_1) = 0.0676U_{14} + 2.43$, $\log(A_2) = 0.959U_{14}^{1/2} - 1.476$, where U_{14} is the wind velocity at 14 m above the mean surface.

Andreas (1992) proposed the sea spray generation function that predicts spume production[4]

$$\frac{dF}{dr_{80}} = \begin{cases} C_1(U_{10})\, r_{80}^{-1} & (15\ \mu\text{m} \le r_{80} \le 37.5\ \mu\text{m}) \quad (8.60) \\ C_2(U_{10})\, r_{80}^{-2.8} & (37.5\ \mu\text{m} \le r_{80} \le 100\ \mu\text{m}) \quad (8.61) \\ C_3(U_{10})\, r_{80}^{-8} & (100\ \mu\text{m} \le r_{80}) \quad (8.62) \end{cases}$$

The coefficient C_1 is determined by the known dF/dr_{80} at $r_{80} = 15\ \mu$m (e.g. using Eq. (8.59)). Equating dF/dr_{80} estimated by Eqs. (8.60) and (8.61) at $r_{80} = 37.5\ \mu$m, C_2 can be calculated. C_3 is also determined to match dF/dr_{80} by Eqs. (8.61) and (8.62) at $r_{80} = 100\ \mu$m.

When large shoaling waves impact on breakwaters and coastal cliffs, sea sprays rising several tens of meters into the air have been often observed in coasts (see Fig. 8.22). In this case, further downwind transport of the sea sprays and longer residence time, depending on the size distributions, from higher source levels may largely affect the atmospheric dynamics in a thicker spray-mediated boundary layer.

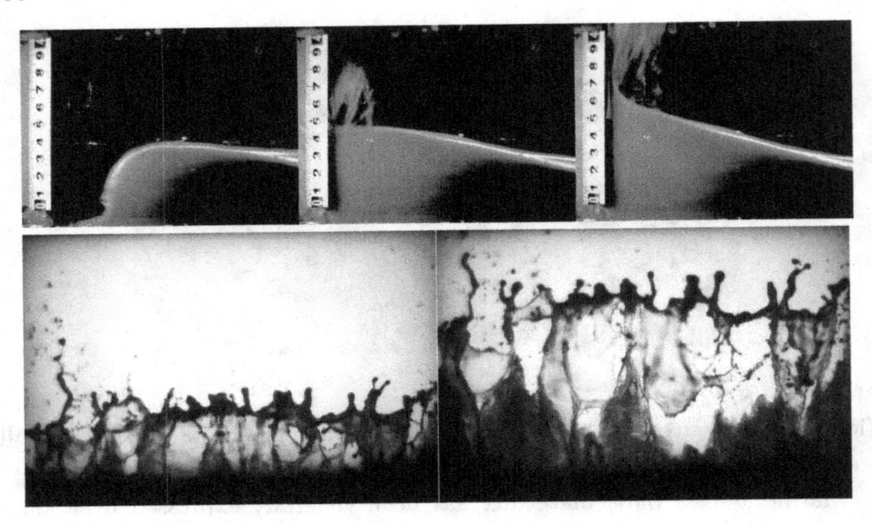

FIGURE 8.23
Sequential images of the shoaling wave face impact at the vertical wall (top); the wall is located at the right edge of the ruler. Transverse backlit images of the ascending flip-through jet (bottom).

According to the free-surface dynamics of overtopping water waves explained in Watanabe and Ingram[121], when a shoaling wave crest approaches a vertical wall, the wall prevents the forward flow of water causing the water level at the wall to rise rapidly. The water surface between the rising wave trough and advancing crest converges rapidly resulting in the violent ejection of a sheet (see Fig. 8.23 top), the so-called flip-through event[15]. As the sheet of water rises, it evolves cusp-like formations which are amplified to producing finger jets and ligaments subsequently fragmented into droplets (see Fig. 8.23 bottom). This breakup process is triggered by two distinct mechanisms caused by the rim instability (Section 6.4) and liquid sheet instability (Section 6.2). Accordingly, initial amplification of undulations in the sheet thickness resulting in the formation of ligaments which fragment into droplets; and transverse deformations of the rim bounding the liquid sheet form finger jets which, due to capillary instability, also break up to form droplets. Watanabe and Ingram[122] assumed Rayleigh-Plateau instability (Section 6.3) results in breakup of the rim-ligament system and related number of the total droplets (n) to the drop radius (r) fragmented from the flip-through jet for $r > 0.4$ mm:

$$n \propto \begin{cases} r^{-5/2} & \text{(in early state)} \\ r^{-2} & \text{(in steady state)} \end{cases} \tag{8.63}$$

They also found the lognormal probability distribution with parameters related to the elapsed time since the initial wave impact describes the whole drop sizes observed in experiments.

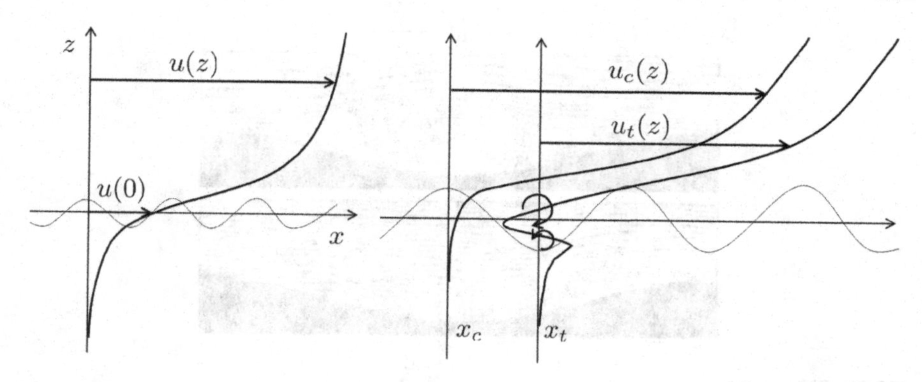

FIGURE 8.24
Schematic illustration of velocity profiles through the boundary layers formed at the both sides of the wave surface in low Reynolds number (left), and in the case of vortices produced in high Reynolds number (right).

8.4 Air–sea interactions

When weak wind blows above a still water surface, a laminar wind boundary layer is produced on the air side of the interface (Section 1.3.4) and the viscous stress at the surface works as friction for the air flow, while the wind-drift layer is simultaneously produced below the surface[127] (Fig. 8.24 left). In the case of high-speed wind blowing large amplitude waves, the air flow patterns behind the wave crests become complex (Fig. 8.24 right), as the transition of the flow states from laminar to turbulence depending on Reynolds number (see Fig. 2.1). Accordingly, the lee vortex produced behind the crest is successively ejected toward downstream, causing pressure fluctuations in the boundary layer (Fig. 8.25 top). In the turbulent regime, shown in Fig. 8.25 (bottom), many smaller vortices are produced and disturbed the flow over the layer, causing local fluctuations of mean velocity field within the layer.

Fig. 8.26 shows the organized vortex structures evolving with the wind velocity. As explained, the lee vortices having uniform cylindrical forms in the direction of wave crest govern the vortex structure under the slow wind (Fig. 8.26*a*), while the cylindrical vortices are stretched in the wind direction and axially deformed at the higher wind velocity (*b*), resulting in a formation of vortex loops. Further increase in wind velocity stretches bends of the loops downstream to create longitudinal vortex structure (*c*). Finally the organized vortices are decomposed into numbers of smaller vortices in the turbulent regime (*d*). In addition to the wind shear, pressure fluctuations caused by the vortices strongly affect the sea surface drag and the momentum transfer.

Wind waves have roles to exchange momentum, heat, gas, and moisture between air and sea across the boundary layer formed at the both sides of

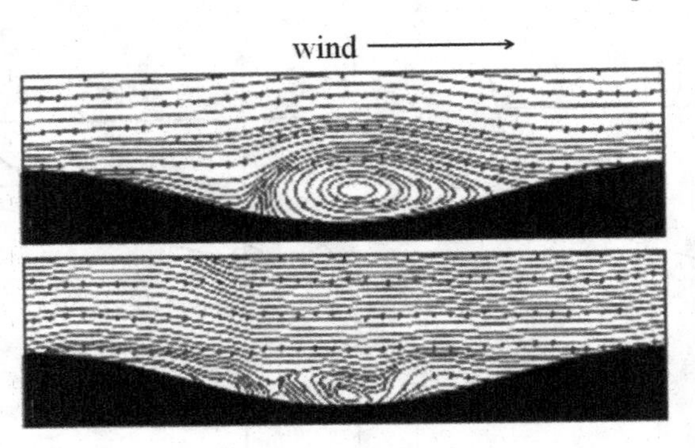

FIGURE 8.25
Typical streamlines of wind above the wave surface (steepness $H/L = 0.05$); wind velocity out of the boundary layer $\overline{u}_{\infty} = 5.5$ m/s (top) and 18.8 m/s (bottom).

the sea surface. As noted, while the small-scale variations of flows define local mechanical balance at the interface and interaction between air and sea, such small-scale quantities in turbulent flows are difficult to obtain in real ocean. If the local effects of turbulence to specific sea surface process can be described by observed macroscopic parameters, the other processes, parameterized by the same macro quantities, can be connected each other; that is, any interactions over the different sea surface processes may be estimated through the parameterization. The quantities observed at specific level (e.g. 10 m above the mean sea surface) are commonly used as the representative bulk parameters. For instance, the sea surface shear is expressed by

$$\tau_w = \rho C_D U_{10}^2 \tag{8.64}$$

where C_D is the sea-surface drag coefficient, and U_{10} is the wind velocity at 10 m above the mean surface. It should be noted that the definition of C_D by Eq. (8.64) is derived from the skin friction coefficient for wall turbulence, Eq. (2.46). The representation of the specific quantity by bulk parameters, like as Eq. (8.64), is called the bulk formula. Numbers of parameterization studies, based on ocean observations, have been performed for bulk modeling of sea-surface drag, gas, and heat transfers. Important achievements of those investigations are introduced in the following sections.

8.4.1 Friction at sea surfaces

Wind field above ocean surfaces is generally described by a logarithmic form of turbulent boundary layer introduced in Section 2.1.4. As the wind generates and develops waves, the boundary layer has rough surface. Therefore a logarithmic velocity profile on a rough wall, Eq. (2.48), describes the winds

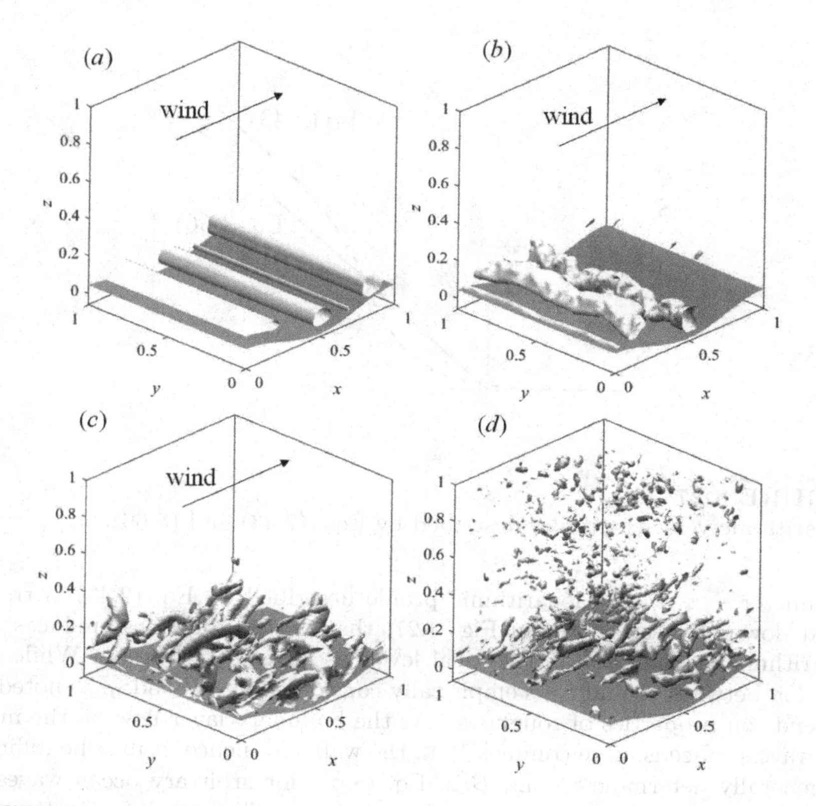

FIGURE 8.26
Typical distributions of the vortex cores above the wave surface ($H/L = 0.10$); (*a*) $\overline{u}_\infty = 5.3$ m/s, (*b*) $\overline{u}_\infty = 6.3$ m/s, (*c*) $\overline{u}_\infty = 6.7$ m/s, (*b*) $\overline{u}_\infty = 17.9$ m/s

above wave surfaces:

$$\frac{\overline{u}}{u_*} = \frac{1}{\kappa} \ln \frac{z}{z_r} + B \tag{8.65}$$

where u_* is the friction velocity, κ is the Kármán constant, z_r is the roughness, and B is a function of roughness Reynolds number ($= u_* z_r/\nu$). The physical modeling of z_r and B for arbitrary roughness wall properties is still challenging, which has been actively studied [16].

For wind boundary layer, the logarithmic wind velocity has been conventionally assumed to take the simpler form, instead of Eq. (8.65):

$$\frac{\overline{u}}{u_*} = \frac{1}{\kappa} \ln \frac{z}{z_0} \tag{8.66}$$

where the roughness parameter z_0 is the virtual origin of the velocity profile (the height to achieve $\overline{u} = 0$) (see Fig. 8.27). With the dimensionless

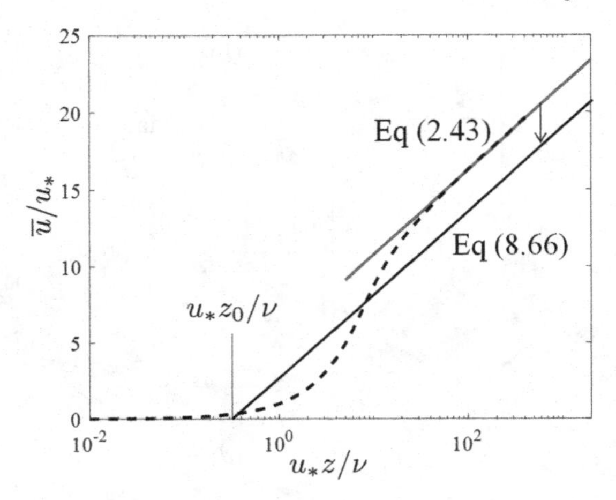

FIGURE 8.27
Logarithmic velocity profiles described by Eqs. (2.43) and (8.66).

parameter $u_* z_0/\nu$, the logarithmic profile described by Eq. (2.43) is translated downward, as shown in Fig. 8.27; that is, the roughness reduces the logarithmic velocity at fixed vertical levels as a result of friction. While the relation between z_r and z_0 is empirically considered as $z_r \sim 30z_0$, as noted, it depends on properties of roughness. As the boundary layer flow on the moving wave surface is more complex than the wall turbulence, it may be difficult to generally determine z_r and B in Eq. (8.65) for arbitrary ocean waves in the identical framework of static wall turbulence. The past investigations to estimate the roughness, as a result of parameterization, are introduced below.

The simplest model for z_0 is the well-known Charnock's formula[31)]:

$$z_0 = \frac{m_0 u_*^2}{g} \tag{8.67}$$

where m_0 is the constant to be empirically determined ($m_0 \approx 0.014 - 0.0185$). Masuda and Kusaba (1987) considered the effect of wave age, c_p/u_* to the roughness:

$$\frac{g z_0}{u_*^2} = m_0 \left(\frac{u_*}{c}\right)^{\beta} \tag{8.68}$$

where β are constant, and c is the wave speed. Eq. (8.68) with $\beta = 0$ corresponds to Eq. (8.67). Kitaigorodskii and Volkov (1965) considered the wind velocity in a reference frame moving with the wave speed and proposed the velocity profile:

$$\frac{\overline{u} - c}{u_*} = \frac{1}{\kappa} \ln \frac{z}{A} \tag{8.69}$$

$$\therefore \frac{\overline{u}}{u_*} = \frac{1}{\kappa} \ln \frac{z}{a \exp\left(-\kappa c/u_*\right)} \tag{8.70}$$

where A is the wave amplitude. While z_0 corresponds to $a \exp(-\kappa c/u_*)$, the later work by Kitaigorodskii (1970) [58] explicitly proposed

$$z_0 = 0.3H_{rms} \exp(-\kappa c_p/u_*) \tag{8.71}$$

where H_{rms} is the root-mean-square height of surface waves, Eq. (7.135), and c_p is the wave speed at the spectrum peak. Anctil and Donel (1996) [21] parameterized z_0 normalized with H_{rms} by the inverse wave age U_{10}/c_p and the root-mean-square wave slope S as

$$\frac{z_0}{H_{rms}} = 2.26 \left(\frac{U_{10}}{c_p}\right)^{1.82} S^{3.83} \tag{8.72}$$

Drennan et al. (2003) [20] estimated the relationship between the dimensionless roughness and the inverse wave age:

$$\frac{z_0}{H_{rms}} = 13.4 \left(\frac{u_*}{c_p}\right)^{3.4} \tag{8.73}$$

Taylor and Yelland (2001) [97] presented z_0, scaled with the significant wave height $H_{1/3}$, parameterized by the wave steepness:

$$\frac{z_0}{H_{1/3}} = 1200 \left(\frac{H_{1/3}}{L_p}\right)^{4.5} \tag{8.74}$$

On the one hand, Eqs. (8.64) and (2.30) give the sea-surface drag:

$$C_D = \left(\frac{u_*}{\overline{u}}\right)^2 \tag{8.75}$$

The substitution of Eq. (8.66) into Eq. (8.75) gives

$$C_D = \left(\frac{1}{\kappa} \ln \frac{z}{z_0}\right)^{-2} \tag{8.76}$$

This expression indicates z_0 is the only parameter to determine C_D. If the Charnock formula, Eq. (8.67), is applied, Eq. (8.77) gives

$$C_D = \kappa^2 \left(\ln \frac{gz}{m_0 u_*^2}\right)^{-2} \tag{8.77}$$

Similarly, the other z_0 models can define C_D.

On the one hand, numbers of bulk formulas of C_D on the basis of parameterization only with U_{10}, regardless of z_0, have been considered, assuming wind-wave equilibrium state. Garratt (1967) [23] proposed two options for estimating C_D;
A power law relation:

$$C_D \times 10^3 = 0.51 U_{10}^{0.46} \quad (4 < U_{10} < 21 \text{ m/s}) \tag{8.78}$$

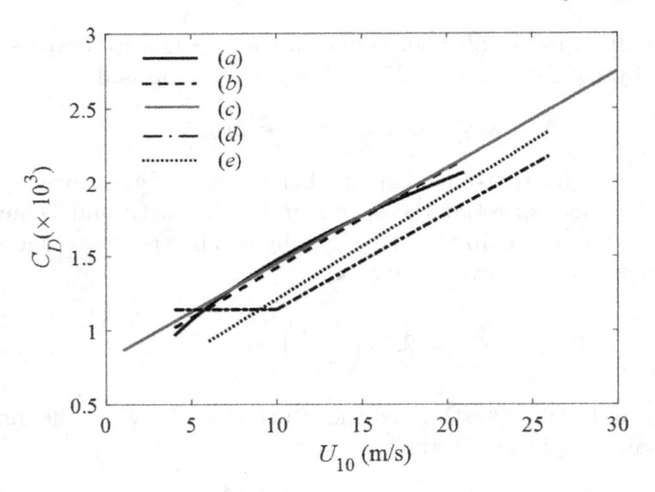

FIGURE 8.28
Drag coefficients as a function of U_{10}; (*a*) Eq. (8.78), (*b*) Eq. (8.79), (*c*) Eq. (8.80), (*d*) Eq. (8.81), and (*e*) Eq. (8.82).

A linear relation:

$$C_D \times 10^3 = 0.75 + 0.067 U_{10} \ \ (4 < U_{10} < 21 \text{ m/s}) \tag{8.79}$$

The other common formulas are listed as:

- Wu (1980)[112)]

$$C_D \times 10^3 = 0.8 + 0.065 U_{10} \ \ (U_{10} > 1 \text{ m/s}) \tag{8.80}$$

- Large and Pond (1981)[50)]

$$C_D \times 10^3 = \begin{cases} 1.14 & (4 < U_{10} \leq 10 \text{ m/s}) \\ 0.49 + 0.065 U_{10} & (10 < U_{10} < 26 \text{ m/s}) \end{cases} \tag{8.81}$$

- Yelland et al. (1998)[126)]

$$C_D \times 10^3 = 0.50 + 0.071 U_{10} \ \ (6 < U_{10} < 26 \text{ m/s}) \tag{8.82}$$

We find all the above conventional models monotonically increase with the wind velocity (see Fig. 8.28). They are unavailable to violent wind state occurring in extreme events, since the previous direct observations have only been available for weak winds ($U_{10} < 26$ m/s). On the one hand, Powell(2003)[83)] analyzed profiles of the strong winds, measured by GPS sondes, in the marine boundary layer associated with tropical cyclones. They provided sensational finding that C_D decreases with the wind velocity in the range of strong winds, $U_{10} > 33$ m/s, which turns the conventional prospect of increasing C_D with winds even in the strong wind states. The investigations of C_D for strong winds have been accelerated since the Powell's pioneer research, and provide new bulk formulas to support the decreasing or saturating C_D in the strong wind range. They are listed as:

- Moon et al. (2007)[70)]

$$z_0 = \begin{cases} \dfrac{0.0185}{g}\left(0.001U_{10}^2 + 0.028U_{10}\right)^2 & (U_{10} \le 12.5 \text{ m/s}) \\ (0.085U_{10} - 0.58) \times 10^{-3} & (U_{10} > 12.5 \text{ m/s}) \end{cases} \tag{8.83}$$

- Andreas et al. (2012)[5)]

$$u_* = 0.239 + 0.0433\left(U_{10} - 8.271 + \left(0.120\left(U_{10} - 8.271\right)^2 + 0.181\right)^{1/2}\right) \tag{8.84}$$

- Zijlema et al. (2012)[130)]

$$C_D \times 10^3 = 0.55 + 2.97\left(\frac{U_{10}}{U_{ref}}\right) - 1.49\left(\frac{U_{10}}{U_{ref}}\right)^2 \tag{8.85}$$

  where U_{ref}=31.5 m/s.

- Holthuijsen et al. (2012)[37)]

$$C_D \times 10^3 = \max\left[0.7, \min\left\{a + b\left(\frac{U_{10}}{U_{ref,1}}\right)^c, d\left(1 - \left(\frac{U_{10}}{U_{ref,2}}\right)^e\right)\right\}\right] \tag{8.86}$$

  where $U_{ref,1} = 27.5$ m/s and $U_{ref,2} = 54$ m/s. For no swell, opposing swell and following swell, the constants $a = 1.05$, $b = 1.25$, $c = 1.4$, $d = 2.3$, $e = 10$. For cross swell, $a = 0.7$, $b = 1.1$, $c = 6$, $d = 8.2$, $e = 2.5$.

- Komori et al. (2018)[45)]

$$C_D \times 10^3 = \begin{cases} 1.0 & (U_{10} < 5.2 \text{ m/s}) \\ 0.44U_{10}^{0.5} & (5.2 \ge U_{10} < 33.6 \text{ m/s}) \\ 2.55 & (U_{10} \ge 33.6 \text{ m/s}) \end{cases} \tag{8.87}$$

The drag coefficients predicted by the above models are shown in Fig. 8.29.

8.4.2 Gas exchanges across sea surfaces

In this section, we consider gas exchanges between air and sea across concentration boundary layers formed at both sides of the sea surface. As introduced

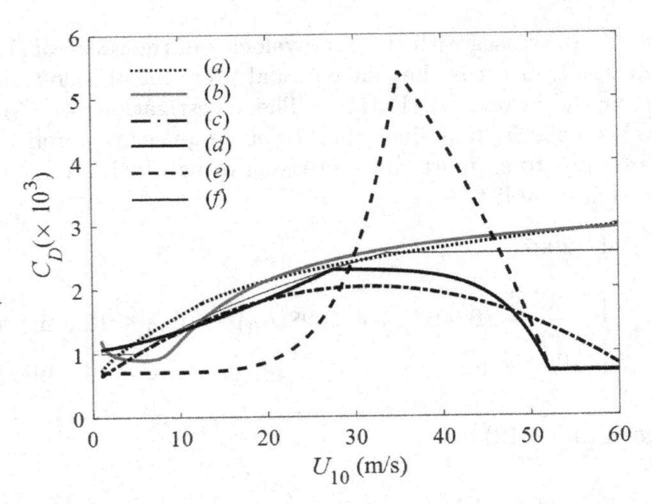

FIGURE 8.29
Drag coefficient as a function of U_{10}; (*a*) Eq. (8.83), (*b*) Eq. (8.84), (*c*) Eq. (8.85), (*d*) Eq. (8.86) (no swell, opposing swell or following swell), (*e*) Eq. (8.86) (cross swell), and (*f*) Eq. (8.87).

in Section 2.2.2, a diffusion flux of the soluble matter across a surface is defined as the product of a transfer velocity k_D and the concentration difference, which generally depends on diffusion coefficient D, the shape of surface, Schmidt number Sc, and Reynolds number Re, as given by Eq. (2.81). The diffusion flux (j) of gases across the atmosphere–ocean interfaces is also described in terms of k_D:

$$j = k_D(C_w - C_s) \tag{8.88}$$

where C_s is the gas concentration at the interface and C_w is the gas concentration in bulk water. C_s may be expressed by the gas concentration in the air phase C_a and solubility α; $C_s = \alpha C_a$.

$$j = k_D(C_w - \alpha C_a) \tag{8.89}$$

k_D is generally expressed in terms of Sc:

$$k_D = (D/n)^n f(v, l) = Sc^{-n} f(v, l) \tag{8.90}$$

where $f(v, l)$ is a function of turbulent velocity, v, and length scale, l. Eq. (8.90) states that the transfer velocities with different diffusivity at identical $f(v, l)$ have a relation:

$$\frac{k_{D1}}{k_{D2}} = \left(\frac{Sc_1}{Sc_2}\right)^{-n} \tag{8.91}$$

While the exponent n varies from 1/2 to 3/2, depending on model and environmental condition, $n = 1/2$ is commonly accepted. Using Eq. (8.91), k_D at arbitrary Sc may be normalized to the one at reference Sc. A common practice is to normalize with $Sc = 600$ (Schmidt number of CO_2 in fresh water at 20 °C, denoted as $k_D(600)$ or $Sc = 660$ (Schmidt number of CO_2 in seawater at 20 °C, denoted as $k(660)$). For instance, k_D for arbitrary gas can be related to $k_D(660)$ in sea water:

$$\frac{k_D(660)}{k} = \left(\frac{660}{Sc}\right)^{-1/2} \tag{8.92}$$

Numbers of the bulk formulas for k_D have been proposed (Wanninkhof et al. 2009 [108]):

- Liss and Merlivat (1986) [56]

$$k_D(600) = \begin{cases} 0.17U_{10} & (U_{10} < 3.6 \text{ m/s}) \\ 2.85U_{10} - 9.65 & (3.6 < U_{10} < 13 \text{ m/s}) \\ 5.9U_{10} & (U_{10} > 13 \text{ m/s}) \end{cases} \tag{8.93}$$

- Wanninkhof (1992) [107]

$$k_D(660) = 0.39U_{10}^2 \tag{8.94}$$

- Nightingale et al. (2000) [77]

$$k_D(600) = 0.333U_{10} + 0.222U_{10}^2 \tag{8.95}$$

- Ho et al. (2006) [36]

$$k_D(600) = 0.266U_{10}^2 \tag{8.96}$$

- Wanninkhof et al. (2009) [108]

$$k_D(660) = 3 + 0.1U_{10} + 0.064U_{10}^2 + 0.011U_{10}^3 \tag{8.97}$$

- Iwano et al. (2013) [38]

$$k_D(600) = \begin{cases} 1.02U_{10}^{1.25} & (U_{10} < 33.6 \text{ m/s}) \\ 5.32 \times 10^{-4}U_{10}^{3.4} & (U_{10} \geq 33.6 \text{ m/s}) \end{cases} \tag{8.98}$$

It is reasonable to explicitly include contributions of bubbles and foams created by breaking waves to gas transfer. Asher and Wanninkhof (1998) [7] introduced the whitecap coverage W (see Section 8.3.4) to predict the contributions of the bubble-mediated transfer and turbulence-driven exchange components. The total transfer velocity k_L is given by the form:

$$k_L = (k_M + W(k_T - k_M)) + Wk_B \tag{8.99}$$

where k_T is the transfer velocity due to turbulence generated by breaking waves, k_M is the one due to turbulence generated by all other processes, and k_B is the bubble-mediated transfer velocity. The parameterization gives

$$k_L = (47U_{10} + W(115200 - 47U_{10}))\,Sc^{-1/2} + W\left(-\frac{37}{H_{rms}} + 6120H_{rms}^{-0.137}Sc^{-0.18}\right) \tag{8.100}$$

for k_L in units of centimeters per hour and U_{10} in meters per second.

A

Appendix

A.1 Vector formulas

Useful vector formulas, involving, for vector $\boldsymbol{A}$, $\boldsymbol{B}$, and scalar ϕ, are listed below.

$$\nabla \cdot (\nabla \times \boldsymbol{A}) = 0 \tag{A.1}$$

$$\nabla \times \nabla \phi = 0 \tag{A.2}$$

$$\nabla \cdot (\phi \boldsymbol{A}) = \boldsymbol{A} \cdot \nabla \phi + \phi \nabla \cdot \boldsymbol{A} \tag{A.3}$$

$$\nabla \cdot (\boldsymbol{A} \otimes \boldsymbol{B}) = (\nabla \boldsymbol{A})\, \boldsymbol{B} + \boldsymbol{A}(\nabla \cdot \boldsymbol{B}) \tag{A.4}$$

$$(\nabla \boldsymbol{A})\, \boldsymbol{B} = \boldsymbol{B} \cdot \nabla \boldsymbol{A} \tag{A.5}$$

$$\nabla^2 \boldsymbol{A} = \nabla \cdot (\nabla \boldsymbol{A}) = \nabla (\nabla \cdot \boldsymbol{A}) - \nabla \times (\nabla \times \boldsymbol{A}) \tag{A.6}$$

$$\begin{aligned} \nabla \times (\boldsymbol{A} \times \boldsymbol{B}) &= \nabla \cdot (\boldsymbol{A} \otimes \boldsymbol{B} - \boldsymbol{B} \otimes \boldsymbol{A}) \\ &= (\nabla \cdot \boldsymbol{B})\, \boldsymbol{A} - (\nabla \cdot \boldsymbol{A})\, \boldsymbol{B} + \boldsymbol{B} \cdot (\nabla \boldsymbol{A}) - \boldsymbol{A} \cdot (\nabla \boldsymbol{B}) \end{aligned} \tag{A.7}$$

A.2 Hyperbolic functions

A.2.1 Definitions

$$\sinh x = \frac{\mathrm{e}^x - \mathrm{e}^{-x}}{2} \tag{A.8}$$

$$\cosh x = \frac{\mathrm{e}^x + \mathrm{e}^{-x}}{2} \tag{A.9}$$

$$\tanh x = \frac{\sinh x}{\cosh x} \tag{A.10}$$

$$\mathrm{csch} x = \frac{1}{\sinh x} \tag{A.11}$$

$$\mathrm{sech} x = \frac{1}{\cosh x} \tag{A.12}$$

$$\mathrm{coth} x = \frac{1}{\tanh x} \tag{A.13}$$

DOI: 10.1201/9781003140160-A

A.2.2 Derivatives

$$\frac{d}{dx}\sinh x = \cosh x \tag{A.14}$$

$$\frac{d}{dx}\cosh x = \sinh x \tag{A.15}$$

$$\frac{d}{dx}\tanh x = 1 - \tanh^2 x = \mathrm{sech}^2 x = \frac{1}{\cosh^2 x} \tag{A.16}$$

$$\frac{d}{dx}\coth x = 1 - \coth^2 x = -\mathrm{csch}^2 x = -\frac{1}{\sinh^2 x} \tag{A.17}$$

$$\frac{d}{dx}\mathrm{sech} x = -\tanh x \; \mathrm{sech} x \tag{A.18}$$

$$\frac{d}{dx}\mathrm{csch} x = -\coth x \; \mathrm{csch} x \tag{A.19}$$

A.2.3 Useful relations

$$\sinh(-x) = -\sinh x \tag{A.20}$$

$$\cosh(-x) = \cosh x \tag{A.21}$$

$$\cosh^2 x - \sinh^2 x = 1 \tag{A.22}$$

$$\sinh(x+y) = \sinh x \cosh y + \cosh x \sinh y \tag{A.23}$$

$$\cosh(x+y) = \cosh x \cosh y + \sinh x \sinh y \tag{A.24}$$

$$\tanh(x+y) = \frac{\tanh x + \tanh y}{1 + \tanh x \tanh y} \tag{A.25}$$

$$\sinh 2x = 2\sinh x \cosh x \tag{A.26}$$

$$\cosh 2x = 2\sinh^2 x + 1 = 2\cosh^2 x - 1 \tag{A.27}$$

$$\tanh 2x = \frac{2\tanh x}{1 + \tanh^2 x} \tag{A.28}$$

A.3 Gauss' divergence theorem

If we consider a control volume V and its surface boundary S in vector field $\boldsymbol{A}$, the divergence theorem is given by

$$\int_V \nabla \cdot \boldsymbol{A} dv = \int_S \boldsymbol{A} \cdot \boldsymbol{n} ds \tag{A.29}$$

where $\boldsymbol{n}$ is the outward unit normal vector.

A.4 Delta function

Dirac's delta function is defined by

$$\delta(\boldsymbol{x}) = \begin{cases} \infty & (\boldsymbol{x} = 0) \\ 0 & (\boldsymbol{x} \neq 0) \end{cases} \tag{A.30}$$

where

$$\int_{-\infty}^{\infty} \delta(\boldsymbol{x}) d\boldsymbol{x} = 1 \tag{A.31}$$

While $\delta(\boldsymbol{x})$ is singular at $\boldsymbol{x} = 0$ and its value is infinite there, Eq. (A.31) provides an important formula of convolution for an arbitrary function $f(\boldsymbol{x})$:

$$f(\boldsymbol{x}) = \int_{-\infty}^{\infty} f(\boldsymbol{\xi}) \delta(\boldsymbol{x} - \boldsymbol{\xi}) d\boldsymbol{\xi} \tag{A.32}$$

A.5 Stokes' theorem

Stokes' theorem for vector $\boldsymbol{f}$ states that

$$\oint_C \boldsymbol{f} \cdot d\boldsymbol{x} = \int_A (\nabla \times \boldsymbol{f}) \cdot \boldsymbol{n} da \tag{A.33}$$

where $d\boldsymbol{x}$ is differential vector along boundary curve C of surface A and $\boldsymbol{n}$ is the unit normal vector.

A.6 Green function

Consider the following Poisson equation:

$$\nabla^2 \psi = -f(\boldsymbol{x}) \tag{A.34}$$

The Green function $G(\boldsymbol{x})$ is introduced to solve the equation:

$$\nabla^2 G(\boldsymbol{x}) = -\delta(\boldsymbol{x}) \tag{A.35}$$

The solution of Eq. (A.34) is given

$$\psi(\boldsymbol{x}) = \int G(\boldsymbol{x} - \boldsymbol{x}') f(\boldsymbol{x}') d\boldsymbol{x}' \tag{A.36}$$

since

$$\nabla^2\psi(\boldsymbol{x}) = \int \nabla^2 G(\boldsymbol{x}-\boldsymbol{x}')f(\boldsymbol{x}')d\boldsymbol{x}' = -\int \delta(\boldsymbol{x}-\boldsymbol{x}')f(\boldsymbol{x}')d\boldsymbol{x}' = -f(\boldsymbol{x}) \tag{A.37}$$

which derives Eq. (A.34). Fourier transform and inverse transform are defined by

$$\hat{f}(\boldsymbol{k}) = \int f(\boldsymbol{x})\mathrm{e}^{-i\boldsymbol{k}\cdot\boldsymbol{x}}d\boldsymbol{x} \tag{A.38}$$

$$f(\boldsymbol{x}) = \frac{1}{(2\pi)^d}\int \hat{f}(\boldsymbol{k})\mathrm{e}^{i\boldsymbol{k}\cdot\boldsymbol{x}}d\boldsymbol{k} \tag{A.39}$$

where d is number of dimensions. When the inverse transform of G and δ

$$G(\boldsymbol{x}) = \frac{1}{(2\pi)^d}\int \hat{G}(\boldsymbol{k})\mathrm{e}^{i\boldsymbol{k}\cdot\boldsymbol{x}}d\boldsymbol{k} \tag{A.40}$$

$$\delta(\boldsymbol{x}) = \frac{1}{(2\pi)^d}\int \mathrm{e}^{i\boldsymbol{k}\cdot\boldsymbol{x}}d\boldsymbol{k} \tag{A.41}$$

are substituted into Eq. (A.35), we find

$$\hat{G}(k) = \frac{1}{k^2} \tag{A.42}$$

The substitution of Eq. (A.42) into Eq. (A.40) gives

$$G(\boldsymbol{x}) = \frac{1}{(2\pi)^d}\int \frac{1}{k^2}\mathrm{e}^{i\boldsymbol{k}\cdot\boldsymbol{x}}d\boldsymbol{k} \tag{A.43}$$

In a spherical coordinate for wave number $\boldsymbol{k}$, (k,θ,φ), as $d\boldsymbol{k} = k^2 dk \sin\theta d\theta d\varphi$, substituting $\cos\theta = \mu$,

$$\begin{aligned} G(r) &= \frac{1}{(2\pi)^3}\int_0^\infty dk \int_{-1}^{1} d\mu \int_0^{2\pi} d\varphi \mathrm{e}^{ikr\mu} \\ &= \frac{1}{2\pi^2}\int_0^\infty \frac{\sin kr}{kr}dk = \frac{1}{2\pi^2 r}\int_0^\infty \frac{\sin t}{t}dt = \frac{1}{4\pi r} \end{aligned} \tag{A.44}$$

since $\int_0^\infty \sin t/t dt = \pi/2$.

A.7 Leibniz rule

The derivative of integral $\int_{\alpha(x)}^{\beta(x)} Q(x,y)dy$ is expressible as

$$\frac{d}{dx}\int_{\alpha(x)}^{\beta(x)} Q(x,y)dy = \int_{\alpha(x)}^{\beta(x)} \frac{\partial}{\partial x}Q(x,y)dy + Q(x,\beta(x))\frac{d\beta(x)}{dx} - Q(x,\alpha(x))\frac{d\alpha(x)}{dx} \tag{A.45}$$

If the interval α and β are constant,

$$\frac{d}{dx}\int_{\alpha}^{\beta} Q(x,y)dy = \int_{\alpha}^{\beta} \frac{\partial}{\partial x} Q(x,y)dy \tag{A.46}$$

A.8 Bessel functions

A.8.1 Bessel function

$$x^2\frac{d^2y}{dx^2} + x\frac{dy}{dx} + \left(x^2 - n^2\right)y = 0 \tag{A.47}$$

where the solution

$$y(x) = AJ_n(x) + BY_n(x) \tag{A.48}$$

Series representation:

$$J_n(x) = \sum_{m=0}^{\infty} \frac{(-1)^m (x/2)^{n+2m}}{m!\,(m+n)!} \tag{A.49}$$

$$Y_n(x) = \frac{J_n(x)\cos(n\pi) - J_{-n}(x)}{\sin(n\pi)} \tag{A.50}$$

Properties:

$$2nJ_n(x) = x\left(J_{n-1}(x) + J_{n+1}(x)\right) \tag{A.51}$$

$$J_{-n}(x) = (-1)^n J_n(x) \tag{A.52}$$

$$J_n(x) \approx \frac{1}{n!}\left(\frac{x}{2}\right)^n \quad x \to 0 \tag{A.53}$$

$$J_n(x) \approx \sqrt{\frac{2}{\pi x}}\cos\left(x - \frac{\pi n}{2} - \frac{\pi}{4}\right) \quad x \to \infty \tag{A.54}$$

$$2nY_n(x) = x\left(Y_{n-1}(x) + Y_{n+1}(x)\right) \tag{A.55}$$

$$Y_{-n}(x) = (-1)^n Y_n(x) \tag{A.56}$$

Derivative:

$$\frac{d}{dx}J_n(x) = \frac{1}{2}\left(J_{n-1}(x) - J_{n+1}(x)\right) \tag{A.57}$$

$$\frac{d}{dx}Y_n(x) = \frac{1}{2}\left(Y_{n-1}(x) - Y_{n+1}(x)\right) \tag{A.58}$$

In particular, from Eqs. (A.57) and (A.52),

$$\frac{d}{dx}J_0(x) = -J_1(x) \tag{A.59}$$

$$\frac{d}{dx}Y_0(x) = -Y_1(x) \tag{A.60}$$

A.8.2 Modified Bessel function

$$x^2 \frac{d^2 y}{dx^2} + x \frac{dy}{dx} - \left(x^2 + n^2\right) y = 0 \tag{A.61}$$

where the solution

$$y(x) = AI_n(x) + K_n(x) \quad \text{or} \quad y(x) = CJ_n(ix) + DY_n(ix) \tag{A.62}$$

Series representation:

$$I_n(x) = \sum_{m=0}^{\infty} \frac{(x/2)^{n+2m}}{m!\,(m+n)!} \tag{A.63}$$

$$K_n(x) = \frac{\pi}{2} \frac{I_{-n}(x) - I_n(x)}{\sin n\pi} \tag{A.64}$$

Properties:

$$2nI_n(x) = x\left(I_{n-1}(x) - I_{n+1}(x)\right) \tag{A.65}$$

$$2nK_n(x) = x\left(-K_{n-1}(x) + K_{n+1}(x)\right) \tag{A.66}$$

$$I_n(-x) = (-1)^n I_n(x) \tag{A.67}$$

$$I_n(x) \approx \frac{1}{n!} \left(\frac{x}{2}\right)^n \quad x \to 0 \tag{A.68}$$

$$I_n(x) \approx \frac{e^x}{\sqrt{2\pi x}} \quad x \to \infty \tag{A.69}$$

Derivative:

$$\frac{d}{dx} I_n(x) = \frac{1}{2}\left(I_{n-1}(x) + I_{n+1}(x)\right) \tag{A.70}$$

$$\frac{d}{dx} K_n(x) = -\frac{1}{2}\left(K_{n-1}(x) + K_{n+1}(x)\right) \tag{A.71}$$

Note:

$$I_n(x) = i^{-n} J_n(ix), \quad I_n(ix) = i^n J_n(x) \tag{A.72}$$

Bibliography

[1] A. Saruwatari, Y. Watanabe, and D.M. Ingram. Scarifying and fingering surfaces of plunging jets. *Coastal Engineering*, 56:1109–1122, 2009.

[2] G. Agbaglah, C. Josserand, and S. Zaleski. Longitudinal instability of a liquid rim. *Physics of Fluids*, 25(2):022103, 2013.

[3] A.N. Kolmogorov. The local structure of turbulence in incompressible viscous fluid for very large Reynolds numbers. *Proceedings of the Royal Society of London. Series A: Mathematical and Physical Sciences*, 434(1890):9–13, 1991.

[4] E. L. Andreas. Sea spray and the turbulent air-sea heat fluxes. *Journal of Geophysical Research*, 97(C7):11429–11441, 1992.

[5] E. L. Andreas, L. Mahrt, and D. Vickers. A new drag relation for aerodynamically rough flow over the ocean. *Journal of the Atmospheric Sciences*, 69(8):2520–2537, 2012.

[6] W.E. Asher, J. Edson, W. McGillis, R. Wanninkhof, D.T. Ho, and T. Litchendorf. Fractional area whitecap coverage and air-sea gas transfer velocities measured during GasEx-98. *Gas Transfer at Water Surfaces, Geophysical Monograph 127*, 199–203, 2002.

[7] W. E. Asher and R. Wanninkhof. The effect of bubble-mediated gas transfer on purposeful dual-gaseous tracer experiments. *Journal of Geophysical Research: Oceans*, 103(C5):10555–10560, 1998.

[8] G.K. Batchelor. *An introduction to fluid dynamics.* Cambridge University Press, first mathematical library edition, 2000.

[9] J.A. Battjes. Surf similarity. Coastal Engineering 1974, American Society of Civil Engineers, 1974.

[10] C.L. Bretschneider. Significant waves and wave spectrum. *Ocean Industry*, 40–46, 1968.

[11] M. Brocchini and D.H. Peregrine. The dynamics of strong turbulence at free surfaces. Part 1. description. *Journal of Fluid Mechanics*, 449:225–254, 2001.

[12] A.H. Callaghan, G.B. Deane, and M.D. Stokes. Observed physical and environmental causes of scatter in whitecap coverage values in a fetch-limited coastal zone. *Journal of Geophysical Research*, 113(C5), 2008.

[13] B. Camenen and M. Larson. Predictive formulas for breaker depth index and breaker type. *Journal of Coastal Research*, 234:1028–1041, 2007.

[14] R. Clift, J.R. Grace, and M.E. Weber. *Bubbles, Drops and Particles*. Academic Press, 1978.

[15] M.J. Cooker and D.H. Peregrine. Pressure-impulse theory for liquid impact problems. *Journal of Fluid Mechanics*, 297:193–214, 1995.

[16] D. Chung, N. Hutchins, M.P. Schultz, and K.A. Flack. Predicting the drag of rough surfaces. *Annual Review of Fluid Mechanics*, 53:439–471, 2021.

[17] R.G. Dean and R.A. Dalrymple. *Water Wave Mechanics for Engineers and Scientists*. World Scientific, 1991.

[18] D.M. Farmer, C.L. McNeil, and B.D. Johnson. Evidence for the importance of bubbles in increasing air-sea gas flux. *Nature*, 361:620–623, 1993.

[19] N. Dombrowski and W.R. Johns. The aerodynamic instability and disintegration of viscous liquid sheets. *Chemical Engineering Science*, 18(3):203–214, 1963.

[20] W.M. Drennan. On the wave age dependence of wind stress over pure wind seas. *Journal of Geophysical Research*, 108(C3), 2003.

[21] F. Anctil and M.A. Donelan. Air-water momentum flux observation sover shoaling waves. *Journal of Physical Oceanography*, 26:1344–1353, 1996.

[22] N. Frössling. On the evaporation of falling drops (in German). *Gerlands Beiträge zur Geophysik*, 52:170–216, 1938.

[23] J.R. Garratt. Review of drag coefficients over oceans and continents. *Monthly Weather Review*, 105(7):915–929, 1977.

[24] C. Garrett, M. Li, and D. Farmer. The connection between bubble size spectra and energy dissipation rates in the upper ocean. *Journal of Physical Oceanography*, 30(9):2163–2171, 2000.

[25] G.B. Deane and M.D. Stokes. Scale dependence of bubble creation mechanisms in breaking waves. *Nature*, 418:839–844, 2002.

[26] Y. Goda. Statistical variability of sea state parameters as a function of a wave spectrum. *Coastal Engineering in Japan*, 31(1):39–52, 1988.

[27] Y. Goda. New wave pressure formulae for composite breakwaters. *Coastal Engineering Proceedings*, 1(14):100, 1974.

[28] Y. Goda. Empirical formulation of sediment pickup rate in terms of wave energy flux dissipation rate. *Coastal Engineering Journal*, 55(4):1350012–1–1350012–17, 2013.

[29] H. Gotoh, A. Okayasu, and Y. Watanabe. *Computational wave dynamics.* World Scientific, 2013.

[30] J.R. Grace, T. Wairegi, and T.H. Nguyen. Shape and velocities of single drops and bubble moving freely through immiscible liquids. *Transactions of the Institution of Chemical Engineers*, 54:167–173, 1976.

[31] H. Charnock. Wind stress on a water surface. *Quarterly Journal of the Royal Meteorological Society*, 81:639–640, 1955.

[32] H. Mitsuyasu and T. Honda. The high frequency spectrum of wind-generated waves. *Journal of the Oceanographical Society of Japan*, 30:185–198, 1974.

[33] M. Hatori. Nonlinear properties of laboratory wind waves at energy containing frequencies. *Journal of the Oceanographical Society of Japan*, 40(1):1–11, 1984.

[34] J.O. Hinze. Fundamentals of the hydrodynamic mechanism of splitting in dispersion processes. *AIChE Journal*, 1(3):289–295, 1955.

[35] J.O. Hize. *Turbulence.* McGraw Hill, 1975.

[36] D.T. Ho, C.S. Law, M.J. Smith, P. Schlosser, M. Harvey, and P. Hill. Measurements of air-sea gas exchange at high wind speeds in the southern ocean: Implications for global parameterizations. *Geophysical Research Letters*, 33(16), 2006.

[37] L.H. Holthuijsen, M.D. Powell, and J.D. Pietrzak. Wind and waves in extreme hurricanes. *Journal of Geophysical Research*, 117:C09003, 2012.

[38] K. Iwano, N. Takagaki, R. Kurose, and S. Komori. Mass transfer velocity across the breaking air–water interface at extremely high wind speeds. *Tellus B: Chemical and Physical Meteorology*, 65(1):21341, 2013.

[39] J. Otsuka, A. Saruwatari, and Y. Watanabe. Vortex-induced suspension of sediment in the surf zone. *Advances in Water Resources*, 110:59–76, 2017.

[40] J. Otsuka and Y. Watanabe. Characteristic time, length and velocity scales of transverse flows in the surf zone. *Coastal Engineering Journal*, 57(2):15500006, 2015.

[41] B. Jahne and H. Haußecker. Air-water gas exchange. *Annual Review of Fluid Mechanics*, 30(1):443–468, 1998.

[42] K. Hasselmann, T.P. Barnett, E. Bouws, H. Carlson, D.E. Cartwright and K. Enke, J.A. Ewing, H. Gienapp, D.E. Hasselmann, P. Kruseman, A. Meerburg, P. Muller, D.J. Olbers, K. Richter, W. Sell, and H. Walden. Measurements of wind-wave growth and swell decay during the Joint North Sea Wave Project (JONSWAP). *Deutschen Hydrographischen Zeitschrift*, Suppl 8. A(12):1–95, 1973.

[43] K.T. Kiger and J.H. Duncan. Air-entrainment mechanisms in plunging jets and breaking waves. *Annual Review of Fluid Mechanics*, 44(1):563–596, 2012.

[44] A.N. Kolmogorov. The local structure of turbulence in incompressible viscous fluid for very large Reynolds' numbers. Proceedings of the Royal Society of London, Series A: Mathematical and Physical Science, 434(1890): 9–13, 1991.

[45] S. Komori, K. Iwano, N. Takagaki, Ryo Onishi, Ryoichi Kurose, Keiko Takahashi, and Naoya Suzuki. Laboratory measurements of heat transfer and drag coefficients at extremely high wind speeds. 48(4):959–974, 2018.

[46] V.N. Kudryavtsev and V.K. Makin. Impact of ocean spray on the dynamics of the marine atmospheric boundary layer. *Boundary-Layer Meteorology*, 140(3):383–410, 2011.

[47] C. Lafon, J. Piazzola, P. Forget, O. Le Calve, and S. Despiau. Analysis of the variations of the whitecap fraction as measured in a coastal zone. *Boundary-Layer Meteorology*, 111(2):339–360, 2004.

[48] E. Lamarre and W.K. Melville. Air entrainment and dissipation in breaking waves. *Nature*, 351(6326):469–472, 1991.

[49] H. Lamb. *Hydrodynamics (6th edition)*. Dover Publication, 1945.

[50] W.G. Large and S. Pond. Open ocean momentum flux measurements in moderate to strong winds. *Journal of Physical Oceanography*, 11(3):324–336, 1981.

[51] M. Larson and N. Kraus. Estimation of suspended sediment trapping ratio for channel infilling and bypassing. Technical report, US Army Corps of Engineers, Technical Notes ERDC/CHL CHETN-IV-34, 2001.

[52] T.G. Leighton. *The Acoustic Bubble*. Academic Press, 1994.

[53] L.J. Leng. Splash formation by spherical drops. *Journal of Fluid Mechanics*, 427:73–105, 2001.

[54] V.G. Levich. *Physicochemical Hydrodynamics.* Prentic-Hall, New York, 1962.

[55] H. Lhuissier and E. Villermaux. Bursting bubble aerosols. *Journal of Fluid Mechanics*, 696:5–44, 2011.

[56] P.S. Liss and L. Merlivat. Air-sea gas exchange rates: Introduction and synthesis. 113–127, 1986.

[57] P. Lubin and S. Glockner. Numerical simulations of three-dimensional plunging breaking waves: Generation and evolution of aerated vortex filaments. *Journal of Fluid Mechanics*, 767:364–393, 2015.

[58] M.A. Donelan. *The Sea.* Wiley-Interscience, 1990.

[59] J. McCowan. XXXIX. on the highest wave of permanent type. *The London, Edinburgh, and Dublin Philosophical Magazine and Journal of Science*, 38(233):351–358, 1894.

[60] C.C. Mei, M. Stiassnie, and D.K.-P. Yue. *Theory and applications of ocean surface waves.* World Scientific, 2005.

[61] W.K. Melville. The role of surface-wave breaking in air-sea interaction. *Annual Review of Fluid Mechanics*, 28(1):279–321, 1996.

[62] M. Miche. Mouvements ondulatoires de la mer en profondeur constante ou décroissante. *Annales de Ponts et Chaussées*, 114:25–406, 1944.

[63] J.H. Michell. XLIV. The highest waves in water. *The London, Edinburgh, and Dublin Philosophical Magazine and Journal of Science*, 36(222):430–437, 1893.

[64] J.W. Miles. On the generation of surface waves by shear flows. *Journal of Fluid Mechanics*, 3(02):185, 1957.

[65] J.W. Miles. On the generation of surface waves by turbulent shear flows. *Journal of Fluid Mechanics*, 7(3):469–478, 1960.

[66] J.W. Miles. On the generation of surface waves by shear flows. part 4. *Journal of Fluid Mechanics*, 13(3):433–448, 1962.

[67] H. Mitsuyasu. On the growth of spectrum of wind-generated waves (2) – spectral shape of wind waves at finite fetch. *Proceedings of the Japanese Conference on Coastal Engineering*, 1–7, 1970.

[68] E.C. Monahan and M. Lu. Acoustically relevant bubble assemblages and their dependence on meteorological parameters. *IEEE Journal of Oceanic Engineering*, 15(4):340–349, 1990.

[69] E.C. Monahan and I. Muircheartaigh. Optimal power-law description of oceanic whitecap coverage dependence on wind speed. *Journal of Physical Oceanography*, 10(12):2094–2099, 1980.

[70] Il-J. Moon, I. Ginis, T. Hara, and B. Thomas. A physics-based parameterization of air-sea momentum flux at high wind speeds and its impact on hurricane intensity predictions. *Monthly Weather Review*, 135(8):2869–2878, 2007.

[71] N. Mori and T. Takahashi. Nationwide post event survey and analysis of the 2011 Tohoku earthquake tsunami. *Coastal Engineering Journal*, 54(1):1250001–1–1250001–27, 2012.

[72] M.S. Longuet-Higgins. Capillary rollers and bores. *Journal of Fluid Mechanics*, 240:659–679, 1992.

[73] K. Nadaoka, M. Hino, and Y. Koyano. Structure of the turbulent flow field under breaking waves in the surf zone. *Journal of Fluid Mechanics*, 204(1):359, 1989.

[74] J.M. Nelson, Y. Shimizu, T. Abe, K. Asahi, M. Gamou, T. Inoue, T. Iwasaki, T. Kakinuma, S. Kawamura, I. Kimura, T. Kyuka, R.R. McDonald, M. Nabi, M. Nakatsugawa, F.R. Simões, H. Takebayashi, and Y. Watanabe. The international river interface cooperative: Public domain flow and morphodynamics software for education and applications. *Advances in Water Resources*, 93:62–74, 2016.

[75] M.M. Nicolson. The interaction between floating particles. *Mathematical Proceedings of the Cambridge Philosophical Society*, 45(2):288–295, 1949.

[76] P. Nielsen. Field measurements of time-averaged suspended sediment concentrations under waves. *Coastal Engineering*, 8(1):51–72, 1984.

[77] P.D. Nightingale, G. Malin, C.S. Law, A.J. Watson, P.S. Liss, M.I. Liddicoat, J. Boutin, and R.C. Upstill-Goddard. In situ evaluation of air-sea gas exchange parameterizations using novel conservative and volatile tracers. *Global Biogeochemical Cycles*, 14(1):373–387, 2000.

[78] C.D. O'Dowd and G. de Leeuw. Marine aerosol production: A review of the current knowledge. *Philosophical Transactions of the Royal Society A: Mathematical, Physical and Engineering Sciences*, 365(1856):1753–1774, 2007.

[79] Y. Okada. Surface deformation due to shear and tensile faults in a half-space. *Bulletin of the Seismological Society of America*, 75(4):1135–1154, 1985.

[80] D.H. Peregrine. The fascination of fluid mechanics. *Journal of Fluid Mechanics*, 106(1):59, 1981.

[81] O.M. Phillips. On the generation of waves by turbulent wind. *Journal of Fluid Mechanics*, 2(05):417, 1957.

[82] W.J. Pierson and L. Moskowitz. A proposed spectral form for fully developed wind seas based on the similarity theory of S. A. Kitaigorodskii. *Journal of Geophysical Research*, 69(24):5181–5190, 1964.

[83] M.D. Powell, P.J. Vickery, and T.A. Reinhold. Reduced drag coefficient for high wind speeds in tropical cyclones. *Nature*, 422(6929):279–283, 2003.

[84] H. Rause. *Fluid mechanics for hydraulic engineers*. Dover, New York, 1961.

[85] L.C. Van Rijn and A. Kroon. Sediment transport by currents and waves. Coastal Engineering 1992, American Society of Civil Engineers, 1993.

[86] F. Risso. Agitation, mixing, and transfers induced by bubbles. *Annual Review of Fluid Mechanics*, 50(1):25–48, 2018.

[87] T. Sarpkaya and P. Suthon. Interaction of a vortex couple with a free surface. *Experiments in Fluids*, 11:205–217, 1991.

[88] L. Schiller and A. Naumann. Uber die grundlegenden berechnungen bei der schwerkraftaufbereitung. *Zeitschrift des Vereins deutscher Ingenieure*, 77:318, 1933.

[89] M.H. Smith, P.M. Park, and I.E. Consterdine. Marine aerosol concentrations and estimated fluxes over the sea. *Quarterly Journal of the Royal Meteorological Society*, 119(512):809–824, 1993.

[90] D.B. Spalding. A single formula for the "law of the wall". *Journal of Applied Mechanics*, 28(3):455–458, 1961.

[91] H.B. Squire. Investigation of the instability of a moving liquid film. *British Journal of Applied Physics*, 4(6):167–169, 1953.

[92] J.J. Stoker. Water waves. John Wiley & Sons, Inc., 1992.

[93] G.G. Stokes. Report on recent researches in hydrodynamics. *16th Report of the British Association for the Advancement of Science*, 1–20, 1846.

[94] P. Sutherland and W. K. Melville. Field measurements and scaling of ocean surface wave-breaking statistics. *Geophysical Research Letters*, 40(12):3074–3079, 2013.

[95] Y. Tanioka and K. Satake. Tsunami generation by horizontal displacement of ocean bottom. *Geophysical Research Letters*, 23(8):861–864, 1996.

[96] G. Taylor. The dynamics of thin sheets of fluid. II. Waves on fluid sheets. *Proceedings of the Royal Society of London. Series A*, 1959.

[97] P.K. Taylor and M.J. Yelland. The dependence of sea surface roughness on the height and steepness of the waves. *Journal of Physical Oceanography*, 31:572–590, 2001.

[98] Y. Toba and M. Koga. A parameter describing overall conditions of wave breaking, whitecapping, sea-spray production and wind stress. 37–47, 1986.

[99] A. Tomiyama, I. Kataoka, I. Zun, and T. Sakaguchi. Drag coefficients of single bubbles under normal and micro gravity conditions. *JSME International Journal Series B*, 41(2):472–479, 1998.

[100] S. Tomotika. On the instability of a cylindrical thread of a viscous liquid surrounded by another viscous fluid. *Proceedings of the Royal Society of London. Series A – Mathematical and Physical Sciences*, 150(870):322–337, 1935.

[101] M.K. Tripathi, K.C. Sahu, and R. Govindarajan. Dynamics of an initially spherical bubble rising in quiescent liquid. *Nature Communications*, 6(1), 2015.

[102] G. Tryggvason. Deformation of a free surface as a result of vortical flows. *Physics of Fluids*, 31(5):955–957, 1988.

[103] F. Ursell. Edge waves on a sloping beach. *Proceedings of the Royal Society A*, 214:79–97, 1952.

[104] S. Vagle and D.M. Farmer. The measurement of bubble-size distributions by acoustical backscatter. *Journal of Atmospheric and Oceanic Technology*, 9(5):630–644, 1992.

[105] P. Valiorgue, N. Souzy, M.E. Hajem, H.B. Hadid, and S. Simons. Concentration measurement in the wake of a free rising bubble using planar laser-induced fluorescence (PLIF) with a calibration taking into account fluorescence extinction variations. *Experiments in Fluids*, 54(4), 2013.

[106] P. Wang, W. Yuan, and L. Min. Measuring and modeling suspended sediment concentration profiles in the surf zone. *Journal of Palaeogeography*, 1(2):172–192, 2012.

[107] R. Wanninkhof. Relationship between wind speed and gas exchange over the ocean. *Journal of Geophysical Research*, 97(C5):7373, 1992.

[108] R. Wanninkhof, W.E. Asher, D.T. Ho, C. Sweeney, and W.R. McGillis. Advances in quantifying air-sea gas exchange and environmental forcing. *Annual Review of Marine Science*, 1(1):213–244, 2009.

[109] Y. Watanabe, Y. Mitobe, A. Saruwatari, T. Yamada, and Y. Niida. Evolution of the 2011 Tohoku earthquake tsunami on the pacific coast of Hokkaido. *Coastal Engineering Journal*, 54(1):1250002–1–1250002–17, 2012.

[110] Y. Watanabe, H. Oyaizu, H. Satoh, and Y. Niida. Bubble drag in electrolytically generated microbubble swarms with bubble-vortex interactions. *International Journal of Multiphase Flow*, 136:103541, 2021.

[111] C.H.K. Williamson. Three-dimensional wake transition. *Journal of Fluid Mechanics*, 328:345–407, 1996.

[112] J. Wu. Wind-stress coefficients over sea surface near neutral conditions—a revisit. *Journal of Physical Oceanography*, 10(5):727–740, 1980.

[113] J. Wu. Variations of whitecap coverage with wind stress and water temperature. *Journal of Physical Oceanography*, 18(10):1448–1453, 1988.

[114] J. Wu. Contributions of film and jet drops to marine aerosols produced at the sea surface. *Tellus B*, 41B(4):469–473, 1989.

[115] J. Wu, J.J. Murray, and R.J. Lai. Production and distributions of sea spray. *Journal of Geophysical Research*, 89(C5):8163, 1984.

[116] Y. Goda. *Directional wave spectrum and its engineering applications*, volume 3. World Scientific, Advances in coastal and ocean engineering edition, 1996.

[117] Y. Goda. *Random seas and design of maritime structures*. World Scientific, 2010.

[118] Y. Niida and Y. Watanabe. Oxygen transfer from bubble-plumes. *Physics of Fluids*, 30:107104, 2018.

[119] Y. Toba. Local balance in the air-sea boundary processes, I. On the growth process of wind waves. *Journal of Oceanographical Society of Japan*, 28:109–120, 1972.

[120] Y. Watanabe, A. Saruwatari, and D.M. Ingram. Free-surface flows under impacting droplets. *Journal of Computational Physics*, 227:2344–2365, 2008.

[121] Y. Watanabe and D.M. Ingram. Transverse instabilities of ascending planar jets formed by wave impacts on vertical walls. *Proceedings of the Royal Society A*, 471:20150397, 2015.

[122] Y. Watanabe and D.M. Ingram. Size distributions of sprays produced by violent wave impacts on vertical sea walls. *Proceedings of the Royal Society A*, 472:20160423, 2016.

[123] Y. Watanabe, H. Saeki, and R.J. Hosking. Three-dimensional vortex structures breaking waves. *Journal of Fluid Mechanics*, 545:218–328, 2005.

[124] Y. Watanabe and N. Mori. Infrared measurements of surface renewal and subsurface vortices in nearshore breaking waves. *Journal of Geophysical Research*, 113:C07015, 2008.

[125] A.L. Yarin. Drop impact dynamics: Splashing, spreading, receding, bouncing. *Annual Review of Fluid Mechanics*, 38(1):159–192, 2006.

[126] M.J. Yelland, B.I. Moat, P.K. Taylor, R.W. Pascal, J. Hutchings, and V.C. Cornell. Wind stress measurements from the open ocean corrected for airflow distortion by the ship. *Journal of Physical Oceanography*, 28(7):1511–1526, 1998.

[127] W.R. Young and C.L. Wolfe. Generation of surface waves by shear-flow instability. *Journal of Fluid Mechanics*, 739:276–307, December 2013.

[128] X. Yu, K. Hendrickson, and D.K.P. Yue. Scale separation and dependence of entrainment bubble-size distribution in free-surface turbulence. *Journal of Fluid Mechanics*, 885, 2019.

[129] D. Zhao and Y. Toba. Dependence of whitecap coverage on wind and wind-wave properties. *Journal of Oceanography*, 57(5): 603–616, 2001.

[130] M. Zijlema, G.Ph. van Vledder, and L.H. Holthuijsen. Bottom friction and wind drag for wave models. *Coastal Engineering*, 65:19–26, 2012.

Index